Sweet Cherries

Crop Production Science In Horticulture

CROP PRODUCTION SCIENCE IN HORTICULTURE SERIES

This series examines economically important horticultural crops selected from the major production systems in temperate, subtropical and tropical climatic areas. Systems represented range from open field and plantation sites to protected plastic and glass houses, growing rooms and laboratories. Emphasis is placed on the scientific principles underlying crop production practices rather than on providing empirical recipes for uncritical acceptance. Scientific understanding provides the key to both reasoned choice of practice and the solution of future problems.

Students and staff at universities and colleges throughout the world involved in courses in horticulture, as well as in agriculture, plant science, food science and applied biology at degree, diploma or certificate level will welcome this series as a succinct and readable source of information. The books will also be invaluable to progressive growers, advisers and end-product users requiring an authoritative, but brief, scientific introduction to particular crops or systems. Keen gardeners wishing to understand the scientific basis of recommended practices will also find the series very useful.

The authors are all internationally renowned experts with extensive experience of their subjects. Each volume follows a common format covering all aspects of production, from background physiology and breeding, to propagation and planting, through husbandry and crop protection, to harvesting, handling and storage. Selective references are included to direct the reader to further information on specific topics.

Sweet Cherries

Crop Production Science In Horticulture

Lynn E. Long

Oregon State University, USA

Gregory A. Lang

Michigan State University, USA

Clive Kaiser

Oregon State University, USA

CABI is a trading name of CAB International

CABI
Nosworthy Way
Wallingford
Oxfordshire OX10 8DE
UK

Tel: +44 (0)1491 832111
Fax: +44 (0)1491 833508
E-mail: info@cabi.org
Website: www.cabi.org

CABI
WeWork
One Lincoln St
24th Floor
Boston, MA 02111
USA
T: +1 (617)682-9015
E-mail: cabi-nao@cabi.org

A catalogue record for this book is available from the British Library, London, UK.

References to Internet websites (URLs) were accurate at the time of writing.

ISBN-13:   9781786398284 (paperback)
           9781786398291 (ePDF)
           9781786398307 (ePub)

Commissioning Editor: Rachael Russell
Editorial Assistant: Emma McCann
Production Editor: Marta Patino

Typeset by Exeter Premedia Services Pvt Ltd, Chennai, India
Printed and bound in the UK by Severn, Gloucester

Disclaimer

Sweet Cherries: Crop Production Science in Horticulture is a work of reference and the information contained herein has been collected and collated from diverse sources, scientific studies, personal observations, textbooks, journal articles, and online resources. While the authors and the publisher have made all reasonable efforts to present reliable data and information, together with the relevant citations, neither the authors nor the publisher can accept any legal responsibility or liability for any errors or omissions that may have been made and do not accept any liability for any damage caused by, or economic loss arising from reliance upon information contained in this publication. The authors and the publishers have included the information contained herein with the belief that it is in accordance with the standard practices at the time of publication and restricted within the scope and space availability. This information is provided strictly as a supplement to a grower's own judgement, relevant manufacturer's instructions and appropriate best practise guidelines. Neither the authors nor the publisher can take responsibility for a reader's own actions. The authors and publishers have attempted to trace the copyright holders of all material reproduced in this publication and apologize to copyright holders if permission to publish in this form has not been obtained. If any copyright material has not been appropriately acknowledged, please write to let us know so we may rectify this in any future reprint.

Contents

Acknowledgements

The authors wish to thank the following individuals for their valuable contributions to this book:

Miljan Cvetkovic, University of Banja Luka, Bosnia and Herzegovina
David Granatstein, Washington State University, USA
Eugeniu Gudumac, Horti Management SRL, Moldova
Denise Neilsen, Agriculture and Agri-Food Canada
Gerry Neilsen, Agriculture and Agri-Food Canada
Tim Smith, Washington State University, USA

Trends in Sweet Cherry Production

1

This is an exciting time to be a sweet cherry producer. Today, growers have access to an unprecedented selection of high quality cultivars, precocious and dwarfing rootstocks, and high yielding innovative training systems. Additionally, increasing recognition of soil biology as an integral component of a healthy, productive orchard system is changing the way that orchards are managed. New technology that facilitates mechanization of some production, harvest and postharvest practices is being developed to respond to the uncertainty and increasing expense associated with the labor supply. Although new challenges, such as invasive pests and a changing climate, also are apparent, research and innovations are expected to continue to develop greater efficiencies, reduced inputs and higher quality fruit as growers adopt new orchard practices.

Sweet cherry research and development, along with market demand, have shown an upward trend since the late 20th century. In the 1990s, new rootstocks and training systems initiated the shift towards high density cherry production for the first time, resulting in earlier yields, greater returns on investment and orchard profitability. Smaller trees and innovative training systems facilitated the development of 'pedestrian' orchards. No longer was it necessary to harvest trees with ladders that were up to 6 m (20 ft) tall, but rather some orchards provided the potential to harvest all of the fruit from the ground, without ladders. At the same time, sweet cherry breeding programs in Canada, Germany, the USA, and elsewhere around the world significantly increased new cultivar fruit size while maintaining fruit firmness for long-distance export.

1.1 Global Production

Since the mid-1980s, cherry growers in Turkey, the USA, Chile and other countries have expanded production to supply greater fresh market exports to Europe and Asia. In fact, with only a few exceptions (mostly in Europe), sweet cherry production has been expanding around the world, increasing from 1.7 million t in 2000 to over 2.6 million t in 2015, a gain of about 54% (O'Rourke, 2016). Increases in production have been especially dramatic in China and Chile. According to the US Department of Agriculture (USDA), China increased production from 11,000 t in 2000–2002 to 380,000 t in 2017, and Chile increased production from 29,333 t to 206,741 t in the same period. This moved China and Chile to the third and fifth largest producers in the world in 2017 (O'Rourke, 2018). Expansion was not limited solely to these two countries, however. Of the top five global producers in 2001 (Turkey, USA, Iran, Italy and Spain), Turkey and the USA have at least doubled their harvested area since the mid-1980s. All of this expansion is being driven by fresh market demand, as the market for processing cherries has remained essentially static or has even decreased over the same period, mirroring similar trends in other processed fruits.

Although past trends are no guarantee of future direction, all of the leading cherry producing countries have the potential to not only expand production area, but to increase yields through orchard renovation to higher density plantings using dwarfing rootstocks and new training systems that maximize both yield and fruit quality. Furthermore, it is not just the leading countries, but also previously minor countries such as Uzbekistan and Bulgaria, that are expanding production rapidly, using new technology to improve yields and fruit quality.

As sweet cherry production continues to expand worldwide, each producer must decide whether to increase production (either through the renovation of old orchards or expansion into new sites), continue at a steady pace or decrease acreage and thus reduce risk but also yields. Since growing cherries is a high risk venture, the best way to proceed will depend on past successes and failures, perceived risks, the possibilities for mitigating those risks and future market potential. With a few notable exceptions, the fresh market has so far been able to absorb increased production through expansion into new territory and increased buying power in developing markets. A good example of this is China, where imports from both the northern and southern hemisphere have increased exponentially, yet so has new domestic production capacity. As recently as 2000, few consumers beyond China's eastern coast cities had ever seen cherries in their markets. As direct imports increased, mainly from the USA, Australia and Chile, and as the buying power of the Chinese middle-class consumer has grown, the Chinese market has absorbed an ever increasing amount of cherries. Future market stability will depend not only on a continued expansion of the Chinese market, but also in strengthening currently weak markets (such as India) and other developing economies. Since cherries are perceived and marketed as a luxury item, most sales are made to middle- and upper-class consumers, so continued growth in these markets is vital to the overall success of the worldwide cherry industry as well as to individual producers.

1.2   Fresh Market Sweet Cherry Production – yield, fruit quality and economics

Expanding acreage is not always the best strategy for every grower. Many producers believe that bigger operations are always better, but economies of scale with regards to equipment investment, labor availability and matching supply to demand are critical components of profitability. Producing more cherries during certain times of the season may actually depress prices, leading to a smaller per unit return, and requiring labor when availability is at its most competitive (and thus costly). US growers often farm relatively large cherry orchards of 40 to 80 ha (100 to 200 acres) or more. However, it is not uncommon for European producers to farm only 5 to 10 ha (12 to 25 acres) of cherries. The average price received by US growers for a unit of cherries often is less than that received by many European producers. In this case, finding a unique market niche, protecting the crop from inclement weather to increase yields per area, or increasing fruit size to take advantage of premium prices, may be a better strategy than expanding orchard size.

Similarly, ideas of expansion of production capacity must be weighed against potential impacts on fruit quality. Producing the largest, highest quality fruit possible may increase profitability in the long run for producers since larger fruit is often more efficient to pick, tends to ship better, arrives better, holds up longer on the market shelf and receives a higher price than smaller fruit of the same cultivar. Most packing houses can sell larger fruit for a premium. As production has increased, many packers have stopped marketing smaller cherries for fresh consumption. It is now rare to see fresh market packages that include 21 mm (12-row) cherries, and more common to see fruit that averages 28 mm or larger (9.5-row and larger). Some training systems, such as the super slender axe (SSA), produce a high proportion of fruit at the base of one-year-old shoots, which is typically larger, firmer and sweeter than fruit produced on spurs (on two-year-old and older branches). Most training systems, however, produce a greater proportion of the crop on spurs, resulting in higher yields per tree but smaller average fruit size. As will be discussed in detail in later chapters, whatever orchard production system is chosen, fruiting site renewal has become a critical part of orchard management, to maintain a balance of new shoots and young (relative to older) spurs, which helps maintain the production of higher quality fruit throughout the life of the orchard.

1.3   Organic Sweet Cherry Production

Growers not only have choices regarding rootstock, variety and canopy training, but also general orchard management, including codified management systems associated with a specific identity for the buyer or end user. Many 'sustainable management' programs delineate specific practices or tools that can or cannot be used. These typically relate to environmental impact, human health impact and treatment of workers. As more institutional buyers, such as supermarket chains, develop their own sustainable management programs, their suppliers (i.e. growers) are required to also show compliance with specific standards or principles. The resulting 'product attributes' may be communicated to the end consumer as labels signifying 'pesticide free', 'GMO free' or 'organic', or may simply be tracked internally as part of sustainability metrics or future liability protection. 'Certified organic' is one of the more widely used codified management systems for fruit crops, and the production area of sweet cherries managed organically has been increasing globally.

Organic production is legally regulated in many countries, and sales of food products labeled 'organic' are highly regulated in the two dominant markets, the European Union and the USA. Therefore, commercial growers considering organic production would need to undertake a formal organic certification process. Many resources (e.g. the USDA National Organic Program, www.ams.usda.gov/about-ams/programs-offices/national-organic-program, accessed 19 June 2020) are available online to help growers understand this process and determine whether it might be a good fit.

The global area of organic (certified and transition) sweet cherries was estimated in 2015 to be 13,282 ha (32,806 acres), which does not include China, a major country for organic tree fruit production (Willer and Lernoud, 2017). Figure 1.1 illustrates the changing production area

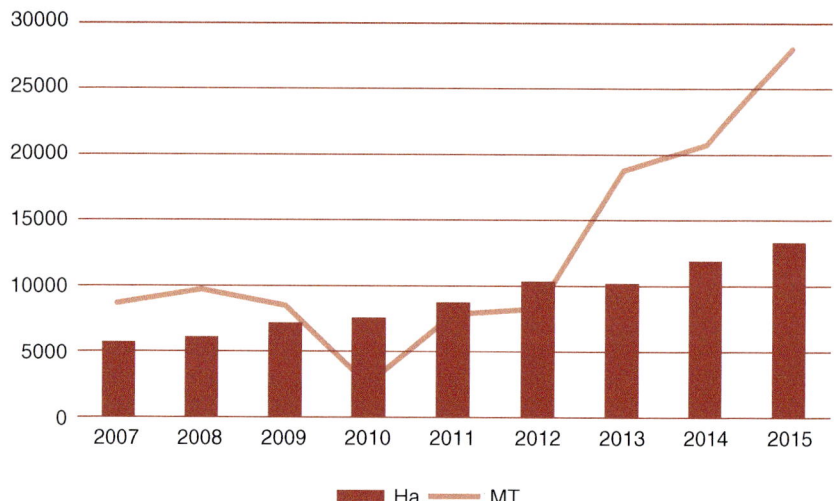

Fig. 1.1. Global organic sweet cherry production area (hectares, ha) and volume (tons, t) (Kirby and Granatstein, 2017; Willer and Lernoud, 2017).

Table 1.1. Major producing countries for organic sweet cherries in 2015, with the top three ranked (in parentheses) by area, production, and estimated yield per orchard area (Kirby and Granatstein, 2017; Willer and Lernoud, 2017).

	Area (ha)	Production (MT)	Yield (MT/ha)
Turkey	3165 (1)	6832 (2)	2.16
Italy	2776 (2)	6035 (3)	2.17
Bulgaria	1618 (3)	879	0.54
USA	1082	8714 (1)	8.05 (1)
Poland	1041	328	0.32
Spain	449	1468	3.27 (2)
Hungary	491	1228	2.50 (3)

(hectares) and estimated global production (tons) of organic sweet cherries over the past decade. When compared with data from the Food and Agriculture Organization (FAO) for all cherries, organic production comprised ~3% of global area and 0.8% of global volume. Organic production area increased slightly from 2012 to 2015 while tonnage increased dramatically. Turkey and Italy had the greatest area of organic production in 2015, while the USA produced the most organic fruit (Table 1.1). A rough estimate of orchard yields (dividing production by area) suggests that organic cherry yields in the USA far surpassed other countries, although these values do not account for the amount of area in transition or new plantings that were not yet producing an organic crop.

Organic cherry prices in Washington State, the leading US producer, were 55–65% higher than conventional prices for 2014–2016, and the year to year patterns have been similar since 2008. Depending on production region, major challenges for organic producers include spotted wing *Drosophila* (SWD), black cherry aphid, plum curculio, cherry powdery mildew, brown rot, bacterial canker and cherry viruses. The recent advances in genetics, biocontrols and horticulture all apply to organic systems. Reliance on one primary organic control option for SWD (i.e. spinosad) is an ongoing concern and research focus.

1.4 Future Production Issues and Innovations

Sweet cherry production, from planting through growing, harvesting and packing, is extremely labor intensive. In fact, labor is the single most expensive input for most cherry growers around the world. Reducing labor costs can increase potential grower profits. However, the labor issue is not just a matter of more profitable production; labor in most production areas of the world is becoming more difficult to source. Profitability and the concern for a sustainable, well-trained labor force are the impetus for both recent production innovations and the research behind future technologies. Dwarfing rootstocks, high density plantings, pedestrian orchards and simple intuitive training systems are all examples of new techniques that have been adopted by growers in recent years to respond to these needs. The future promises even greater efficiencies as scientists and growers work together to mechanize such labor-intensive operations as pruning, harvest and delivery of foliar pesticide, nutrient and growth regulator applications.

Several canopy training systems have the potential for at least partially mechanized pruning. Trees trained to the Kym Green bush (KGB) can be topped at a desired height with a sickle bar or rotary blade hedger in late summer or early fall, reducing some of the labor needed for summer pruning, but dormant pruning requires hand labor. Other systems, however, are more conducive to mechanical pruning and additional labor savings may be possible. Narrow fruiting wall orchards using trees trained to the upright fruiting offshoots (UFO), SSA, bi-axis SSA or espalier (ESP) canopy architectures can be mechanically hedged for both summer and dormant pruning, with some follow-up hand labor required to complete dormant pruning. Even spindle type canopy architectures such as the tall spindle axe (TSA) can be mechanically pruned in summer or winter, with follow-up hand pruning. For most uses of hedging, shoots and branches are cut mechanically at a given distance from the tree axis in the tractor alley and then finished with hand pruning in the row between the trees. This can reduce labor inputs by at least half. Such trees are usually pruned in this way for one or two years, and then fully hand pruned for one year before returning to the hedging cycle.

Another approach to reducing input costs has been the study of solid set canopy delivery (SSCD) spray systems to apply pesticides and growth regulators through microsprinklers, located above or in the canopy, rather than tractor-based airblast sprayers. In theory, SSCD systems can reduce application time, labor and energy consumption, as well as reduce soil compaction and spray drift, and improve worker safety. However, SSCD systems have very high installation costs, relatively low operating pressures that affect spray droplet size and distribution, and to date, spray coverage for sweet cherries becomes increasingly inadequate as the season progresses due to the growth of new shoots and leaves that can block uniform spray penetration into the canopy.

Trees on highly productive rootstocks, such as Gisela 3, 5, 6 and 12, and occasionally even mature trees on Mazzard and Colt, have the potential to set excessive crops relative to the available leaf area for supporting fruit growth. This imbalance in fruit load and leaf area can result in not only smaller fruit size for that season's crop, but also less shoot growth and smaller shoot leaves that season, as well as lingering effects for the next year manifested as smaller spur leaves and more flower buds. Thus, excessive cropping one season creates a negative cycle that can increase the severity of imbalance in subsequent seasons. In parts of the world where labor is relatively inexpensive, workers may manually remove flowers or young fruit by hand to rebalance the leaf area-to-fruit ratio. However, hand thinning is becoming cost-prohibitive for most growers. Chemical thinning of blossoms or fruitlets using caustic agents such as ammonium thiosulfate, fish oil and lime sulfur, and vegetable oil emulsion have been researched, but in these cases, results have been too variable to be reliable and larger fruit have not always resulted from such treatments. It is clear that additional thinning research needs to be conducted before commercial recommendations can be made. At present, the best technique available for balancing crop loads with leaf area is through annual pruning to remove a portion of the developing flower buds and to stimulate additional leaf area through new shoot formation, as will be discussed in later chapters.

The partial or full automation of harvest is also a research area of high interest, but not yet commercially viable. At present, all mechanized harvest techniques require a stem free fruit. Consumer surveys, as well as marketing trials, have shown that stem free cherries are not an obstacle to consumer sales. To avoid the use of abscission-inducing chemicals (such as ethephon) to reduce the pedicel–fruit retention force (**PFRF**) necessary for stem free harvest, cultivars with a genetically low **PFRF**, such as Selah or Skeena, must be selected. Due to the sweet cherry's bearing habit – clusters of relatively small fruit (compared to individual apples or peaches) – and large leaves that can obscure the identification of cherries via machine vision, fully automated (robotic) harvest technologies are likely far into the future. Human-operated 'shake and catch' self-propelled harvesters designed in tandem with specialized planar tree canopies are at advanced stages of testing, and have shown greater harvest labor efficiencies compared to a full crew of human pickers. Hand-held harvest assist shake and catch equipment also has been studied; this requires more pickers shaking individual limbs of purpose designed tree canopies compared to self-propelled machine harvesters, but it greatly improves picker efficiency and requires a lower investment in specialized harvest equipment. Training systems should be chosen with specific labor and mechanical efficiencies in mind. The upright fruiting offshoots in a Y- or V-trellis (UFO-Y or UFO-V) canopy architecture is ideal for mechanical shake and catch harvesters since cherries can drop freely from branches grown as a single plane at an inclined angle to the catch basin without contacting other branches that might cause impact damage. Hand-held harvest assist limb shakers have been used successfully with several canopy architectures, including UFO, KGB and steep leader.

References

Kirby, E. and Granatstein, D. (2017) Recent trends in certified organic tree fruit. Washington State University, Wenatchee, WA. Available at: www.tfrec.cahnrs. wsu.edu/organicag/wp-content/uploads/sites/9/2017/03/Org_Tree_Fruit_ Trends_2016.pdf (accessed 19 June 2020).

O'Rourke, D. (2016) *World Sweet Cherry Review.* Belrose, Inc., Pullman, WA.

O'Rourke, D. (2018) *World Sweet Cherry Review.* Belrose, Inc., Pullman, WA.

Willer, H. and Lernoud, J.(eds.) (2017) The world of organic agriculture. Statistics and emerging trends 2017. Research Institute of Organic Agriculture (FiBL), Frick, and IFOAM – Organics International, Bonn. Available at: www.shop.fibl. org/CHen/mwdownloads/download/link/id/785/?ref=1 (accessed 19 June 2020).

Cherry Flowering, Fruiting and Cultivars

2

2.1 Flowering and Fruiting Habit

Cherry trees produce only simple buds, i.e. one (solitary) bud per node, and each bud is either vegetative or floral. In spring, vegetative buds on one-year-old shoots either elongate into new shoots or develop a whorl of 5–8 leaves to become non-fruiting spurs. In subsequent years, these spurs may elongate into new shoots, or develop another whorl of 5–8 leaves with flowers (fruiting spurs) or without (non-fruiting spurs) (Fig. 2.1). In other words, fruiting spurs are only formed on wood that is two years of age or older. Spurs can remain productive for up to ten years if light interception is adequate. Floral buds may contain up to five flowers per bud. Floral buds develop either from an axillary (lateral) bud at the base of one-year-old shoots, or from an axillary bud of a spur leaf. When axillary buds at the base of one-year-old shoots become floral, the subsequent flowers will either abscise (if not pollinated successfully) or produce fruit (if pollinated and the ovary fertilized) (Fig. 2.2). In either case, this will result in a blind node in subsequent years. Where numerous axillary buds at the base of one-year-old shoots become floral, the entire basal section will ultimately exhibit blind wood that produces neither leaves nor additional fruit. Cultivars can vary significantly in the extent of their flowering spur or basal shoot flower formation. Some cultivars, such as Tieton, Lapins and Sweetheart, have a greater propensity for the development of significant portions of shoot 'blind wood' than other cultivars.

Many of the new precocious rootstocks (Chapter 3) modulate fruiting habit, resulting in increased flower density and greater numbers of spurs or basal shoot flowers, the latter of which can result in increased blind wood.

Fig. 2.1. This fruiting spur contains four floral buds and an elongated vegetative bud located in the center. Each floral bud contains 2–5 flower primordia (courtesy of C. Kaiser).

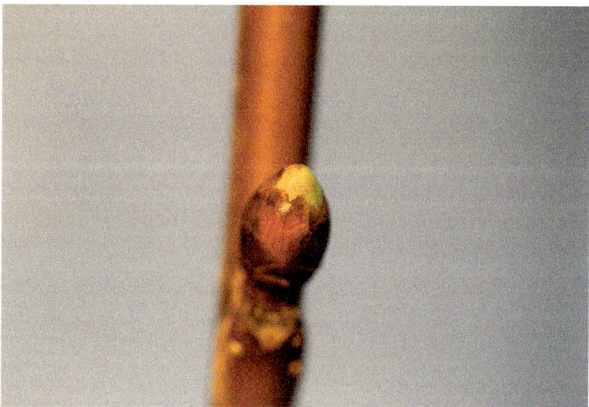

Fig. 2.2. This single floral bud is borne at the base of one-year-old wood (courtesy of C. Kaiser).

2.2 Cultivar Pollen Compatibility, Pollination and Fertilization

Sweet cherry cultivars differ in productivity, which is partly due to genetic factors (such as number of flower buds, viable flower longevity and compatibility of the pollen parent [known as the pollinizer]) and partly due

Fig. 2.3. European honeybees are the most common pollinators for sweet cherries, but bumblebees and native solitary bees can be more efficient in some severe climatic conditions such as cool temperatures and moderate winds (courtesy of G.A. Lang).

to conditions for successful pollination (such as weather during bloom, pollinator [bee] activity and bloom overlap with compatible pollinizers). Pollination is the transfer of pollen grains from the male reproductive parts of the pollinizer cultivar flower (known as the anthers) to the receptive surface (known as the stigma) of the female reproductive part of the fruiting cultivar flower (known as the pistil). Unlike many forest trees with nondescript flowers (like oaks and conifers) that have dry, light pollen for pollination by the wind, the showy flowers of sweet cherries produce relatively heavy, sticky pollen most suitable for insect pollination. Sweet cherry flowers also produce nectar, which helps to attract insects.

The vector that transfers the pollen grains from the pollinizer cultivar to the fruiting cultivar is known as the pollinator. In cherries, the most common pollinators are European honeybees, bumblebees, solitary bees, and to a lesser extent, non-bee insects like syrphid flies (Fig. 2.3). Once pollination has occurred, the pollen grains germinate on the stigma and the emerging pollen tubes must grow down the style to the ovary, and release the male gamete (sperm cell) that then fertilizes the female gamete (ovule) to produce an embryo, thus completing fertilization. The ovary then develops into a fleshy fruit and the embryo develops into a seed.

There are many obstacles to successful pollination, fertilization and fruit development, including asynchronous flowering between the fruiting cultivar and the pollinizer cultivar, insufficient pollen, low pollen viability, insufficient pollinator activity, pollen incompatibility, poor stigma receptivity, ovule degeneration before fertilization can take place and fruit abortion due to various types of stress.

Pollen viability may be affected by a number of factors. First, temperatures during winter may adversely affect pollen viability. Low temperature injury of pollen grains depends on their stage of development and the cold acclimation status of the buds. Critical temperatures at which buds are injured depends on the cultivar, the precise temperature (cold severity) as well as the length of exposure to that temperature. When buds begin acclimation, they typically achieve a baseline hardiness of about $-21°C$ ($-6°F$). If temperatures remain below freezing, bud hardiness improves by about $2.5°C$ ($4°F$) per day, reaching a potential maximum hardiness at about $-34°C$ ($-30°F$). Once temperatures start to warm above $0°C$ ($32°F$), buds begin de-acclimating and cold hardiness may be quickly lost. Indeed, opening buds can be killed when subjected to 6 h at $-8°C$ ($17.6°F$), whereas exposure to $-2°C$ ($28°F$) for 24 h may not cause any damage.

Pollen viability is governed by both nutrient status and carbohydrate reserves. Boron, in particular, plays a major role in pollen germination and pollen tube growth, and close attention should be paid to this element during tissue nutrient analysis. Stigma receptivity also affects fruit set. If the stigmatic surface is not sticky and wet, pollen grains will not adhere to it. Papillate cells, which secrete the sticky solution, are usually only viable for 1 to 2 days, so pollination must take place during that time. Low temperatures or strong winds during the period of stigma receptivity can hinder bee activity. Sites with strong winds should be avoided or ameliorated with suitable windbreaks.

The effective pollination period (EPP) is the difference between the period of time required for pollen tube growth to the ovule and that of ovule longevity (i.e. when the ovule is no longer receptive). The longer the EPP, the greater the chance of adequate fertilization and seed development. Studies of sweet cherry trees have found that the EPP may be as long as 9 days, depending on the cultivar, but the EPP is reduced markedly when temperatures are $>22°C$ ($>72°F$). The growth of the pollen tube and eventual fertilization of the embryo is largely dependent on temperature and its relationship to the EPP. Cool temperatures ($<13°C$ [$<55°F$]) for protracted periods of time will result in slow pollen tube growth and carbohydrate reserves in the pollen grain may be exhausted due to respiration before the pollen tube reaches the ovary sac, resulting in no fertilization.

Conversely, hot temperatures $>25°C$ ($>77°F$) during spring can result in carbohydrate reserves of the pollen grain being exhausted before fertilization occurs. High humidity ($>75\%$) at high temperatures also tends to reduce pollen tube longevity. Even if ovules are fertilized, there is still

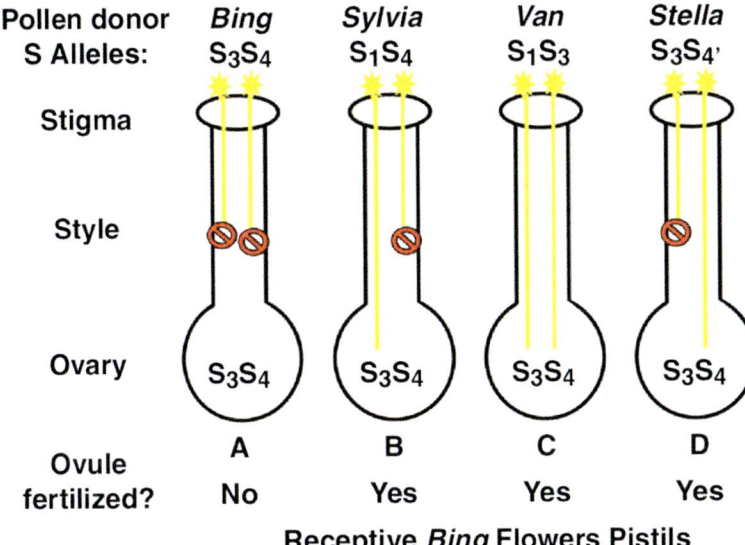

Pollen donor S Alleles:	**Bing** S_3S_4	**Sylvia** S_1S_4	**Van** S_1S_3	**Stella** $S_3S_{4'}$
Stigma				
Style				
Ovary	S_3S_4	S_3S_4	S_3S_4	S_3S_4
Ovule fertilized?	**A** **No**	**B** **Yes**	**C** **Yes**	**D** **Yes**

Receptive *Bing* Flowers Pistils

Fig. 2.4. Sweet cherry pollen compatibility. Bing flowers pollinated with pollen grains from four cultivars: Bing, Sylvia, Van and Stella. (A) Since Bing is self-incompatible, it cannot be fertilized with pollen from another Bing flower; at least one pollen grain that germinates on the stigma must have an allele that differs from S_3 or S_4 for successful fertilization. (B) Sylvia pollen grains with the S_4 allele will not fertilize Bing, but pollen grains with the S_1 allele are compatible. (C) Both alleles of Van are different from, and therefore compatible with, those of Bing for successful fertilization. (D) Pollen grains with the $S_{4'}$ allele do not elicit an incompatibility reaction regardless of the S alleles of the pollinated flower, allowing 'self-compatible' cultivars like Stella to serve as a universal pollen donor to successfully fertilize the ovules of all cultivars as well as themselves (courtesy of G.A. Lang, based on a figure from N. Ibuki, Summerland Variety Corporation).

a chance that embryo abortion may take place due to other physiological conditions, e.g. lack of water, lack of nutrients or available carbohydrate reserves, or high temperatures.

Most cherry cultivars, including such commercial standards as Bing and Regina, are self-incompatible and require cross-pollination to set fruit. Pollen compatibility in each sweet cherry cultivar is governed by two S alleles (a small part of a gene), which are designated S_1, S_2, S_3... S_{38}. Flower pistils contain each S allele in the pair, while pollen grains contain one or the other S allele (Fig. 2.4). The paired combinations of S alleles specific to each cultivar were originally classified into pollen incompatibility groups I, II, III, etc., but now that more than three dozen S alleles have been identified, it is much easier to compare specific S allele pairs than attempt to keep track of the potential combinations as incompatibility groups (currently up to 60). Common examples of S allele pairs are Bing

(S_3S_4), Regina (S_1S_3) and Van (S_1S_3). In these cultivars, the S_3 allele is common, so none of the S_3 pollen grains, regardless of pollen cultivar, will grow down the styles of pollinated flowers of any of these cultivars. Regina and Van also share the S_1 allele, meaning they are completely incompatible for pollination of each other. However, Bing pollen grains with the S_4 allele can successfully pollinate and fertilize Regina and Van flowers, and pollen grains from those cultivars with the S_1 allele can successfully pollinate and fertilize Bing flowers. So, when pollen cultivars and fruiting cultivars share no S alleles, 100% of the pollen is compatible; when one allele is shared, 50% of the pollen is compatible; and when both alleles are shared, none of the pollen is compatible (the cultivars are incompatible).

In 1954, X-rays were used to cause pollen mutations for the S_3 and S_4 alleles that eliminated their incompatible reaction, resulting in the new self-compatible 'prime' S alleles of $S_{3'}$ and $S_{4'}$ (Lewis and Crowe, 1954). Pollen with either of these prime S alleles can successfully pollinate and fertilize the flowers of cultivars with any S allele pair, including even the pollen cultivar itself (Fig. 2.4). This has led to the breeding of many new self-compatible (or 'self-fertile') cultivars, primarily with the $S_{4'}$ allele, such as Lapins, Skeena, Santina $(S_1S_{4'})$, Sir Hans $(S_2S_{4'})$, Sweetheart, Staccato, Sonata $(S_3S_{4'})$, BlackGold (S_4S_6), Benton, Early Star, Grace Star (S_4S_9), and Sir Don (S_4S_{13}). Recent genetic studies have also identified several additional sources of self-compatibility, including an $S_{5'}$ allele and natural mutations in a few unusual cultivars, such as Cristobalina (S_3S_6).

Beyond pollen compatibility, it is also important to consider the time of flowering when matching pollinizer cultivars with fruiting cultivars. Both pollen S alleles and relative bloom timing are listed in Table 2.1 for over 130 standard and new cultivars. Depending on climate, flowering of various cultivars may extend over a period of several weeks, and in temperate regions, may constitute up to five flowering periods (early, early–mid, mid-season, mid–late, late). Every cultivar requires a distinct amount of time at chilling temperatures during winter to complete their dormant period; that is, with insufficient chilling, warm temperatures will not cause flowering, or flowering and growth will be weak and abnormal. Once adequate chilling has occurred to complete dormancy, every cultivar requires a distinct amount of time at warm temperatures to complete the final stages of flower bud development and bloom. Thus, fruiting cultivars and pollinizer cultivars with relatively similar bloom periods, as well as chilling and warmth requirements, will tend to have good bloom overlap year after year. Some cultivars may exhibit a good bloom overlap in some years, but not in others, indicating that they likely have different chilling or warmth requirements that shift their relative bloom periods due to year-to-year differences in late winter or early spring temperatures.

Regions where winter chilling is sometimes insufficient (such as warm winter regions in Australia, California, Chile, South Africa, Spain, etc.) have a higher risk of asynchronous or weak flowering and bloom overlap.

Table 2.1. Relative bloom timing and pollen S alleles for 130+ standard and new international sweet cherry cultivars. Bloom timing is a compilation of relative data from several growing regions and may vary somewhat across different climates. The use of hydrogen cyanamide and other dormancy-breaking agents may alter relative bloom times.

S alleles	Bloom timing				
	Early	Early–mid	Mid	Mid–late	Late
S_1S_2		Black Tartarian, Starking Hardy Giant	Tulare	Canada Giant, Summit	
S_1S_3		Samba, Royal Lee, Vera	Black Star, Coral Champagne, Early Robin, Prime Giant (Giant Red or Giant Ruby), Satin, Van	Areko (Hamid), Cristalina, Lala Star, Oktavia, Olympus, Royal Ansel (Royal Bailey), Sonnet	Regina
S_1S_4	Royal Lynn		Bada, Black Republican, Black York, Ebony Pearl, Garnet, Rainier, Sweet Gabriel	Sylvia	Hudson
S_1S_6	Cheery Blush	BlushingGold, Cheery Glow, Henriette, Hertford, Nugent		Vanda	
S_1S_9		Bellise, Earlise (Early Lori), Sweet Early	Brooks, Rocket	Tamara	
S_1S_{13}		Cheery Treat		Giorgia, Radiance Pearl	
$S_1S_{undocumented}$					
S_2S_3	Nimba	Cavalier		Vega	

Continued

Table 2.1. Continued

S alleles	Bloom timing				
	Early	Early–mid	Mid	Mid–late	Late
S_2S_4		Ferprime (Primulat)	Royalton	Suite Note	Sam
S_2S_9					
S_3S_4	Somerset	Royal Dawn, Sweet Lorenz	Bing, Burgundy Pearl, Emperor Francis, Kristin, Napoleon (Royal Ann), Sweet Valina, Ulster	Karina, Lambert	
S_3S_5	Cheery Crunch	Hartland	Emma	Andersen, Kordia (Attika), Starks Gold, Techlovan	Hedelfingen
S_3S_6					Duroni 3, Ferdiva, Fertard, Gold
S_3S_9		Burlat	Chelan, Tieton	Walter	Linda
S_3S_{12}				0900 Ziraat, Ferrovia, Germersdorfi, Noir de Meched, Nordwunder, Schneiders	
S_3S_{14}		Durone di Verona			
S_4S_5	Royal Hazel, Royal Tenaya (Royal Marie)				
S_4S_6				Carmen	Irena

Continued

Table 2.1. Continued

S alleles	Bloom timing				
	Early	Early–mid	Mid	Mid–late	Late
S_4S_9		Merchant	Cheery Moon	Kiona	
S_4S_{12}			Black Pearl	Katalin	
S_4S_{13}	Rita				
S_5S_{22}					
S_6S_9		Folfer	Cheery Grand		Penny
S_6S_{13}		Durona di Vignola			
$S_{9}S_{9\ undocumented}$		Cheery Burst			
Self-compatible					
$S_1S_{3'}$			Alex (Axel)		
$S_3S_{3'}$					
$S_1S_{4'}$	Royal Tioga	Frisco	Lapins, Royal Edie, Royal Helen, Santina, Symphony	Celeste, Skeena, Stardust	
$S_3S_{4'}$	Sweet Aryana	Marysa	Index, Selah, Sentennial, Staccato, Stella, Sweetheart	Sandra Rose, Sonata, Sovereign, Sunburst	
S_4S_6	Pacific Red		Blaze Star		
S_4S_9			Early Star, Grace Star	Benton, Cashmere, Glacier	
$S_4S_{unknown}$		Sabrina	Big Star	WhiteGold	BlackGold

Under these conditions, the use of dormancy-breaking growth regulators such as hydrogen cyanamide (Dormex®) or nitrogen-based products (Erger), have been effective in promoting the transition from dormancy to growth and improving synchronous flowering. Another alternative is to operate orchard covers to provide shade on sunny winter days or use over-tree sprinklers to evaporatively cool trees on such days, thereby improving the winter chilling needed to naturally alleviate dormancy. However, judicious selection of cultivars with low chilling requirements is the best solution for such locations. As sweet cherry production has expanded over the past 20 years, efforts to breed low chilling requirement cultivars has increased. Some standard cultivars, such as Van and cultivars with Van in their pedigrees, such as Lapins, Rainier, Brooks and Coral Champagne, have performed relatively well in moderately low chilling regions.

2.3   Major Sweet Cherry Cultivars and their Characteristics

Breeding and selection of improved sweet cherry cultivars has kept pace in recent decades with the expansion of sweet cherry production around the world. Longstanding breeding programs in North America (particularly in Summerland, Canada, Cornell University in New York, and Washington State University at Prosser) and throughout Europe (e.g. Czech Republic, France, Germany, Hungary, Italy, Spain, Ukraine, UK) have been joined by new breeding programs in Turkey and Chile, as well as breeding programs focused on developing low chilling varieties (e.g. International Fruit Genetics (IFG), and Zaiger Genetics in California). Surprisingly, other than the breeding of cultivars specifically adapted to particular growing conditions (such as low chilling regions), many of the world's major cultivars grow well in a wide variety of locations (e.g. Lapins). On the other hand, the most planted variety in the USA, Bing, was a chance seedling that was selected in Oregon in the 1800s and does well in the Pacific Northwest (PNW) of the USA, but grows poorly in many other cherry production regions due to sensitivity to rain-induced cracking, cold damage and bacterial canker.

Fifty-nine major and promising new fresh market sweet cherry cultivars are discussed briefly below; there are at least 100 more from around the world that could be discussed (see Bargioni, 1996 and Quero-Garcia *et al.*, 2017 in the References). Many cultivar traits can vary depending on climate, rootstock, training system and orchard management, so the specific traits noted are relative, primarily based on evaluations in the PNW. Harvest timing is relative to Bing, which is considered to be a typical mid-season cultivar. Suggestions for appropriate rootstocks are generally related inversely to cultivar productivity, i.e. high yielding cultivars are usually matched with moderately productive rootstocks, and low yielding

Table 2.2. Rootstock effects on sweet cherry productivity; these should be taken into account when matching the productivity of specific cultivars.

Low productivity	Moderate productivity	High productivity	Very high productivity
Colt	CAB 6P	Gisela 5	Cass
Mahaleb	Krymsk 5	Gisela 6	Clare
Mazzard	Krymsk 6	Gisela 12	Clinton
	MaxMa 14	Gisela 13	Crawford
	MaxMa 60	Weigi 1	Gisela 3
	Gisela 17	Weigi 2	Lake
		Weigi 3	

cultivars are usually matched with highly productive rootstocks. The effect of rootstock on productivity is generalized in Table 2.2. Relative fruit flesh firmness categories are based on available force displacement (g/mm) measurements (FirmTech, Bioworks, Kansas), with a value of 275 g/mm being the minimum for US export fruit quality: marginal (<250), good (250–299), excellent (300–349), extraordinary (≥350). Most of the fruit firmness rankings listed are based on the commercial use of preharvest gibberellic acid application, which enhances firmness and size.

2.3.1 Key attributes of sweet cherry cultivars for fresh markets

Most consumers prefer large, flavorful, sweet, firm cherries. For the cherry grower, fruit size and firmness are two of the most important attributes. The most obvious reason for this is that growers frequently receive a premium for larger fruit, since consumers generally prefer larger fruit and will pay more for it. In addition, there is a relationship between the largest fruit of a given variety and fruit firmness. The firmest fruit will have the highest resistance to pitting (small sunken depressions on the fruit surface that appear during storage) and typically perform best during the severe conditions of shipping. For these reasons, packing houses prefer to export the largest fruit. In most cases, exported fruit receives higher returns than fruit sold to the domestic market. This is true for fruit grown in most countries with an established export market.

Cherry firmness varies due to both orchard management techniques (which will be discussed in later chapters) and cultivar genetics. Some cultivars are naturally more resistant to pitting and other shipping disorders than others. For example, Bing is a cherry of moderate firmness, but is usually more resistant to pitting than Sweetheart, which is considered to be a firm cherry. However, the firmest Bing or Sweetheart fruit will be more resistant to shipping disorders than a less firm cherry of the same cultivar.

Similar to firmness, cherry flavor is a function of both the genetics of the cultivar and orchard management techniques. For example, a fruit grown with a good balance between the number of leaves and fruit will be sweeter, more flavorful and firmer than a cherry that is grown on a tree with too many fruit for the leaf area to fully support. Correct balance of the leaf to fruit ratio will also produce larger fruit and should therefore be an important goal for all cherry growers, regardless of cultivar.

Harvest timing is of critical importance, as it has a direct effect on a number of fruit quality attributes, including fruit colour, size, firmness, flavor, pitting and shelf life. Most growers outside the USA will selectively harvest an orchard block multiple times to make sure that only fruit of optimum maturity is harvested. Skin color is the most convenient indicator of ripeness since most quality attributes are directly related to skin color. Generally, as fruit persist on the tree, skin and flesh color darken while size and sugar levels increase. In addition, firmness and stem retention force decrease. The goal for cherry growers is to find the optimal balance of these quality attributes. For most dark-fleshed cultivars, this is a light mahogany to mahogany skin color. However, there are some cultivars, such as Regina, that are naturally darker in color when mature. These fruit are generally best harvested when skin color reaches mahogany to dark mahogany.

2.3.2   Red-fleshed sweet cherries for the fresh market (listed in approximate order of ripening)

2.3.2.1   *Cheery Crunch*

Harvest timing: Very early season (14–17 days before Bing)
Color when ripe: Mahogany skin and flesh
Bloom time: Early season
Pollen S alleles: S_3S_6
Suggested rootstock productivity: Low to moderate
Average fruit size: 24 to 25 mm (11- to 10.5-row)
Average firmness: Extraordinary
Fruit-cracking susceptibility: High

Cheery Crunch (Fig. 2.5) has an extremely low chilling requirement (~150 h). For this reason, it has the potential to open up entirely new production areas, where it has been impossible to grow cherries before. Fruit size is smaller than most fresh market cherries, but cherries that ripen for early market niches can be successful without large size. Fruit size is larger on upright fruiting wood, so upright fruiting offshoot (UFO) and Kym Green bush (KGB) systems may be the best choice to maximize fruit size. Since the tree is very productive, aggressive pruning and crop thinning may be necessary in some years. Fruit flavor is excellent.

Fig. 2.5. Cheery Crunch (courtesy of L.E. Long © 2017 International Fruit Genetics, LLC, California, USA).

2.3.2.2   Nimba

Harvest timing: Very early season (14–17 days before Bing or Van)
Color when ripe: Light mahogany skin with red flesh
Bloom time: Early season
Pollen S alleles: S_2S_3
Suggested rootstock productivity: Moderate
Average fruit size: 30 mm (9-row)
Average firmness: Good
Fruit-cracking susceptibility: Moderate

Nimba (Fig. 2.6) is of interest to cherry growers in early production areas mainly due to its early ripening characteristic and large size. The flavor is sweet but rather weak and texture is only moderately firm. The tree is precocious and productive.

2.3.2.3   Sweet Aryana

Harvest timing: Very early season (14 days before Bing)
Color when ripe: Mahogany skin and flesh
Bloom time: Early season
Pollen S alleles: S_3S_4 (self-compatible)
Suggested rootstock productivity: Low to moderate
Average fruit size: 30 mm (9-row)

Fig. 2.6. Nimba (courtesy of L.E. Long).

Average firmness: Excellent
Fruit-cracking susceptibility: Moderate

Sweet Aryana fruit are rounded, cordate and sweet. The trees are vigorous, spreading and highly productive.

2.3.2.4   Cheery Treat

Harvest timing: Very early season (12–14 days before Bing)
Color when ripe: Light mahogany skin and flesh
Bloom time: Early–mid season
Pollen S alleles: $S_1S_{unnamed}$
Suggested rootstock productivity: Moderate to high
Average fruit size: 30 mm (9-row)
Average firmness: Good to excellent
Fruit-cracking susceptibility: Low to moderate

Cheery Treat (Fig. 2.7) is a high quality, very early ripening cherry with large fruit and good flavor. Natural cracking rates in test plots indicate good to moderate rain-cracking resistance. A large percentage of the fruit is borne on annual wood, similar to Regina, so training systems that remove one-year-old wood, such as KGB and UFO, should be avoided. One of Cheery Treat's two alleles is very rare and has never been documented in the literature, so nearly any early–mid blooming variety will work as a pollinizer. Chill level is low at 300–500 chill hours, so it should do well in warmer, early production sites.

Fig. 2.7. Cheery Treat (courtesy of L.E. Long © 2017 International Fruit Genetics, LLC, California, USA).

2.3.2.5   Pacific Red

Harvest timing: Very early season (12–14 days before Bing or Van)
Color when ripe: Red to light mahogany and red flesh
Bloom time: Early season
Pollen S alleles: $S_4 S_9$ (self-compatible)
Suggested rootstock productivity: Moderate
Average fruit size: 28–30 mm (9.5- to 9-row)
Average firmness: Excellent
Fruit-cracking susceptibility: Low to moderate

Pacific Red (Fig. 2.8) is a high quality, early ripening cherry that has the potential for the export market. The flavor is pleasant, but only moderately strong and the flesh is firm. The tree is precocious and productive, semi-vigorous and semi-erect in form.

2.3.2.6   Royal Dawn

Harvest timing: Very early season (10–12 days before Bing)
Color when ripe: Red to light mahogany skin and flesh
Bloom time: Early–mid season
Pollen S alleles: $S_3 S_4$

Fig. 2.8. Pacific Red (courtesy of L.E. Long).

Suggested rootstock productivity: Low to moderate
Average fruit size: 28 mm (9.5-row)
Average firmness: Good
Fruit-cracking susceptibility: High

Due to its early ripening, Royal Dawn (Fig. 2.9) has been successful in early production areas, such as Chile and Spain. In other regions, its success has been limited by fruit storage/shipping for longer than 2 weeks (soft fruit, a dull finish and brown, dehydrated stems) and rain-cracking susceptibility. It is best grown under rain covers due to cracking susceptibility. Flavor and fruit quality are best when harvested at 18 to 20°Brix.

2.3.2.7   Cheery Grand™

Harvest timing: Very early season (10–12 days before Bing)
Color when ripe: Red to light mahogany skin and flesh
Bloom time: Mid-season
Pollen S alleles: S_6S_9
Suggested rootstock productivity: Moderate to high
Average fruit size: 30–32 mm (9- to 8.5-row)
Average firmness: Good
Fruit-cracking susceptibility: Moderate

Fig. 2.9. Royal Dawn (courtesy of L.E. Long).

Cheery Grand fruit is very large (Fig. 2.10), ripens early and ships well to distant markets. It has good flavor, but can be a bit acidic if harvested before it is fully ripe. The tree is open and crops consistently. Gibberellic acid and calcium treatments may be needed to enhance firmness. Pedicels are moderately long and retain their green color during storage.

2.3.2.8 Frisco

Harvest timing: Very early season (10–12 days before Bing)
Color when ripe: Red to light mahogany skin and flesh
Bloom time: Early–mid season
Pollen S alleles: $S_1S_{4'}$ (self-compatible)
Suggested rootstock productivity: Low to moderate
Average fruit size: 28 mm (9.5-row)
Average firmness: Good to excellent
Fruit-cracking susceptibility: Moderate

Frisco fruit has a somewhat kidney shape, but a very nice flavor. The tree is medium to high in vigor and very productive, with an open and readily branching canopy. The pedicel is thick and medium in length.

Fig. 2.10. Cheery Grand (courtesy of L.E. Long © 2017 International Fruit Genetics, LLC, California, USA).

2.3.2.9   Chelan

Harvest timing: Early season (10–12 days before Bing)
Color when ripe: Light mahogany to mahogany skin and flesh
Bloom time: Mid-season
Pollen S alleles: S_3S_9
Suggested rootstock productivity: Low to moderate (may be incompatible with Mahaleb)
Average fruit size: 25 mm (10.5-row)
Average firmness: Excellent
Fruit-cracking susceptibility: Low to moderate

Chelan (Fig. 2.11) is a reasonably good, early season cultivar that is very firm, ships well, and has genetic resistance to powdery mildew. Tree vigor is moderate to low. The flavor is somewhat mild. Oversetting can be a problem on most rootstocks, resulting in smaller than average fruit sizes, so careful crop load management is needed to achieve suitable fruit size. Chelan has been accepted well by commercial buyers in the USA and is widely grown in the PNW region, but growers continue to look for a higher quality variety for this harvest niche. Due to modest fruit size and flavor, there is only limited production outside the PNW.

Fig. 2.11. Chelan (courtesy of C. Denby and Oregon State University).

2.3.2.10 Brooks

Harvest timing: Early season (10 days before Bing)
Color when ripe: Red to light mahogany
Bloom time: Mid-season
Pollen S alleles: S_1S_9
Suggested rootstock productivity: Moderate
Average fruit size: 28 mm (9.5-row)
Average firmness: Excellent
Fruit-cracking susceptibility: High

Brooks (Fig. 2.12) has become the standard for early season cherries in California. Fruit flavor is excellent, and the fruit ships very well to distant markets. It is best grown under rain covers due to cracking susceptibility.

Fig. 2.12. Brooks (courtesy of L.E. Long).

2.3.2.11   *Tulare*

Harvest timing: Early season (10 days before Bing)
Color when ripe: Bright red
Bloom time: Mid-season
Pollen S alleles: S_1S_2
Suggested rootstock productivity: Low to moderate
Average fruit size: 27 mm (10-row)
Average firmness: Good to excellent
Fruit-cracking susceptibility: Moderate

Tulare trees are semi-upright with medium vigor and good productivity.

2.3.2.12   *Royal Hazel*

Harvest timing: Early season (10 days before Bing)
Color when ripe: Light mahogany skin and flesh
Bloom time: Early season
Pollen S alleles: S_4S_6
Suggested rootstock productivity: Low to moderate

Fig. 2.13. Royal Hazel (courtesy of L.E. Long).

Average fruit size: 30 mm (9 row)
Average firmness: Excellent
Fruit-cracking susceptibility: Low to moderate

Royal Hazel (Fig. 2.13) has a moderately low chilling requirement (~500 h), tends to be very productive and sets very heavy crops, even in years with some frost. The flavor and fruit quality are very good, especially for such an early ripening cherry.

2.3.2.13   Tieton

Harvest timing: Early season (6–9 days before Bing)
Color when ripe: Light mahogany to mahogany skin and flesh
Bloom time: Mid-season
Pollen S alleles: S_3S_9
Suggested rootstock productivity: High to very high (may be incompatible with Mahaleb)
Average fruit size: 30 mm (9-row)
Average firmness: Good
Fruit-cracking susceptibility: High

A glossy, mahogany red finish, thick stems and very large fruit make Tieton (Fig. 2.14) an eye-catching cherry. However, it has a propensity for doubling where summer temperatures are high, occasionally causing a soft texture, low productivity and bland flavor. In recent years, acreage in the USA has been decreasing, but Tieton has been one of the most planted new varieties in China.

Fig. 2.14. Tieton (courtesy of C. Denby and Oregon State University).

2.3.2.14 *Blaze Star*

Harvest timing: Early to mid-season (8 days before Bing and Van)
Color when ripe: Mahogany
Bloom time: Mid-season
Pollen S alleles: S_4S_6 (self-compatible)
Suggested rootstock productivity: Low to moderate
Average fruit size: 28 mm (9.5-row)
Average firmness: Marginal to good
Fruit-cracking susceptibility: Low

Blaze Star (Fig. 2.15) fruit flavor is very good with a strong, balanced sweet/acid flavor. Pitting potential is higher than average.

2.3.2.15 *Black Pearl*

Harvest timing: Early season (7 days before Bing)
Color when ripe: Mahogany to very dark mahogany skin and flesh
Bloom time: Mid-season
Pollen S alleles: S_4S_{13}
Suggested rootstock productivity: Low to moderate
Average fruit size: 30 mm (9-row)
Average firmness: Extraordinary
Fruit-cracking susceptibility: Low to moderate

Fig. 2.15. Blaze Star (courtesy of L.E. Long).

Black Pearl (Fig. 2.16) is a large, very firm cherry for the early season market, with a good flavor that is rare for this harvest timing. The tree is rather upright and can be very productive on Gisela rootstocks. Aborted fruit do not drop cleanly from the tree but can be easily sorted in the packing house; ripe fruit can hang on the tree for a long time with no loss of firmness. Fruit pitting incidence is lower than average in laboratory tests.

2.3.2.16   Coral Champagne (aka Coral)

Harvest timing: Early season (7 days before Bing)
Color when ripe: Pink to bright red skin, dark red flesh
Bloom time: Mid-season
Pollen S alleles: S_1S_3
Suggested rootstock productivity: Low to moderate
Average fruit size: 30 mm (9-row)

Fig. 2.16. Black Pearl (courtesy of L.E. Long).

Average firmness: Excellent
Fruit-cracking susceptibility: Moderate

Coral Champagne (Fig. 2.17) fruit are very sweet with an impression of low acid, although still quite flavorful. The skin color is light red. It has proven

Fig. 2.17. Coral Champagne (courtesy of L.E. Long).

Fig. 2.18. Santina (courtesy of C. Denby and Oregon State University).

suitable for lower chill areas, becoming one of the most widely produced cherries in California, but planting is also increasing in the PNW US region. Stem pull force is weak. The tree is precocious and productive.

2.3.2.17   Santina

Harvest timing: Early season (7 days before Bing)
Color when ripe: Mahogany skin, light mahogany flesh
Bloom time: Mid-season
Pollen S alleles: S_1S_4 (self-compatible)
Suggested rootstock productivity: Moderate to high (may be incompatible with CAB 6P)
Average fruit size: 31 mm (8.5-row)
Average firmness: Excellent
Fruit-cracking susceptibility: Moderate to high

Planting of Santina (Fig. 2.18) has increased significantly in Chile, with most of the production exported to China. US production is low, but there is some interest in the PNW. Fruit size is large to very large, but the flavor is mild with low acid, and stylar end fruit cracking can occur even in the absence of rain. The tree form is spreading to pendent.

2.3.2.18 Samba

Harvest timing: Early season (7 days before Bing)
Color when ripe: Mahogany skin, light mahogany flesh
Bloom time: Early season
Pollen S alleles: S_1S_3
Suggested rootstock productivity: High
Average fruit size: 29 mm (9.5-row)
Average firmness: Good
Fruit-cracking susceptibility: Moderate

Samba was released as a mid-season cherry, but some growers in Australia and Oregon have harvested it early, when the skin color is pink to light mahogany, although flavor is impacted. Reports from Europe indicate that when allowed to fully ripen, stem retention can occasionally be problematic. The tree form is upright and spurry.

2.3.2.19 Big Star

Harvest timing: Early to mid-season (5 days before Bing or Van)
Color when ripe: Mahogany
Bloom time: Mid-season
Pollen S alleles: $S_4S_?$ (unknown at this time but self-compatible)
Suggested rootstock productivity: High
Average fruit size: 32 mm (8.5-row)
Average firmness: Good
Fruit-cracking susceptibility: Low

Big Star (Fig. 2.19) fruit has very good flavor with a moderately low pitting potential. Productivity has been low.

2.3.2.20 Suite Note

Harvest timing: Early mid-season (5 days before Bing)
Color when ripe: Mahogany skin, light mahogany flesh
Bloom time: Mid–late season
Pollen S alleles: S_2S_4
Suggested rootstock productivity: Moderate to high
Average fruit size: 31 mm (8.5-row)
Average firmness: Excellent
Fruit-cracking susceptibility: Moderate

Suite Note (Fig. 2.20) plantings have been limited mostly to Oregon in the PNW region. Productivity has been moderate to low, even on Gisela rootstocks. The fruit has good to excellent flavor and stores well.

Fig. 2.19. Big Star (courtesy of L.E. Long).

2.3.2.21 Burgundy Pearl

Harvest timing: Early mid-season (4–5 days before Bing)
Color when ripe: Mahogany to dark mahogany skin and flesh
Bloom time: Mid-season
Pollen S alleles: S_3S_4
Suggested rootstock productivity: Moderate to high
Average fruit size: 31 mm (8.5-row)
Average firmness: Extraordinary
Fruit-cracking susceptibility: Moderate

Burgundy Pearl (Fig. 2.21) fruit have a crunchy firmness and a mild but good flavor. The tree reportedly has some resistance to bacterial canker; in some regions, flower buds have been reported to be more susceptible to cold damage than other cultivars.

2.3.2.22 Benton

Harvest timing: Mid-season (2 days before Bing)
Color when ripe: Mahogany skin and flesh
Bloom time: Mid–late season
Pollen S alleles: S_4S_9 (self-compatible)
Suggested rootstock productivity: High to very high

Fig. 2.20. Suite Note (courtesy of L.E. Long).

Average fruit size: 31 mm (8.5-row)
Average firmness: Excellent
Fruit-cracking susceptibility: Moderate

Benton (Fig. 2.22) fruit have an excellent flavor as well as other high quality attributes. However, in spite of being self-compatible, productivity has been low to moderate even on productive rootstocks, resulting in limited commercial success in the PNW region.

2.3.2.23 *Bing*

Harvest timing: Mid-season
Color when ripe: Light mahogany skin and flesh
Bloom time: Mid-season
Pollen S alleles: S_3S_4
Suggested rootstock productivity: Moderate to high
Average fruit size: 26 to 28 mm (10.5- to 9.5-row)
Average firmness: Good
Fruit-cracking susceptibility: High

Bing (Fig. 2.23) has been the standard cultivar for PNW fresh market producers for more than a century. Excellent flavor, texture and long-term storage capability that allows fruit to be shipped to distant markets have

Fig. 2.21. Burgundy Pearl (courtesy of L.E. Long).

made Bing among the world's leading cherries due to exports from the USA and Chile. However, new cultivars with various improved traits (i.e. larger size, later bloom, less cracking, less susceptibility to bacterial canker) have reduced Bing's market share in recent years.

2.3.2.24 Black Star

Harvest timing: Mid-season (ripens with Bing and Van)
Color when ripe: Dark mahogany to black
Bloom timing: Mid-season
Pollen S alleles: S_1S_3 (some reports indicate self-compatible)
Suggested rootstock productivity: Moderate to high
Average fruit size: 30 mm (9-row)
Average firmness: Good
Fruit-cracking susceptibility: Low to moderate

Fig. 2.22. Benton (courtesy of G.A. Lang).

Fig. 2.23. Bing (courtesy of C. Denby and Oregon State University).

Fig. 2.24. Black Star (courtesy of L.E. Long).

Black Star (Fig. 2.24) fruit have good flavor. Under European conditions, the tree is reported to be very productive, but in the PNW region of the USA on Gisela 6, productivity has been moderate.

2.3.2.25   Ebony Pearl

Harvest timing: Mid-season (with Bing)
Color when ripe: Mahogany to dark mahogany skin and flesh
Bloom time: Mid-season
Pollen S alleles: S_1S_4
Suggested rootstock productivity: High to very high
Average fruit size: 32 mm (8.5-row)
Average firmness: Excellent
Fruit-cracking susceptibility: Low

Ebony Pearl (Fig. 2.25) fruit have an excellent flavor, better size and better resistance to rain cracking than Bing. The tree is open, spreading and reportedly less susceptible to bacterial canker than Bing. Productivity is low, so Ebony Pearl should be grown on highly productive rootstocks.

Fig. 2.25. Ebony Pearl (courtesy L.E. Long).

2.3.2.26   Sandra Rose

Harvest timing: Mid-season (with Bing)
Color when ripe: Mahogany skin and flesh
Bloom time: Mid–late season
Pollen S alleles: $S_3S_{4'}$ (self-compatible)
Suggested rootstock productivity: Moderate
Average fruit size: 32 mm (8.5-row)
Average firmness: Good
Fruit-cracking susceptibility: Moderate

Sandra Rose (Fig. 2.26) fruit have an excellent flavor, and the tree is vigorous, spreading and productive. Consistency in firmness has limited its planting in the PNW region.

2.3.2.27   0900 Ziraat

Harvest timing: Late mid-season (5 days after Bing)
Color when ripe: Mahogany skin, light mahogany flesh
Bloom time: Mid–late season
Pollen S alleles: S_3S_{12}
Suggested rootstock productivity: High

Fig. 2.26. Sandra Rose (courtesy of C. Denby, Oregon State University).

Average fruit size: 30 mm (9-row)
Average firmness: Good
Fruit-cracking susceptibility: Moderate

0900 Ziraat (Fig. 2.27) is the foremost cultivar produced for export by growers in Turkey. Large quantities of 0900 Ziraat are exported to Europe each year. Productivity is moderate, so 0900 Ziraat should be grown on productive rootstocks. The flavor is good. There is some evidence that 0900 Ziraat may be genetically the same as Schneiders Späte Knorpel in Germany, Germersdorfi in Hungary, Ferrovia in Italy, Belge in France, and Noir de Meched in Iran, due to the same rare S allele combination and other similarities.

2.3.2.28   Kordia (Attika)

Harvest timing: Late mid-season (5–7 days after Bing)
Color when ripe: Mahogany skin and flesh
Bloom time: Mid–late season
Pollen S alleles: S_3S_6
Suggested rootstock productivity: Moderate to high
Average fruit size: 30 mm (9-row)
Average firmness: Excellent
Fruit-cracking susceptibility: Moderate

Fig. 2.27. 0900 Ziraat (courtesy of C. Denby and Oregon State University).

Kordia (known as Attika in the USA) (Fig. 2.28) is a high quality cherry that is widely planted in Europe due to its excellent flavor and ability to serve as a pollinizer for Regina. From both Chile and Oregon, it has the reputation of arriving in distant markets in excellent condition due to its low susceptibility to impact damage. In the PNW region and Europe, yields are adequate, but in areas with low winter chill accumulation, such as Chile, fruit set can be poor. Kordia reportedly has a higher than normal susceptibility to late season frosts, but objective data has yet to confirm this trait.

2.3.2.29   Oktavia

Harvest timing: Mid- to late season (5–7 days after Bing or Van)
Color when ripe: Mahogany
Bloom timing: Mid–late season

Fig. 2.28. Kordia (Attika) (courtesy of C. Denby and Oregon State University).

Pollen S alleles: S_1S_3
Suggested rootstock productivity: Moderate
Average fruit size: 26 to 29 mm (10- to 9.5-row)
Average firmness: Good to excellent
Fruit-cracking susceptibility: Moderate

Oktavia is a spreading tree of moderate vigor and good productivity.

2.3.2.30   *Tamara*

Harvest timing: Mid- to late season (6–8 days after Bing or Van)
Color when ripe: Mahogany
Bloom timing: Mid–late season
Pollen S alleles: S_1S_9

Fig. 2.29. Tamara (courtesy of L.E. Long).

Suggested rootstock productivity: Low to moderate
Average fruit size: 32 mm (8.5-row)
Average firmness: Excellent
Fruit-cracking susceptibility: Low

Tamara (Fig. 2.29) fruit are very high quality with excellent flavor and very low pitting potential. Stem retention may be a problem during harvest or postharvest handling. Proper harvest timing is essential as fruit quickly becomes overripe.

2.3.2.31   Karina

Harvest timing: Mid- to late season (9–11 days after Bing or Van)
Color when ripe: Mahogany
Bloom timing: Mid–late season
Pollen S alleles: S_3S_4
Suggested rootstock productivity: Moderate
Average fruit size: 27 to 28 mm (10-row)
Average firmness: Good
Fruit-cracking susceptibility: Moderate

Karina serves primarily as a pollinizer for Regina. The tree has very strong growth and is upright.

Fig. 2.30. Lapins (courtesy of C. Denby and Oregon State University).

2.3.2.32 *Lapins*

Harvest timing: Late season (10 days after Bing)
Color when ripe: Mahogany skin, light mahogany flesh
Bloom time: Mid-season
Suggested rootstock productivity: Low to moderate (may be incompatible with Mahaleb)
Average fruit size: 30 mm (9-row)
Average firmness: Excellent
Fruit-cracking susceptibility: Moderate

Although widely planted in the PNW region, Canada and New Zealand, Lapins (Fig. 2.30) has fallen from favor in hot climate areas of the PNW due to poor adaptation leading to low packouts and a propensity for pitting in storage. Lapins crops heavily and tends to form tight fruit clusters that are difficult to harvest or penetrate with fungicides. To reduce this tendency, it is important to head prune all new shoots by one-third each year to eliminate the portion of the branch where the most dense clusters form. In addition, careful handling during picking and packing can reduce the potential for pitting. The tree form is very upright, making it difficult to manage as a central leader but very suitable for KGB or UFO training.

2.3.2.33   Selah

Harvest timing: Late season (10–12 days after Bing)
Color when ripe: Mahogany skin, light mahogany flesh
Bloom time: Mid-season
Pollen S alleles: S_3S_4, (self-compatible)
Suggested rootstock productivity: Moderate to high
Average fruit size: 31 mm (8.5-row)
Average firmness: Excellent
Fruit-cracking susceptibility: High

Selah (Fig. 2.31) bears fruit in loose clusters with excellent flavor and quality, though pedicel retention can be low, making stem-on harvest unfeasible in some years. However, this trait may make it a good candidate for mechanical or mechanical-assist harvest. The growth habit of the tree is spreading and yields are moderate in spite of the variety being self-compatible.

2.3.2.34   Skeena

Harvest timing: Late season (11 days after Bing)
Color when ripe: Mahogany to dark mahogany skin, mahogany flesh
Bloom time: Mid–late season
Pollen S alleles: S_1S_4, (self-compatible)
Suggested rootstock productivity: Low to moderate
Average fruit size: 31 mm (8.5-row)
Average firmness: Extraordinary
Fruit-cracking susceptibility: High

Skeena (Fig. 2.32) has increasingly been adopted by some growers in place of Lapins, having firmer, slightly larger fruit, less dense clusters, and thick stems. Developing fruit are negatively impacted by sustained temperatures of 38°C (100°F) and higher. The tree is upright and open, and much easier to manage as a central leader tree than Lapins, but pendant branches are sensitive to sunburn and must be removed.

2.3.2.35   Regina

Harvest timing: Late season (12–15 days after Bing or Van)

Fig. 2.31. Selah (courtesy of C. Denby and Oregon State University).

Color when ripe: Mahogany to very dark mahogany skin and flesh
Bloom time: Late season
Pollen S alleles: S_1S_3
Suggested rootstock productivity: High to very high
Average fruit size: 31 mm (8.5-row)
Average firmness: Excellent
Fruit-cracking susceptibility: Low

Regina (Fig. 2.33) is widely planted in the late cherry production districts of Europe due to its large, high quality fruit and resistance to rain cracking. Production is also increasing in Chile as the mild but pleasantly flavored fruit is highly sought after in the Chinese export market. Although Regina tends to have low productivity, this has been partially overcome with the use of productive rootstocks, higher than normal levels of late-blooming

Fig. 2.32. Skeena (courtesy of C. Denby and Oregon State University).

pollinizer cultivars, and use of ethylene biosynthesis-inhibiting growth regulators.

2.3.2.36 Royal Edie

Harvest timing: Late season (13 days after Bing or Van)
Color when ripe: Mahogany skin and flesh
Bloom time: Mid-season
Pollen S alleles: S_1S_4 (though some reports indicate self-incompatible)
Suggested rootstock productivity: Moderate to high
Average fruit size: 32 mm (8.5-row)
Average firmness: Excellent to extraordinary
Fruit-cracking susceptibility: Moderate to high

Fig. 2.33. Regina (courtesy of C. Denby and Oregon State University).

Fig. 2.34. Royal Edie (courtesy of L.E. Long).

Fig. 2.35. Royal Helen (courtesy of L.E. Long).

Royal Edie (Fig. 2.34) fruit have an excellent sweet/acid flavor and long, thick stems that are securely attached. The tree growth habit is rather upright.

2.3.2.37   Royal Helen

Harvest timing: Late season (13 days after Bing)
Color when ripe: Mahogany skin and flesh
Bloom time: Mid-season
Pollen S alleles: $S_1S_{4'}$ (though some reports indicate self-incompatible)
Suggested rootstock productivity: Moderate to high
Average fruit size: 32 mm (8.5-row)
Average firmness: Excellent
Fruit-cracking susceptibility: Moderate

Royal Helen (Fig. 2.35) fruit have a pleasant flavor, but pitting incidence is higher than average in laboratory tests. Although it ripens in the PNW region around Regina timing, harvest in California is generally about a week after Bing. The tree growth habit is upright.

2.3.2.38   Sweetheart

Harvest timing: Late season (18–20 days after Bing)

Fig. 2.36. Sweetheart (courtesy of C. Denby and Oregon State University).

Color when ripe: Red to light mahogany skin and flesh
Bloom time: Mid-season
Pollen S alleles: $S_3S_{4'}$ (self-compatible)
Suggested rootstock productivity: Low to moderate
Average fruit size: 27 mm (10-row)
Average firmness: Excellent
Fruit-cracking susceptibility: Moderate to high

Sweetheart (Fig. 2.36) fruit have an agreeable flavor, but pitting incidence has been higher than average for exported fruit to distant markets. As with Lapins, to reduce the potential for pitting, it is important to handle the fruit carefully and to head prune shoots to prevent trees from oversetting, even on rootstocks of moderate productivity. Trees are spreading in growth

Fig. 2.37. Staccato (courtesy of N. Ibuki).

habit and very susceptible to powdery mildew, so timely application of fungicides throughout the season is critical for disease-free fruit.

2.3.2.39 Staccato

Harvest timing: Very late season (21–25 days after Bing)
Color when ripe: Mahogany skin and flesh
Bloom time: Mid-season
Pollen S alleles: $S_3S_{4'}$ (self-compatible)
Suggested rootstock productivity: Moderate to high
Average fruit size: 31 mm (8.5-row)
Average firmness: Excellent
Fruit-cracking susceptibility: Low

Staccato (Fig. 2.37) fruit have a moderately intense, sweet/acid flavor and are sensitive to extended periods of high heat. Elimination of pendant branches can help reduce sunburn and other heat related symptoms. The tree is also very susceptible to powdery mildew, so timely application of fungicides throughout the season is critical for disease-free fruit.

2.3.2.40 Sentennial

Harvest timing: Very late season (30 days after Bing or Van)
Color when ripe: Mahogany skin and flesh

Fig. 2.38. Sentennial (courtesy of N. Ibuki).

Bloom time: Mid-season
Pollen S alleles: S_3S_4' (self-compatible)
Suggested rootstock productivity: Low to moderate
Average fruit size: 30 mm (9-row)
Average firmness: Extraordinary
Fruit-cracking susceptibility: Low to moderate

Sentennial (Fig. 2.38) fruit have a moderately sweet flavor. The tree is productive, open, spreading and susceptible to powdery mildew, so timely application of fungicides throughout the season is critical for disease-free fruit.

2.3.2.41 Sovereign

Harvest timing: Very late season (33 days after Bing)
Color when ripe: Mahogany skin and flesh
Bloom time: Mid–late season
Pollen S alleles: S_3S_4' (self-compatible)
Suggested rootstock productivity: Moderate to high
Average fruit size: 30 mm (9-row)
Average firmness: Extraordinary
Fruit-cracking susceptibility: Low to moderate

Fig. 2.39. Sovereign (courtesy of N. Ibuki).

Sovereign (Fig. 2.39) fruit have a good, moderately sweet flavor with some acidity. The tree is moderately vigorous with open, flat branches.

2.3.3 Yellow-fleshed blush sweet cherries for the fresh market (listed in order of ripening)

2.3.3.1 Cheery Blush

Harvest timing: Very early season (14–18 days before Rainier)
Color when ripe: Yellow skin with red blush, yellow flesh
Bloom time: Early season
Pollen S alleles: S_1S_6
Suggested rootstock productivity: Low to moderate
Average fruit size: 26 mm (10.5-row)
Average firmness: Excellent
Fruit-cracking susceptibility: Unknown at this time

Cheery Blush (Fig. 2.40) has an extremely low chilling requirement (~150 h). For this reason, it has the potential to open up entirely new production areas. Fruit flavor is sub-acid and very good, similar to Rainier, and the pedicels are moderately long. The tree is very productive, so aggressive pruning and crop thinning may be necessary in some years.

Fig. 2.40. Cheery Blush (courtesy of N. Esch © 2017 International Fruit Genetics, LLC, California, USA; all rights reserved).

2.3.3.2 *Early Robin*

Harvest timing: Early season (5–8 days before Rainier)
Color when ripe: Yellow skin with red blush, yellow flesh
Bloom time: Mid-season
Pollen S alleles: S_1S_3
Suggested rootstock productivity: High to very high
Average fruit size: 30 mm (9-row)
Average firmness: Excellent
Fruit-cracking susceptibility: High

The overall appearance and flavor of Early Robin (Fig. 2.41) fruit are very similar to Rainier, the industry standard for yellow-fleshed cherries with a reddish blush on yellow skin. This trait for light colored skin and flesh readily shows bruises, so harvest and packing costs are typically higher than for red-fleshed cultivars, but returns tend to be higher also. Early Robin fruit have an intense sweet flavor, with a subtle acid balance, and an attractive

Fig. 2.41. Early Robin (courtesy of C. Denby and Oregon State University).

red blush covering 50% or more of the fruit. Productivity tends to be low, and trees are very susceptible to bacterial canker.

2.3.3.3   Radiance Pearl

Harvest timing: Mid-season (1 to 2 days before Rainier)
Color when ripe: Yellow skin with red blush, yellow flesh
Bloom time: Late mid-season
Pollen S alleles: S_1S_{13}
Suggested rootstock productivity: Low to moderate
Average fruit size: 31 mm (8.5-row)
Average firmness: Extraordinary
Fruit-cracking susceptibility: Moderate

Radiance Pearl (Fig. 2.42) fruit have a very good eating quality and a pitting incidence lower than average in laboratory tests; the fruit are very firm.

Fig. 2.42. Radiance Pearl (courtesy of L.E. Long).

The tree is open and spreading, with the potential to overset unless crop load is properly managed.

2.3.3.4 Rainier

Harvest timing: Late mid-season (5–7 days after Bing or Van)
Color when ripe: Yellow skin with red blush, yellow flesh
Bloom time: Mid-season
Pollen S alleles: S_1S_4
Suggested rootstock productivity: Moderate to high
Average fruit size: 32 mm (8.5-row)
Average firmness: Good
Fruit-cracking susceptibility: High

Rainier (Fig. 2.43) fruit can achieve very high sugar levels and, in North American markets, Rainier commands a premium compared with dark red sweet cultivars harvested at the same time. All blush cherries must be

Fig. 2.43. Rainier (courtesy of C. Denby and Oregon State University).

handled carefully at harvest and packing to prevent bruising and unsightly brown marks. The tree is open and highly susceptible to powdery mildew. As with all blush cultivars, pruning for good light distribution throughout the canopy is required to enable formation of the red blush on the skin. Preharvest gibberellic acid should be applied at 10 ppm or less so that normal blush develops.

2.3.3.5   *Stardust*

Harvest timing: Late season (12–14 days after Rainier)
Color when ripe: Yellow skin with reddish-orange blush, white flesh
Bloom time: Late season
Pollen S alleles: $S_1S_{4'}$ (self-compatible)

Fig. 2.44. Stardust (courtesy of C. Denby and Oregon State University).

Suggested rootstock productivity: Low to moderate
Average fruit size: 31 mm (8.5-row)
Average firmness: Good
Fruit-cracking susceptibility: Moderate

Stardust (Fig. 2.44) fruit are moderately sweet. All blush cherries must be handled carefully during harvest to prevent bruising and unsightly brown marks. The tree is very productive, vigorous and upright to spreading.

2.3.4 Sweet cherry cultivars for processing (listed alphabetically)

Sweet cherries can be processed in several ways: brining (for use as maraschino cherries), freezing (for use in ice cream, yogurt, etc.), canning, juicing or drying. Certainly, before planting cherries for the processing industry, a processor must be identified to determine their anticipated needs, specifications for the fruit they process (e.g. most brine cherries must be less than 18°Brix and equal to or smaller than 25 mm in diameter), pit size requirements and shape limitations for mechanical pitting machines, etc. Key attributes for processing cherry cultivars tend to be high yields (as prices for processed fruit are much lower than for fresh market fruit), modest size, the ability to retain firmness during processing, and ease of mechanical pit removal (i.e. minimal adherence of flesh to the pit). Growers of cherries for processing may choose cultivars strictly for processing or dual purpose cultivars that could be grown for fresh or processing markets. In the latter case, growers might divert cherries originally intended for the fresh market to the processing industry depending on crop quality or market opportunities. Fruit for juicing is often cull fruit that could serve no other intended use.

Thirteen standard or promising sweet cherry cultivars for processing are discussed briefly below; the processing market is not nearly as dynamic as the fresh market, so adoption of new cultivars is much more constrained. Some cultivars that may be grown as pollinizers, such as Rainier, often set excessive crops, producing smaller cherries that cannot be sold fresh but can be processed. In some years, entire blocks, portions of blocks or individual trees in a fresh block may overset and be picked for processing. Also, fresh market cherries with rain-cracking damage may be used as low grade brine fruit and processed as pieces, for uses such as in ice cream or yogurt. In some years, unfavorable fresh market supply and demand volumes may make processing more attractive. It is important to anticipate market direction early, because brine cherries must be harvested at a more immature stage then fresh cherries. Cherries for canning are usually harvested at a fully ripe stage to achieve higher sugar levels.

2.3.4.1   *Andersen*

Harvest timing: Late season (15–18 days after Bing or Van)
Color when ripe: Yellow–white with a comprehensive pink–red blush
Bloom time: Mid–late season
Pollen S alleles: S_3S_6
Suggested rootstock productivity: Moderate to high
Average fruit size: 23 mm (11.5-row)
Average firmness: Excellent
Fruit-cracking susceptibility: Moderate

Andersen fruit have an acidic flavor. They tend to bear in singles and strongly retain their long, thick stems for excellent mechanical harvest as a higher value stem-on maraschino brining cherry. The tree is vigorous and widely spreading to weeping in growth habit, cold hardy and less susceptible to bacterial canker.

2.3.4.2 Bing

See description of traits under Section 2.3.2: 'Red-fleshed sweet cherries for the fresh market'.

When Bing is overset with small fruit, it may be used for brining if it is harvested earlier than its normal ripening window, between 16 to 18°Brix. For canning, the harvest window is similar to that of fruit for fresh market to achieve high sugar levels. If the fruit are rain-cracked, they may be used for low grade brining as pieces for use in ice cream, yogurt and baking.

2.3.4.3 Black York

Harvest timing: Mid-season (similar to Bing or Van)
Color when ripe: Purple–red skin and flesh
Bloom time: Mid-season
Pollen S alleles: S_1S_4
Suggested rootstock productivity: Moderate
Average fruit size: 21–23 mm (12 to 11.5-row)
Average firmness: Excellent
Fruit-cracking susceptibility: Low to moderate

Black York fruit have an excellent flavor and are very productive. The tree is moderately vigorous and spreading to moderately upright in growth habit, cold hardy and less susceptible to bacterial canker and brown rot.

2.3.4.4 BlushingGold

Harvest timing: Mid- to late season (5–7 days after Bing or Van)
Color when ripe: Yellow skin with a reddish blush, yellow flesh
Bloom time: Early–mid season
Pollen S alleles: S_1S_6
Suggested rootstock productivity: Moderate
Average fruit size: 23–27 mm (11.5- to 10-row)
Average firmness: Moderate
Fruit-cracking susceptibility: Low to moderate

BlushingGold fruit have a very good flavor and are very productive. The tree is moderately vigorous and spreading to moderately upright in growth habit, cold hardy and less susceptible to bacterial canker and brown rot.

2.3.4.5 Emperor Francis

Harvest timing: Late mid-season (7–8 days after Bing or Van)
Color when ripe: Yellow with a red blush
Bloom time: Mid-season
Pollen S alleles: S_3S_4

Suggested rootstock productivity: Moderate
Average fruit size: 21 mm (12-row)
Average firmness: Excellent
Fruit-cracking susceptibility: Low to moderate

Similar to Napoleon, but the fruit are less susceptible to cracking and have more of a red blush. The tree is vigorous, cold hardy, less susceptible to bacterial canker and productive.

2.3.4.6   Gold

Harvest timing: Late season (12–14 days after Bing or Van)
Color when ripe: Yellow with no blush
Bloom time: Late season
Pollen S alleles: S_3S_6
Suggested rootstock productivity: Moderate
Average fruit size: ≤20 mm (12.5-row)
Average firmness: Excellent
Fruit-cracking susceptibility: Low to moderate

Gold fruit are pure yellow; flavor is poor for fresh eating but is suitable for brining. The tree is vigorous, cold hardy, less susceptible to bacterial canker and productive. Gold is a leading brining variety in Michigan, USA.

2.3.4.7   Napoleon (aka Royal Ann)

Harvest timing: Early season for brining (10–12 days before Bing or Van); Late mid-season for canning (5 days after Bing or Van).
Color when ripe: Yellow with a pink blush
Bloom time: Mid-season
Pollen S alleles: S_3S_4
Suggested rootstock productivity: Moderate
Average fruit size: 21 mm (12-row)
Average firmness: Excellent
Fruit-cracking susceptibility: Moderate

Napoleon is a high quality processing cherry developed in France but grown in the USA as Royal Ann. In the USA it is one of the principal cultivars used for maraschino cherries, but it can also be canned. The fruit has a thin skin, medium–long stem, and a moderately pointed shape, traits which are associated with high quality cocktail-style cherries. Harvesting early helps stem retention during processing as well as reduces the risk of rain crack-ing. If used for brining, the proper harvest window is at 14–18°Brix; if used for canning, the proper harvest window is a minimum of 20°Brix. The light colored skin readily shows brown discoloration from bruising, especially in hot weather. Trees are relatively slow to come into production, even on precocious rootstocks such as Gisela 6.

2.3.4.8   Nugent

Harvest timing: Late season (12–14 days after Bing)
Color when ripe: Yellow with no blush
Bloom time: Early–mid season
Pollen S alleles: S_1S_6
Suggested rootstock productivity: Moderate
Average fruit size: 21 mm (12-row)
Average firmness: Excellent
Fruit-cracking susceptibility: Low

Nugent fruit are pure yellow, achieve a sugar level of 20°Brix, and could be a potential alternative to Gold for brining. The tree is spreading, cold hardy, less susceptible to bacterial canker and very productive.

2.3.4.9   Rainier

See description of traits under Section 2.3.3: 'Yellow-fleshed blush sweet cherries for the fresh market'.

When Rainier is used for brining, it is harvested 7–10 days before its normal ripening window. For canning, the harvest window is similar to that for fresh market fruit to achieve high sugar levels.

2.3.4.10   Sam

Harvest timing: Early mid-season (4 days before Bing) for brining
Color when ripe: Very dark mahogany
Bloom time: Late season
Pollen S alleles: S_2S_4
Suggested rootstock productivity: Moderate to high
Average fruit size: 24 mm (11-row)
Average firmness: Good
Fruit-cracking susceptibility: Low

Sam fruit have very dark flesh and poor flavor until fully ripe, though flavor is not a problem when harvested early for brining. Tight fruit clusters can be a problem for brown rot control. The tree is cold hardy, less susceptible to bacterial canker and is often planted as one of several late blooming pollinizers for Regina.

2.3.4.11   Skeena

See description of traits under Section 2.3.2.

When Skeena is overset with small fruit, it may be used for brining if it is harvested earlier than its normal ripening window, between 16 to 18°Brix. Stem retention may be a problem for more mature Skeena fruit.

2.3.4.12   Sweetheart

See description of traits under Section 2.3.2.

Sweetheart fruit make a high quality brined product if harvested before fully ripe (16–18°Brix), providing an alternative market if the tree crop load is excessive and fruit size is too small for the fresh market.

2.3.4.13   Ulster

Harvest timing: Mid-season (with Bing or Van)
Color when ripe: Dark mahogany
Bloom time: Mid-season
Pollen S alleles: S_3S_4
Suggested rootstock productivity: Moderate to high
Average fruit size: 27 mm (10-row)
Average firmness: Good
Fruit-cracking susceptibility: Moderate

Ulster fruit have very dark flesh and excellent flavor. Ulster can be harvested for fresh market if pruned for crop load management, or for processing if allowed to crop more heavily. The tree is cold hardy and less susceptible to bacterial canker. Ulster is a leading variety for canning and freezing in Michigan, USA.

2.3.4.14   Van

Harvest timing: Mid-season (with Bing), slightly earlier for canning
Color when ripe: Red to mahogany
Bloom time: Mid-season
Pollen S alleles: S_1S_3
Suggested rootstock productivity: Moderate
Average fruit size: 21–24 mm (11- to 12-row)
Average firmness: Good
Fruit-cracking susceptibility: Moderate

In the PNW region of the USA, Van is typically used as an excellent pollinizer for both Royal Ann and Bing. Although it has excellent flavor, since it is prone to pitting it is not suitable for fresh market production, and as a pollinizer it generally sets excessive crops resulting in small fruit size. Such fruit can be harvested at 20°Brix for canning or 21°Brix for brining or freezing. The trees are vigorous, hardy and productive, but very susceptible to bacterial canker.

2.4 Advantages of Cultivar Diversification

Planting new, largely untested cultivars is always a risk; however, there are several good reasons to diversify and expand the number of cultivars produced. When growing multiple cultivars, bloom times may be staggered, reducing the risk that the entire crop will be affected by poor weather during final flower development and pollination. In addition, producing cultivars that range from early to late ripening times lengthens the growing season, spreading the risk of fruit cracking during ripening, and potentially providing greater financial returns during market windows when supplies are limited. An extended harvest season also spreads the demand for labor and improves labor efficiency as fewer pickers are needed and, once trained, they can be employed over a longer period of time. Cultivar diversification helps improve cropping consistency, and the risk of rain-cracked fruit can be reduced further by choosing cultivars with moderate to high resistance to rain cracking. On the other hand, a longer growing season can be problematic if labor availability is itinerant or if populations of certain pests, like spotted wing *Drosophila* (SWD) fruit flies, increase as the season progresses.

Acknowledgements

Portions of this chapter were adapted with permission from Oregon State University. *PNW 604 Sweet cherry cultivars for the fresh market*, copyright 2020. Oregon State University Extension and Experiment Station Communications, Corvallis, Oregon. http://catalog.extension.oregonstate.edu/pnw604 (accessed 19 June 2020). *EM 9056 Sweet cherry cultivars for brining, freezing and canning in Oregon*, copyright 2013. Oregon State University Extension and Experiment Station Communications, Corvallis, Oregon. www.catalog.extension.oregonstate.edu/em9056 (accessed 19 June 2020).

References

Bargioni, G. (1996) Sweet cherry scions: characteristics of the principal commercial cultivars, breeding objectives, and methods. In: Webster, A.D. and Looney, N.E. (eds) *Cherries: Crop Physiology, Production, and Uses*. CAB International, Wallingford, UK, pp. 73–112.

Lewis, D. and Crowe, L.K. (1954) Structure of the incompatibility gene. *Heredity* 8(3), 357–363.

Quero-Garcia, J., Schuster, M., López-Ortega, G. and Charlot, G. (2017) Sweet cherry varieties and improvement. In: Quero-Garcia, J., Iezzoni, A., Pulawska, J. and Lang, G.A. (eds) *Cherries: Botany, Production and Uses*. CAB International, Wallingford, UK, pp. 60–94.

Sweet Cherry Rootstocks

3

All commercial sweet cherry trees are either budded or grafted. The part of the tree above the graft/bud union is known as the 'scion' and the part below the graft/bud union is known as the 'rootstock'. Rootstocks are used for several purposes: (i) ease for propagating and producing more trees of a superior cultivar; (ii) better adaptation to particular soil or site characteristics; and (iii) the potential improvement of production due to additional traits like precocious flowering, higher productivity, and greater or reduced scion vigor as appropriate.

Sweet cherry scion cultivars have been selected and clonally propagated over millennia for many reasons, but over the last century, breeding programs have concentrated mainly on achieving improved characteristics such as yield, taste, fruit size, fruit firmness, fruit color, precocity, and resistance to diseases and fruit cracking. In contrast, up until only the past couple of decades, rootstocks have been dominated by genetically non-uniform seedlings propagated from seed. Indeed, it is believed that Mazzard sweet cherry (*P. avium*) seedlings were first used as rootstocks more than 2400 years ago by early Greek and Roman horticulturists. The fact that Mazzard continues to be used throughout many of the leading commercial sweet cherry growing areas of the world is testimony to the widespread availability and success of this seedling rootstock.

Over the last few decades, several new rootstock choices have gained prominence, offering important attributes lacking in Mazzard. An early attempt to develop genetically uniform, clonally propagated rootstocks focused on eliminating performance variability, resulting in the Mazzard clone F.12/1. By the late 1950s, the East Malling Research Station, UK, released Colt (a hybrid of *P. avium* and *P. pseudocerasus*), the first so-called size-controlling rootstock as it was considered to be semi-dwarfing. Besides size control, modern cherry rootstocks may induce precocity, impart some disease resistance, increase productivity and enable growers to harvest premium quality fruit from high density orchards. Furthermore, full production now may be achieved on precocious rootstocks within 5 or 6 years;

similar trees on Mazzard may take up to 12 years. The choice of rootstock also directly influences pruning, training, tree support and labor management decisions.

3.1 Growth Habits and Tree Size

The dwarfing nature of modern rootstocks imparts numerous advantages over full-size rootstocks. Small trees are easier and cheaper to manage, are more precocious and productive, and with proper management, will produce fruit of equal or greater quality. In addition, dwarf trees more easily allow for the installation of rain and hail covers, as well as bird and insect netting. The challenge with trees on dwarfing rootstocks is to manage them properly to prevent overcropping and promote adequate annual vigor to maintain a good proportion of young fruiting sites in the canopy.

The degree to which a tree is dwarfed depends not only on the rootstock but also scion cultivar selection, soils, pruning severity and training system choice. When grafted with Bing, Gisela 12 can be more dwarfing than Gisela 6. However, a Regina/Gisela 12 combination can produce a tree that is approximately 10% larger than Regina/Gisela 6. Other cultivars can exhibit similar relative variations, but the complex interactive factors that also include site and management, make broad conclusions difficult.

Orchard site or location can significantly influence relative tree size. For example, in the Eastern USA, trees on Gisela 6 irrigated largely by natural rainfall may be only 60% of the size of a standard tree, whereas in the irrigated Pacific Northwest (PNW) region, Gisela 6 produces a vigorous tree growing to 90% or more of full size. Soil type, daily solar radiation and cloudiness, daily air temperatures and soil moisture management probably play a role in this discrepancy. Likewise, MaxMa 14 grows more vigorously in the rich soils of the PNW than in the calcareous soils of southern France. Similarly, Colt, which was released in Britain as a semi-dwarfing rootstock under non-irrigated conditions, produces a full-sized tree in the irrigated orchards of California, Oregon and Washington.

Gisela 5 is the most common commercially grown, dwarfing to semi-dwarfing rootstock in the USA and Europe, reducing tree size to approximately 50% of trees on standard Mazzard. Research undertaken in the 'NC140 Coordinated Regional Rootstock Trials' across North America has reported mature tree sizes relative to Mazzard, of 35–45% for Gisela 3; 45–60% for Gisela 5; and 80–100% for Gisela 6 and Gisela 12.

Knowing that a rootstock can decrease tree size to 50% or 90% of a standard tree can be helpful to know; however, additional factors can further modulate tree size. For example, with proper pruning and an appropriate training system, a tree on Gisela 6 can easily be maintained at a height of only 2.4 m (8 ft). Closer planting distances to achieve higher densities also can introduce root competition between trees, further reducing tree vigor, such that trees on Gisela 6 at 0.75 m (2.5 ft) spacing may be similar in vigor to trees on Gisela 5 spaced at 1.5 m (5 ft) apart. When pruned severely, trees on many of the productive rootstocks, such as the Gisela and Krymsk series, respond with moderate, controlled growth, whereas trees on Mazzard will be invigorated through the growth of water sprouts. This moderate response, in conjunction with naturally wider branch angles, makes trees on many of these productive rootstocks much easier to manage.

3.1.1   Precocity, yield and fruit quality

One of the main advantages of most of the new rootstocks developed during the past few decades is their promotion of precocious flower formation compared to traditional rootstocks. Indeed, research results from the co-ordinated NC140 trials showed significantly earlier production on several Gisela rootstocks compared to Mazzard (Perry *et al.*, 1998). After 7 years at Washington State University, Bing on Gisela 6 were most productive, yielding between 13 and 31% more than on Gisela 5, and 212 to 657% more than on Mazzard, depending on the year. Both Gisela rootstocks improved precocity, bearing a significant amount of fruit in year 3 compared to year 5 for Mazzard. Subsequent research has shown Gisela 3 to be even more precocious. The effect of rootstock genotype on fruit quality has been shown to be minimal for most traits, with fruit size in particular being influenced more by management of the crop load and canopy leaf health than by rootstock effects on vigor or precocity.

3.1.2   Graft compatibility

Incompatibility is occasionally a problem with certain scion–rootstock combinations. When the rootstock is *P. avium* (the same species as the cultivar), there is no incompatibility issue. However, most modern rootstocks are hybrid combinations that include various *Prunus* species other than *P. avium*, such as *P. cerasus*, *P. canescens*, *P. pseudocerasus* and *P. mahaleb*. It is this genetic diversity that imparts the desired traits that growers seek in a rootstock, but the various genetic origins of rootstocks also lead to occasional issues of compatibility with some sweet cherry cultivars.

  One example of incompatibility occurs with seedling Mahaleb (*P. mahaleb*) rootstocks. By the mid-19th century, Mahaleb had become popular in the USA and by the early 1900s, it was the most popular cherry rootstock. This was due mainly to its ease of propagation from seed, its cold hardiness and its resistance to some diseases, compared to Mazzard. However, by the mid-1920s, Mahaleb was observed to be incompatible with several scion cultivars, resulting in premature tree death. More recently, Lapins, Chelan, and Tieton, cultivars commonly grown in the PNW region, have exhibited symptoms of incompatibility with Mahaleb. Consequently, production of such cultivars has often defaulted to Mazzard as the rootstock of choice.

Other rootstocks that have shown incompatibility include Weiroot 13 with several scion cultivars, and Colt with Sam or Van. At this time, there have been no reports of scion incompatibility with the commercially available Gisela rootstocks including 3, 5, 6 and 12, nor with Krymsk 5 or 6.

The symptoms of graft incompatibility vary, but it is possible for incompatibility to show up in the nursery as a low percentage of bud and graft takes, or as weak growth or breakage at the graft due to a weak union. When incompatibility is apparent in the nursery, it can create a significant loss to the nursery; however, delayed incompatibility is a costlier issue for the grower. In this case, trees may perform well for 6 years or more before beginning to decline. Symptoms may be expressed through small, yellowish leaves, excessive bloom on increasingly weak vegetative growth, excessive root suckering and ultimately tree death.

Scion–rootstock combinations that would typically be incompatible, may be possible with the use of a compatible interstem (or 'interstock'). Two interstems commercially available include Adara, a selection of myrobalan or cherry plum (*P. cerasifera*), and the presumed interspecific hybrid ZeeStem®. These allow the grafting of sweet cherry scions on to a range of rootstocks that even might include plum (*P. salicina*), apricot (*P. armeniaca*), peach (*P. persica*) and almond (*P. dulcis*) genetic backgrounds.

3.1.3   Site/soil adaptation, anchorage and root suckers

Rootstocks can be important tools for adapting the production of high quality scion cultivars to a wide range of sites and soils. Mahaleb rootstocks tend to be superior for sweet cherry performance in sandy soils and poor choices for heavy soils. They also tend to do better in calcareous soils with higher pH values. The first Gisela rootstocks (Gisela 5 and Gisela 6) have tended to perform poorly in hot, dry climates, but there are reports that Gisela 12 does better in such sites, and the recent releases of Gisela 13 and Gisela 17 are reported to be more suitable to such stresses.

Although typically grown without support, trees on Gisela 6 and sometimes Gisela 5 have been known to tilt away from the prevailing winds. This is especially true for more top-heavy, central or multiple leader trees. Tree support with stakes or a trellis system may be beneficial with these rootstocks. Gisela 12 is more deeply rooted than Gisela 6, but occasionally can have anchorage problems as well. Anchorage seems to be adequate for most other rootstocks.

Most commercial cherry rootstocks used in the PNW region produce limited to no root suckers. Occasionally, depending on the soil type and growing conditions, Mazzard and Colt can show low levels of suckering. This has also been observed with Krymsk 5 and 6, several of the Weiroot clones such as Weiroot 158, and several of the new Michigan State University (MSU) rootstock series.

3.1.4   Cold hardiness, frost and bloom

Cold hardiness is a complex physiological attribute (see Chapter 8, 'Managing the Orchard Environment'), and findings from cherry rootstock research trials around the world are inconsistent. Indeed, one of the main confounding problems is that damage to scion budwood in fall is due to incomplete cold acclimation of the trees. This, in turn, results in poor winter hardiness as well as potential damage in early spring when trees break dormancy too early. That said, multiple studies have shown that many scion cultivars grafted on to Mahaleb will acclimatize earlier in fall/winter than on other rootstocks. The parent material of the Gisela rootstocks, *P. cerasus* and *P. canescens*, are both hardier than Mazzard and impart that hardiness to the Gisela rootstocks currently available commercially. Colt is perhaps the only rootstock used in the USA that is less hardy than Mazzard.

Although trees on highly productive rootstocks are at least as sensitive to frost as those on standard rootstocks, such trees often produce more fruit after a severe frost due to higher initial flower numbers. Furthermore, in situations with no frost protection measures, the flowers closest to ground level are worse affected than those furthest from the ground. Consequently, where spring frosts occur during flowering, and artificial heating of the orchards does not take place, fruit set on dwarfing or semi-dwarfing rootstocks will be affected more severely than on standard vigor rootstocks, simply because the latter trees are taller and the flowers are further from the coldest air that settles near the ground. Given the fact that cherries are such a high value crop, it is advisable to install heaters (usually propane) and/or frost fans ('wind machines') to protect the flowers during spring frosts.

3.1.5   Bacterial canker, *Phytophthora* and virus sensitivity

Bacterial canker, caused by *Pseudomonas syringae*, is a pathogen of sweet cherries found in all cherry production areas around the world. Infection rates of 50 to 80% have been reported in some of the wetter regions of the PNW such as the Willamette and Hood River valleys in Oregon. Even in the drier regions of Central Washington and Oregon, infection and mortality rates can approach 10% or more in some years. Young trees seem to be particularly susceptible and can exhibit significant mortality.

The *P. avium* clone F.12/1 has shown tolerance to this pathogen. Therefore, it is used in the Willamette valley as a high-budded rootstock in order to slow down or stop a branch infection before the pathogen infects the trunk and threatens the entire tree. In this situation, the stock is grown up to the point desired for branching and scion wood is budded on to the rootstock branches.

Reports from the literature (Spotts *et al.*, 2010) as well as limited grower experience suggest that trees on Colt rootstock show greater tolerance to the disease than Mazzard, while trees on Krymsk 5 may show slightly greater tolerance than Mazzard. Additionally, trees on Gisela 6 are less tolerant than Mazzard and have shown considerable infection levels in canker-prone areas such as Hood River. Growers in Chile report that some tolerance to bacterial canker is imparted to the scion by both MaxMa 14 and MaxMa 60, and greater susceptibility is imparted by CAB 6 P and Gisela 6.

Sweet cherry crown and root rot diseases are caused by several species of the fungus *Phytophthora*. Infections are usually initiated where orchard drainage is poor or soils are periodically flooded for an extended period of time. Mahaleb rootstocks tend to be more prone to crown and root rots than Mazzard, Colt or other common cherry rootstocks.

Prune dwarf virus (PDV) and Prunus necrotic ringspot virus (PNRSV) are transmitted via pollen and are commonly found in mature orchards throughout the world. Most strains of these two viruses show few, if any, symptoms when trees on Mazzard, Mahaleb or Colt are infected. However, some of the newer rootstocks, such as Gisela 7 and Weiroot 158, show varying degrees of sensitivity to one or both of these viruses when inoculated in controlled trials (Lang *et al.*, 1998). In this same trial, Gisela 5, 6 and 12 were shown to generally exhibit tolerance to these two viruses, with only a slight reduction in vigor when infected.

A number of rootstocks from the Gisela rootstock breeding program were found to be hypersensitive to PDV and PNRSV, such as Gisela 1, 4, 10 and 11 (all of which were subsequently dropped from commercial release). Following this, two Russian rootstocks, Krymsk 5 and 6, also were found to be sensitive or hypersensitive to these viruses (Howell and Lang, 2001). Much debate has taken place over the past decade concerning the practical importance of these findings to orchard longevity. Since hypersensitive trees die quickly when infected, some scientists believe that hypersensitivity ultimately may be beneficial to an orchard block since infected trees may die before the virus can be transmitted to surrounding trees. Trees on rootstocks sensitive to one or both pollen-borne viruses, on the other hand, may become infected through pollination and then decline over several years, potentially serving as a source of infected pollen each year until they die. However, a limited number of trees on the sensitive (to PNRSV) rootstock Gisela 7 were planted by PNW growers in the mid-1990s, which did not subsequently lead to reports of widespread mortality. Still, it would be wise to avoid trees on hypersensitive rootstocks in an interplant situation among mature trees or close to blocks of older trees that may be infected with one or both of these viruses.

3.2 Choosing the Right Rootstock

Growers now have a wide range of rootstocks to choose from. These include vigorous rootstocks such as Mazzard, Mahaleb and Colt, to full

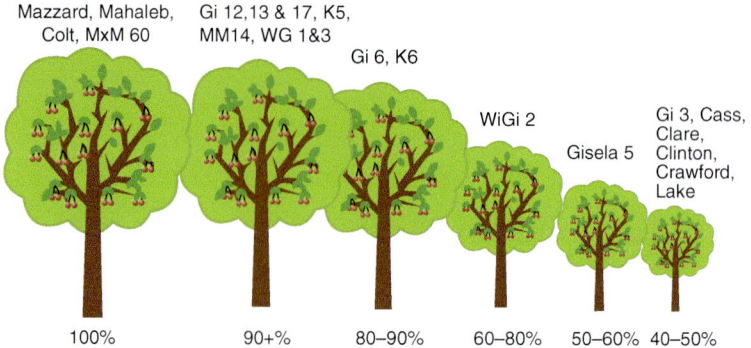

Mazzard, Mahaleb, Colt, MxM 60

Gi 12,13 & 17, K5, MM14, WG 1&3

Gi 6, K6

WiGi 2

Gisela 5

Gi 3, Cass, Clare, Clinton, Crawford, Lake

100%   90+%   80–90%   60–80%   50–60% 40–50%

Fig. 3.1. Relative sizes of sweet cherry rootstocks (L.E. Long).

dwarfing rootstocks, such as Gisela 3 or the Corette series. In between, there are many others ranging in size from semi-vigorous to semi-dwarfing (Fig. 3.1). Several factors must be taken into account when selecting rootstocks for a new orchard. Soil fertility, scion cultivar choice and desired training system are some of the more important aspects. For sites with soils that are shallow or low in fertility and water- or nutrient holding capacity, the most dwarfing rootstocks, such as Gisela 3, Gisela 5, and the Corette rootstocks should be avoided. More vigorous rootstocks, such as Gisela 6, Gisela 12, Krymsk 5, Krymsk 6, and MaxMa 14 may perform better on these sites, but may need to be planted at higher densities than usual since their growth on poor soils is likely to be reduced. Similarly, Mahaleb performs well in light soils, but should be avoided in heavy soils since it is particularly susceptible to *Phytophthora* in wet soils and does poorly under these conditions, to the point of potentially dying out. On the other hand, Krymsk 5 and 6, while not adapted to wet soils, will survive in heavier soil conditions than Mahaleb or even Mazzard.

Full-size rootstocks, such as Mazzard, Mahaleb, Colt and MaxMa 60, are best suited to standard density orchards of 300 to 400 trees/ha (120 to 160 trees/acre). However, when trained as a Spanish bush (SB) or Kym Green bush (KGB), all three of these rootstocks may be grown at densities of between 740 to 840 trees/ha (300 to 340 trees/acre). Commercial plantings using these training systems in the PNW, as well as in Australia and Spain, have proven successful at these higher densities. Due to the high vigor of these rootstocks, a spindle or central leader training system usually is not recommended since vigor tends to be greatest in the top of the canopy, leading to tall trees and heavier upper growth creating shade problems lower in the canopy. Furthermore, scion cultivars of lower productivity, such as Regina, Benton and Tieton, tend to produce unsatisfactory yields on these rootstocks.

Modern semi-vigorous to vigorous rootstocks, such as Gisela 6, Gisela 12, Krymsk 5, Krymsk 6, WeiGi 1, WeiGi 2 and WeiGi 3, result in trees ranging in size between those on Gisela 5 and the previous industry standards like Mazzard, Mahaleb and Colt. These rootstocks tend to perform well across a range of soil properties and with most variety combinations. Highly productive varieties such as Sweetheart and Lapins can be a challenge to produce balanced crop loads and good fruit quality with any precocious, productive rootstock, but growers have been successful at consistently growing high quality fruit with such varieties on semi-vigorous rootstocks. The key is to pay close attention to pruning principles for balanced crop loads (see Chapter 6, 'Sweet Cherry Pruning Fundamentals') and to perform required procedures in a timely fashion. While growing productive varieties on highly productive rootstocks can be challenging, matching them with moderately productive rootstocks such as MaxMa 14 and Krymsk 5 also are good choices to confer some precocity while more easily maintaining good leaf-to-fruit ratios.

Dwarfing to semi-dwarfing rootstocks, such as Gisela 3, Gisela 5, W720, and the MSU rootstocks (Corette series) result in small trees for high- to very high-density plantings. These rootstocks tend to perform well on more fertile soils, requiring more intensive management of water and nutrients, and with most varieties of average to low productivity. Usually, highly productive varieties such as Sweetheart and Lapins should be avoided due to the challenges of maintaining suitable leaf-to-fruit ratios for quality fruit. Cultivars of low productivity, such as Early Robin, Regina, Benton and Tieton, can be improved by matching them with highly productive rootstocks across the vigor range, such as Cass, Gisela 3, 5, 6 or 12.

Table 3.1 provides suggested planting densities, soil types and scion material best suited for cherry rootstocks commonly used around the world.

The following is a list of 25 important standard, relatively recent hybrid and new rootstocks that are currently in use or being evaluated in major cherry production areas around the world. Rootstock characteristics can vary based on both soil and climate; therefore, site specific evaluations should be made before rootstock potential becomes known for a given growing region. The following information is based on data and reports from a variety of North American and European locations.

3.2.1　Vigorous to semi-vigorous rootstocks

3.2.1.1　CAB 6 P (P. cerasus)

CAB 6 P is a clonal selection of sour cherry that makes a vigorous, somewhat precocious rootstock for sweet cherry. Trees have good anchorage. Currently, it is used mainly in Italy and Chile, although in recent years it has fallen out of favor in Chile due to high sensitivity to bacterial canker. Relative to apple rootstocks, CAB 6 P is comparable to EMLA 111.

Table 3.1. Planting parameters for various commercially available cherry rootstocks (L.E. Long).

	Best for super high density	Best for moderately high density[a]	Best for low density[a]	Best for shallow or poor soils	Best with low productive varieties	Best with highly productive varieties
Cass	Yes	Yes	No	High density	Yes	No
Clare	Yes	Yes	No	High density	Yes	No
Clinton	Yes	Yes	No	No	Yes	No
Colt	No	No	Yes	Yes	No	Yes
Crawford	Yes	Yes	No	High density	Yes	No
Gisela 3	Yes	Yes	No	No	Yes	No
Gisela 5	Yes	Yes	No	No	Yes	No
Gisela 6	No	Yes	No	High density	Yes	With proper management
Gisela 12	No	Yes	No	High density	Yes	With proper management
Gisela 13	No	Yes	Yes	Yes	Yes	With proper management
Gisela 17	No	Yes	Yes	Yes	Yes	With proper management
Krymsk 5	No	Yes	No	High density	Yes	With proper management
Krymsk 6	On poor soils	Yes	No	High density	Yes	With proper management
Lake	Yes	Yes	No	High density	Yes	No

Continued

Table 3.1. Continued

	Best for super high density	Best for moderately high density[a]	Best for low density[a]	Best for shallow or poor soils	Best with low productive varieties	Best with highly productive varieties
Mahaleb	No	No	Yes	Avoid heavy soils	No	Yes
MaxMa 14	No	Yes	No	Avoid heavy soils	Yes	With proper management
Mazzard	No	No	Yes	Yes	No	Yes
WeiGi 1	No	Yes	Yes	Yes	Yes	With proper management
WeiGi 2	Yes	Yes	No	High density	Yes	With proper management
WeiGi 3	No	Yes	Yes	Yes	Yes	With proper management

[a]See text

3.2.1.2   Colt (P. avium x P. pseudocerasus)

Colt was released in England in the 1970s as a semi-dwarfing clonal root-stock. However, in the irrigated orchards of the PNW, it produces a vigor-ous tree that is similar in size to Mazzard with similarly low precocity. In addition, Colt is sensitive to droughty soils and to cold winter temperatures. Colt grows well in a wide range of soils from sandy to clay and can tolerate short periods with high water tables. Trees have good anchorage. It is easily propagated, but also is prone to suckering in the orchard. Colt has been planted widely in California due to its resistance to cherry stem pitting, a debilitating virus readily found in that state. It also has shown some resist-ance to *Phytophthora* root rot, bacterial canker and gopher damage, but is susceptible to crown gall (*Agrobacterium tumefaciens*). In one Oregon or-chard, Colt performed well in a replant situation where cherries followed cherries on a non-fumigated site (Long, 1995).

3.2.1.3   Gisela 6 (P. cerasus x P. canescens)

Gisela 6 is widely planted in many production regions around the world. Even though it is semi-vigorous to vigorous under irrigated conditions, it is productive, precocious and easy to manage. Relative to apple rootstocks, Gisela 6 is comparable to EMLA 106 or M7. Recommended planting densi-ties are 750–1250 trees/ha (300–500 trees/acre). It begins producing har-vestable crops by the third leaf, with full production possible by the fifth leaf. Due to these high production levels, trees on Gisela 6 need to be properly pruned from an early age with crop load management in mind to maintain fruit size and quality. Premium fruit quality is possible with culti-vars of moderate to low productivity such as Bing and Regina, but more dif-ficult with very productive cultivars like Lapins. As with all size-controlling rootstocks, it is imperative to maintain adequate levels of vigor to produce high quality fruit. The promotion of new shoots is much easier to achieve with Gisela 6 than with Gisela 5 and is one of the reasons for the popular-ity of this rootstock as growers transition to higher density, earlier yielding orchards.

Gisela 6 tends to advance flowering and fruit ripening only slightly compared to Mazzard. Trees are open and spreading with good branch-ing. Anchorage can be a problem, especially on windy sites, although most growers in the PNW do not provide support. Gisela 6 is well suited to a wide range of soil types from light to heavy; however, good drainage is essential. Trees grown on Gisela 6 have not been prone to suckering and scion com-patibility has been good, although with some cultivars, the scion diameter above the graft union sometimes grows much larger than the rootstock diameter below the union. Bacterial canker has been reported to be more prevalent in trees on Gisela 6 than on other rootstocks of similar vigor.

Fig. 3.2. Regina is more vigorous on Gisela 12 (right) than on Gisela 6 (left) (Oregon) (courtesy of L.E. Long).

3.2.1.4 *Gisela 12* (P. cerasus x P. canescens)

Trees on Gisela 12 are variably semi-vigorous depending upon cultivar. A 10-year trial in The Dalles, Oregon, and Prosser, Washington, indicated that Bing on Gisela 12 produced a tree intermediate in size to Gisela 5 and 6. However, Regina on Gisela 12 has produced a tree about 10% larger than Gisela 6 in some PNW orchards (Fig. 3.2). For this reason, some growers prefer Regina/Gisela 12 as they find it easier to maintain shoot growth and ultimately fruit size. Relative to apple rootstocks, Gisela 12 is comparable to EMLA 106 or M7.

Gisela 12 is both precocious and productive, producing early heavy crops, with full production possible by the fifth leaf. Good fruit size and quality is possible with proper pruning. Gisela 12 is adapted to a wide range of soils, resists suckering and is well anchored. It has been reported to perform better in the hot climate of California's central valley than other Gisela rootstocks. The tree structure is open and spreading, and new branches form readily. Scion compatibility has not been a problem.

3.2.1.5 *Gisela 13* (P. cerasus x P. canescens)

Gisela 13 is a recently commercialized clonal rootstock just introduced in the USA and thus in need of widespread site evaluation. Sweet cherry trees on Gisela 13 are semi-vigorous, precocious, productive, develop horizontal branch angles, have good anchorage and no suckering. Relative to apple rootstocks, Gisela 13 is comparable to EMLA 106 or M7. To date, there have been no reports of incompatibility problems. Gisela 13 is reported to perform well in less fertile soils and better than Gisela 6 in hot climates.

3.2.1.6 *Gisela 17* (P. canescens x P. avium)

Gisela 17 is a recently commercialized semi-vigorous to vigorous clonal rootstock just introduced in the USA and thus in need of widespread site evaluation; however, it was tested from 1998 to 2007 in the coordinated NC140 trials. Initial interest was limited due to the focus on identifying dwarfing rootstocks, but sweet cherry trees on Gisela 17 were found to be vigorous, precocious and productive. In Prosser, Washington, trees on Gisela 17 were slightly more vigorous than on Gisela 6, but of similar vigor to Gisela 6 at Summerland, British Columbia. Relative to apple rootstocks, Gisela 17 is comparable to EMLA 111.

Trees on Gisela 17 develop horizontal branch angles, have good anchorage, no suckering, and are adaptable to a wide range of soil and climatic conditions, possibly being suitable for replant sites. The yield potential at Prosser was less than on Gisela 5 or 6, but intermediate between Gisela 5 and 6 in Corvallis, Oregon. Compatibility with cultivars has been good.

3.2.1.7 *Krymsk 7* (P. serrulata var. lannesiana)

Krymsk 7 is a recently commercialized clonal rootstock just introduced in the USA and thus in need of widespread site evaluation. It is the most vigorous of the Krymsk rootstocks to date, producing a tree slightly smaller than or similar in size to Mahaleb. Relative to apple rootstocks, Krymsk 7 is comparable to EMLA 111. Precocity is moderate, with production beginning in the third or fourth leaf. Krymsk 7 has good anchorage, minimal suckering, and is tolerant to PDV and PNRSV.

3.2.1.8 *Krymsk 5* (P. fruticosa x P. serrulata var. lannesiana)

Krymsk 5 is a precocious, semi-vigorous clonal rootstock that results in trees comparable in size to Gisela 12, though with somewhat less precocity and yield. Relative to apple rootstocks, Krymsk 5 is comparable to EMLA 106 or M7. Due to its intermediate level of productivity, it is particularly well suited for cultivars of higher productivity, such as Lapins, Sweetheart and Chelan.

Krymsk 5 is adapted to a wide range of soil types, with reports that it will grow well in heavier soils than Mazzard. Bred in Russia, it is well adapted to cold as well as dry climates. In addition, it is also reported to perform well in hotter climates. In California and the hottest regions of the PNW, leaves remain turgid in extreme heat and do not show the characteristic cupping of Gisela 6 trees. Trees are well anchored and do not need support. Suckering from the crown is low to moderate, especially in heavy soils.

Suckers growing from the roots of the tree row are minimal. The tree form is excellent, with wide branch angles and good shoot formation throughout the tree. Krymsk 5 is hypersensitive to PDV and PNRSV, but orchard losses to such infections have thus far been minimal.

3.2.1.9   *Krymsk 6* (P. cerasus x [P. cerasus x P. maackii])

Krymsk 6 produces a tree that is semi-vigorous, with a similar level of vigor as Gisela 6. Relative to apple rootstocks, Krymsk 6 is comparable to EMLA 106 or M7. Although Krymsk 6 is not as productive as Gisela 5, 6 or 12, it is more productive than Krymsk 5 and crops well with Regina. Like Krymsk 5, Krymsk 6 rootstocks seem to be adapted to both cold and hot climates as well as heavier soils. Trees are well anchored, but there is low to moderate root suckering that in some soils can become excessive. Tree form is good, with wide crotch angles and good branching. Krymsk 6 is hypersensitive to PDV and PNRSV, but orchard losses to such infections have thus far been minimal. Nevertheless, it is best to avoid planting new blocks of Krymsk 6 next to older blocks of trees that may be infected.

3.2.1.10   *Mahaleb* (P. mahaleb)

Mahaleb is usually a seed propagated rootstock that is slightly more precocious and slightly less vigorous than Mazzard. It is cold hardy and is one of the most drought tolerant cherry rootstocks, having deep roots. Mahaleb is, however, extremely sensitive to waterlogged soils as well as poorly drained soils that may be anaerobic for a short time during winter, as it is susceptible to *Phytophthora* root rot, oak root fungus (*Armillaria*), and root knot. Mahaleb is suited best to deep, well-drained loams and sands as well as the calcareous soils typical of Spain, southern France, southern Italy and Hungary; in similar locales around the world, Mahaleb is the preferred rootstock. In the PNW, Mahaleb generally is used only in light, sandy-loam soils, since trees readily die in ravines and other low-lying areas where water collects.

Some sweet cherry cultivars, such as Chelan, Tieton and Lapins, can be incompatible with Mahaleb, with decline appearing up to 6 years after planting. In addition, Mahaleb is attractive to gophers, and control measures must be pursued with diligence. There are a number of selections of Mahaleb that have been clonally propagated for specific traits and adaptations to local conditions. These include St. Lucia 64, Pontaleb, Bogdány and CEMA, but US growers mainly have access to seedling Mahaleb.

3.2.1.11   *MaxMa 14* (P. mahaleb x P. avium)

MaxMa 14 is a semi-vigorous clonal rootstock, producing a tree intermediate in size between Mazzard and Gisela 6. Relative to apple rootstocks,

MaxMa 14 is comparable to EMLA 106. For this reason, it is not recommended for super high density training systems such as super slender axe (SSA), but could do well with KGB or upright fruiting offshoots (UFO). It has been widely planted in France due to its precocity, moderate level of vigor and resistance to iron chlorosis on calcareous soils. In Chile, growers use MaxMa 14 due to its moderate resistance to bacterial canker; it is also less susceptible to crown gall.

MaxMa 14 has good scion compatibility and a broad adaptation to soil types and environmental conditions, although trees do best on sandier, lighter soils (though it requires irrigation during droughts) and will die in areas with high water tables. Trees have good anchorage and very little suckering. Tree form is open, but branching has been inadequate at times. As with Mahaleb, gophers prefer MaxMa 14 over other cherry rootstocks, so diligent control measures are necessary.

3.2.1.12   *MaxMa 60* (P. mahaleb x P. avium)

MaxMa 60 is a vigorous clonal rootstock grown in limited areas around the world, producing large well-anchored trees similar in size to Mazzard, but with greater precocity and excellent fruit size. It is used by growers in Chile due to its moderate resistance to bacterial canker. Some resistance to *Phytophthora*, crown gall and replant disease has also been reported. Adaptability to a range of soil types has been good.

3.2.1.13   *Mazzard* (P. avium)

Mazzard is a seed propagated rootstock that is well adapted to PNW soils, is winter hardy and graft compatible with all cultivars. In addition, due to high vigor and moderate productivity, premium fruit quality can be obtained with only moderate inputs in pruning and management. However, it lacks precocity, often not coming into production until the fifth or sixth leaf, or achieving full production in low density orchards until the twelfth leaf. Vigorous growth makes it difficult to control in high density plantings, and the large tree size greatly limits picker efficiency and increases worker hazards associated with harvest. Mazzard does well in a wide range of soils from sandy-loam to clay-loam, though, as with other cherry rootstocks, it performs poorly in slow draining or wet soils. Tree anchorage is excellent, with occasional root suckers.

3.2.1.14   *Mazzard F.12/1* (P. avium)

F.12/1 is a clonally propagated, vigorous selection of Mazzard used in many locations around the world instead of seedling Mazzard. F.12/1 is less susceptible to bacterial canker than Mazzard. It performs best on fertile, irrigated soils and should not be planted on light or heavy soils. Many nurseries, however, prefer not to grow F.12/1 due to its high susceptibility to crown gall. F.12/1 is more vigorous than Mazzard in many locations.

3.2.1.15 *WeiGi 1* (P. cerasus x O.P.)

WeiGi 1 is a recently commercialized, semi-vigorous clonal rootstock just introduced in the USA and thus in need of widespread site evaluation. WeiGi 1 is comparable to Gisela 6 in vigor, precocity and productivity, with no suckering, and is reportedly adaptable to hot, dry conditions. Relative to apple rootstocks, WeiGi 1 is comparable to EMLA 106 or M7.

3.2.1.16 *WeiGi 3* (P. cerasus x [P. avium x P. canescens])

WeiGi 3 is a recently commercialized, semi-vigorous clonal rootstock just introduced in the USA and thus in need of widespread site evaluation. WeiGi 3 is comparable to Gisela 6 and Gisela 12 in vigor, precocity and productivity, with no suckering, and is reportedly adaptable to hot, dry conditions. Relative to apple rootstocks, WeiGi 3 is comparable to EMLA 106.

3.2.2 Semi-dwarfing to dwarfing rootstocks

3.2.2.1 *Corette Series* (P. cerasus hybrids)

Five new hybrid, dwarfing clonal rootstocks have recently been released by Michigan State University. Cass and Clare are complex hybrids with *P. avium*, *P. cerasus* and *P. fruticosa* parentage; Clinton and Crawford have *P. canescens* and *P. cerasus* parentage; and Lake has *P. avium* and *P. fruticosa* parentage. In evaluations conducted in Oregon and Washington states, tree vigors on these rootstocks were all smaller than Gisela 5, more in line with Gisela 3 size. Relative to apple rootstocks, the Corette rootstock series ranges in vigor from Bud 9 or M.9T337.

In an earlier Washington trial, precocity and productivity (number of flowering spurs) was very high, with Lake, Cass and Clinton exceeding that for Gisela 5. Due to this prolific flowering, fruit grown on these rootstocks may need to be thinned to achieve acceptable fruit size and quality. However, this may enable growers to more consistently produce optimal yields. This Washington trial found Lake, Cass and Clare advanced in ripening by 4 days compared to Gisela 5.

3.2.2.2 *Gisela 3* (P. cerasus x P. canescens)

Gisela 3 is a recently commercialized clonal rootstock still in need of widespread site evaluation, but of significant interest because trees on Gisela 3 are dwarfing, very precocious, very productive, develop horizontal branch angles, have good anchorage and no suckering. To date, there have been no reports of incompatibility problems. Currently, Gisela 3 is among the weakest cherry rootstocks available, probably best suited for very high density training systems like SSA. Relative to apple rootstocks, Gisela 3 is comparable to B.9 or M.9T337.

3.2.2.3 Gisela 5 (P. cerasus x P. canescens)

Gisela 5 was the first widely successful semi-dwarfing rootstock in the USA and Europe. In the PNW, Gisela 5 is known to reduce vigor by up to 50% or more compared to Mazzard. Relative to apple rootstocks, Gisela 5 is comparable to EMLA 26. The medium–low vigor of this rootstock coupled with very high fruit production has led to fruit size and quality issues when leaf-to-fruit ratios become unbalanced. This problem is accentuated when Gisela 5 is combined with productive cultivars such as Lapins and Sweetheart. When properly pruned and grown on deep, fertile soils, it may be suitable for very high density plantings of 1000–2000 trees/ha (400–800 trees/acre) and is probably best suited to spindle systems such as the Vogel central leader (VCL), tall spindle axe (TSA) or the SSA.

Gisela 5 has become the most popular rootstock in Germany and other parts of northern Europe, and is widely grown in Michigan and the northeast US, where summer temperatures are relatively moderate and humid. Gisela 5 has failed to gain widespread acceptance in hotter cherry growing regions like the US Pacific Coast states, Europe's Mediterranean countries, and Chile. The weak growth that growers in these regions experience with this rootstock is thought to be due to shutting down of photosynthesis resulting from high temperature-related water stress.

Gisela 5 produces trees that are open and spreading with wide branch angles, but branching may be sparse. Anchorage is usually adequate, but some growers have taken the precaution to support the trees. Gisela 5 does not perform well in heavy soils and needs good drainage. Trees on Gisela 5 are sensitive to replant disease and thus should only be planted on virgin sites, where the soil has been properly treated with fumigants prior to planting, or planted at higher densities to compensate for weaker growth. Suckering is not usually a problem. Trees on Gisela 5 rootstock have shown good winter hardiness and scion compatibility. Gisela 5 tends to advance both flowering and fruit ripening by up to 2 to 4 days, a potential advantage for early ripening cultivars where an early harvest window provides higher returns. However, this may be disadvantageous in a frost susceptible site or when later ripening is desired for marketing purposes.

3.2.2.4 W720 (P. cerasus)

A recently commercialized, dwarfing clonal rootstock just introduced in the USA and thus in need of widespread site evaluation. W720 is precocious, productive and reportedly adaptable to a range of soil types. Relative to apple rootstocks, W720 is comparable to EMLA 9.

3.2.2.5   *WeiGi 2* (P. cerasus x [P. avium x P. canescens])

WeiGi 2 is a recently commercialized, dwarfing clonal rootstock just introduced in the USA and thus in need of widespread site evaluation. WeiGi 2 is slightly more vigorous than Gisela 5 in vigor, precocity and productivity, and is reportedly adaptable to hot dry conditions. In tests conducted in Provence, France, WeiGi 2 produced a tree about the same size as Gisela 5. Relative to apple rootstocks, WeiGi 2 is comparable to EMLA 26.

Table 3.2 provides attributes common for many of the commercially available rootstocks found around the world.

3.3   Summary

Unfortunately, no one rootstock can satisfy all the requirements for consistently producing high yields of large, firm fruit of premium quality. Growers are advised to consider carefully the effects of each specific scion–rootstock combination as a function of environmental and cultural practices when replanting an orchard (Table 3.1). Selecting the proper rootstock depends not only on the management skills of the grower, but also on the scion cultivar, training system, and site climate and soil selected for the orchard. For more information about matching the attributes of specific cultivars to rootstocks, see Chapter 2, 'Cherry Flowering, Fruiting and Cultivars'.

Dwarfing, semi-dwarfing and even semi-vigorous rootstocks have major economic advantages over full-size rootstocks (Seavert and Long, 2007). Precocity and higher density plantings mean earlier high yields and a faster return on investment. The development of these new, precocious rootstocks has been almost as significant to the sweet cherry industry as to the apple industry several decades ago. When compared to Mazzard, Colt and even Mahaleb, size-controlling rootstocks have allowed sweet cherry growers an opportunity to plant high density, pedestrian orchards that become profitable more quickly, are more readily protected with orchard covering systems, and promote greater labor efficiency, easier management, and a safer and more productive work environment.

Table 3.2. Attributes of various commercially available cherry rootstocks (from L.E. Long).

	Tree size (% full size)	Precocious	Advance bloom/harvest	Compatibility	Root suckers	Anchorage
Cass	40–50	Yes	4 days	?	Low	Good
Clare	40–50	Yes	4 days	?	Moderate	Good
Clinton	40–50	Yes	0–1 day	?	No	Fair
Colt	100	No	No	Good	No	Good
Crawford	40–50	Yes	0–1 day	?	No	Good
Gisela 3	40–50	Yes	0–2 days	Good	No	Fair–poor
Gisela 5	50–60	Yes	2–4 days	Good	No	Fair–good
Gisela 6	80–85	Yes	0–1 day	Good	No	Fair
Gisela 12	85–90	Yes	No	Good	No	Fair–good
Gisela 13	85–90	Yes	No	Good	No	Good
Gisela 17	85–90	Yes	No	Good	No	Good
Krymsk 5	85–90	Moderate	No	Good	Moderate	Good
Krymsk 6	75–85	Moderate	No	Good	Moderate	Good
Lake	50	Yes	No	?	Moderate	Good
Mahaleb	90–95	Slight	No	Fair–good	No	Good
MaxMa 14	85–90	Moderate	No	Good	No	Good
Mazzard	100	No	No	Good	Low	Good
WeiGi 1	80–85	Yes	No	Good	No	Good
WeiGi 2	60–80	Yes	No	Good	No	Good
WeiGi 3	85–90	Yes	No	Good	No	Good

For More Information

Hrotkó, K. and Rospara, E. (2017) Rootstocks and improvement. In: Quero-Garcia, A., Iezzoni, A., Pulawska, J. and Lang, G. (eds) *Cherries: Botany, Production and Uses*. CAB International, Wallingford, UK, pp. 117–139.

Long, L.E. (2003) *PNW 543 Cherry training systems, selection and development*. Oregon State University Extension and Experiment Station Communications, Corvallis, Oregon.

Long, L.E. (2007) *PNW 592 Four simple steps to pruning cherries on Gisela and other productive rootstocks*. Oregon State University Extension and Experiment Station Communications, Corvallis, Oregon.

Long, L.E., Whiting, M. and Nunez-Elisea, R. (2007) *PNW 604 Sweet cherry cultivars for the fresh market*. Oregon State University Extension and Experiment Station Communications, Corvallis, Oregon.

Acknowledgements

Portions of this chapter were adapted with permission from Oregon State University. *PNW 619 Sweet cherry rootstocks for the PNW*, copyright 2020. Oregon State University Extension and Experiment Station Communications, Corvallis, Oregon. http://catalog.extension. oregonstate.edu/pnw619 (accessed 19 June 2020).

References

Howell, W.E. and Lang, G.A. (2001) Virus sensitivity of new sweet cherry rootstocks. *Compact Fruit Tree* 34(3), 78–80.

Lang, G., Howell, W. and Ophardt, D. (1998) Sweet cherry rootstock/virus interactions. *Acta Horticulturae* 468, 307–314. DOI: 10.17660/ActaHortic.1998.468.36.

Long, L.E. (1995) Colt rootstock may be answer for cherry replant disease. *Good Fruit Grower* 46(4).

Perry, R., Lang, G., Andersen, R., Anderson, L., Azarenko, A. *et al.* (1998) Performance of the NC-140 cherry rootstock trials in North America. *Acta Horticulturae* 451, 225–229.

Seavert, C. and Long, L.E. (2007) Financial and economic comparison between establishing a standard and high density sweet cherry orchard in Oregon, USA. *Acta Horticulturae* 732, 501–504. DOI: 10.17660/ActaHortic.2007.732.76.

Spotts, R.A., Wallis, K.M., Serdani, M. and Azarenko, A.N. (2010) Bacterial canker of sweet cherry in Oregon: infection of horticultural and natural wounds, and resistance of cultivar and rootstock combinations. *Plant Disease* 94(3), 345–350. DOI: 10.1094/PDIS-94-3-0345.

Planning a New Cherry Orchard

4

Growing fruit trees is a long term financial commitment and establishing an orchard is an expensive enterprise. Selecting the right orchard site is arguably the most important decision that a sweet cherry grower will make and it will affect all other managerial decisions related to the orchard. Choosing the wrong site can affect both fruit yield and quality, and ultimately the financial viability of the operation. On the other hand, choosing a good site will reduce expenses and increase profits. Properly evaluating a prospective orchard location, determining available resources and establishing the orchard takes careful planning. Due to the expense and long term commitment involved in establishing an orchard, it is important to consider all factors carefully before proceeding, including the initial costs of land, trees and infrastructure, as well as site selection, labor availability and the target cherry market(s).

4.1 Orchard Economics

4.1.1 Researching market and production information

The first step in determining whether to start an orchard enterprise is to learn as much as possible about the potential market opportunities, since anticipated market returns will guide how much can be invested in orchard establishment and orchard management. Successful, superior orchards result only when market criteria are well understood and exacting standards are met and maintained during cherry production. By developing a thorough understanding of cherry markets and production variables, many basic mistakes can be avoided during tree establishment.

In most tree fruit production areas in the USA, the Extension Service associated with a state land-grant university often has publications and bulletins on tree fruit production that can be a tremendous asset for planning and developing an orchard. In addition, classes, seminars, workshops and tours are often sponsored by the Extension Service. American growers can contact their local Extension office and ask to be placed on postal or electronic mailing lists to receive announcements and updates. In areas without Extension expertise in fruit growing, expert help may be sought from private orchard consultants and/or other growers to provide guidance during orchard establishment.

4.1.2 Developing a business plan

Orchard establishment is costly and it will take at least three years before there is any significant return on investment. Clearly, it is important to procure adequate financial resources to sustain the operation until that time. In determining potential return, it is important to understand that the price received for fruit depends not only on fruit quality, but also other market considerations such as harvest timing, market volumes and other vagaries of the target marketing system.

The cost of planting, developing, and maintaining cherry orchards can be higher than that of many other perennial crops due to the intensive nature of production, which requires relatively large, periodic amounts of skilled manual labor. In addition, due to the delicate nature of the fruit and inherent traits of sweet cherry trees (i.e. early bloom when frost risks are higher, and susceptibility to significant diseases and insects like bacterial canker, brown rot and fruit flies), cherry production can be a financially risky venture.

In the USA, the Extension Service associated with the land-grant university in each major sweet cherry production area often has enterprise budgets and cost-and-return studies that can help identify and evaluate the financial obligations, potential risks and anticipated returns to consider when investing in an orchard operation. Often these will cover a wide range of scenarios, including intensive (high density) and standard orchards, and the returns on investment that are typical in the area. Extension Service-based small farms programs are helpful for developing a business plan, determining crop production costs and returns, and identifying potential funding sources. It is critical to confirm the availability of financing before committing to buy land or establish an orchard. Agricultural loan opportunities should be explored with a local bank's agricultural officers or with representatives of the Farm Loan Programs at the US Department of Agriculture (USDA) Farm Service Agency.

The Natural Resource and Conservation Service of the USDA (USDA–NRCS) provides a partnership effort to help private land owners and managers conserve their soil, water and other natural resources through voluntary conservation programs that provide environmental, societal, financial and technical benefits. Of particular interest to prospective cherry growers is the Environmental Quality Incentives Program (EQIP) that promotes agricultural production and environmental quality. EQIP offers financial and technical help to assist eligible participants install or implement structural and management practices on suitable agricultural land. More information can be obtained from local NRCS offices, as each region has different initiatives.

4.1.3   Determining target market(s) and production region infrastructure

Cherry marketing options should be considered very carefully. Small quantities of fruit can be sold at fruit stands or farm markets. However, production of more than one or two acres (0.5–0.75 ha) will need to be marketed through a packing house. Very early in the orchard planning process, prospective growers should visit packing house personnel in the area to determine if they are accepting fruit from new growers and if they are willing to work with you. Establishing a relationship with a packing house field staff member early on will help in growing fruit of the quality needed by the packing house.

Also consider access and distance to the packing house. The delicate nature of sweet cherries means that road surfaces should be smooth, as transport over rough roads can damage the carefully picked fruit before it reaches the packing house. In addition, packing facilities should be located close enough so that harvested fruit can be cooled to 10°C (50°F) within 2 to 4 h.

Furthermore, wholesale price returns are related to harvest timing, relative to peaks and changes in industry supply. In the USA, cherry supply usually peaks throughout July, corresponding with lower prices. Cherries that are harvested earlier than the 4th of July holiday or later than the beginning of August tend to return the highest prices, so cultivar selection is critical with respect to harvest date. The range of good quality cherry cultivars can now provide a continuous harvest for 1.5 months or more from the same orchard site. If the potential orchard location is in an early harvest area, it may be advantageous to grow primarily early ripening cultivars to take advantage of higher market prices often received for fruit that precedes the supply peak. If the site is at a higher elevation or more northerly latitude, late cultivars may provide a higher return.

4.1.4   Determining labor needs and sources

Fresh market sweet cherry production is a labor intensive operation, requiring permanent labor for certain routine tasks and temporary labor for intensive periods, such as harvest. Although small-scale growers may be able to manage most of the operation by themselves for much of the year, even the smallest of operations will likely need extra employees during harvest. It is estimated that approximately one full time worker is needed for every 2 ha (5 acres) of cherries, as well as an additional 6 to 8 temporary laborers to help with dormant pruning, spring pruning, flower thinning and harvesting. Additional temporary labor will be required if the operation includes packing the fruit (versus delivery to a central packing house).

Long before the trees reach maturity, it is wise to determine how this labor will be sourced. Contract pruning and picking services may be available in some production areas, which may be a viable alternative to employing contract labor. The coordination of picking labor requires recruitment of pickers from reliable local or distant sources, management of picker credentials and payroll, and often provision of housing for migrant workers. One current source of temporary agricultural workers in the US is the H-2A program administered by the US Citizenship and Immigration Services department (www.uscis.gov/working-united-states/temporary-workers/h-2a-temporary-agricultural-workers, accessed 19 June 2020). Harvest timing that is not during the peak of production may allow sourcing of pickers from a neighboring orchard, though this must certainly be worked out well in advance. These are only a few of the issues that must be resolved prior to the first harvest. Visiting other growers in the area can help identify labor sources and options that have worked well for others.

If migrant labor housing will need to be built, start by contacting the local (i.e. county) planning office to make sure that all appropriate permissions are obtained in a timely manner. The USDA and the US Department of Housing and Urban Development (HUD) supply federal requirements, as well as potential grants and loans, pertaining to migrant labor housing on their websites (USDA: www.rd.usda.gov/programs-services; HUD: www.hud.gov/groups/hudprograms; both accessed 19 June 2020). States must abide by these regulations, but can be more restrictive; therefore, it is a good idea to check with state policies as well before starting a migrant worker housing project.

4.2 Orchard Site Evaluation

Whether the planned orchard will be developed from existing farmland or from purchased land currently under another use, it is important to evaluate several crucial site suitability issues, including climatic conditions, topography, soil characteristics and available water quality.

4.2.1 Site climatic conditions

Climate plays a critical role in determining whether a site is suitable for producing premium sweet cherries. The climatic conditions that have the greatest impact on cherry production are temperature, precipitation and wind. When possible, several years of recent weather data should be acquired and thoroughly researched. Site-specific weather data may be available where a regional agricultural weather station network has been established (see 'For More Information' at the end of this chapter). Lacking this, municipal weather data may be available, but this has major limitations as microclimates in an agricultural area may be significantly different from even small urban areas. If no other data are available, consider installing a temperature data logger at the potential orchard site and talk to neighbors who may be able to provide a history of the local weather conditions.

4.2.1.1   Temperature

Cherry floral and vegetative bud development begins in the summer and goes dormant in the fall in response to shorter days and cooler temperatures. When fully dormant (termed 'endodormant'), the buds will not grow in response to warm temperatures until they have been exposed to a cumulative period of cold temperatures (termed the 'chilling requirement'), which varies from cultivar to cultivar. Temperatures that are considered to be 'chilling promotive' range from just above freezing (~1°C or ~34°F) to ~12°C (54°F). Temperatures below this range contribute little to achieving the chilling requirement, and temperatures above 16°C (61°F) can have a negative effect on previous chilling. Most cherry cultivars require from 900 to 1200 h of chilling temperatures, though some require less, and the development of cultivars with low chilling requirements (600 h or less) is increasingly being pursued in breeding programs to extend production potential into regions with milder winter climates.

The chilling requirement helps to protect the trees by keeping them from breaking dormancy and beginning growth during warm spells in the winter. Once the trees receive the required amount of chilling, the buds transition from 'endodormant' to 'ecodormant', a state in which they will respond to warm weather and begin to grow and develop normally. If, however, the chilling requirement is not met before temperatures begin warming irrevocably in spring, such as can happen in more Mediterranean-type climates, vegetative and floral budbreak may be weak, delayed and/or extended, resulting in asynchronous growth, possible bud abortion, reduced fruit set and uneven ripening. This chilling requirement should not be a problem in production regions with more temperate climates like the Pacific Northwest (PNW) and north-central Europe, but in California, Spain, the production regions of central Chile, and other areas with mild winters, the application of dormancy-breaking treatments, such as hydrogen cyanimide or certain nitrogen-based compounds, may be required to improve bloom and fruit set. However, these treatments have limitations (see Chapter 5, 'Orchard Establishment and Production'). Other alternatives include partial shading with nets or plastic covers, or pulsed evaporative cooling with microsprinklers, during winter on sunny days to enhance or preserve chilling unit accumulation by minimizing the potential for buds to heat up from solar radiation even as air temperatures are cool.

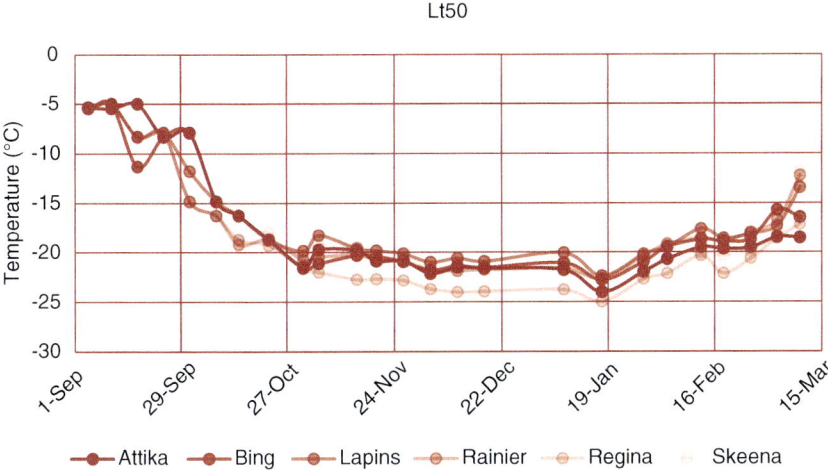

Fig. 4.1. This chart shows the lethal temperature to kill 50% of flower buds (Lt50) of several cultivars at the Oregon State University Mid-Columbia Agricultural Research and Extension Center (courtesy of D. Gibeaut).

As discussed above, low temperatures are necessary during winter to transition from endodormancy to ecodormancy and begin spring growth. These temperatures are also necessary for cold acclimatization, the process by which plant tissues (including flower buds) become progressively more resistant to freeze damage in the fall. Cold damage in sweet cherries can occur in fall, winter or spring, depending on the severity of the low temperature and when it occurs in relation to the dynamics of cold acclimatization and de-acclimatization during fall and winter, and on the stage of budbreak and flowering during spring. Damage in fall usually occurs when temperatures remain relatively warm late into fall, then drop quickly to well below freezing, before the tree has fully cold-acclimatized. Typical damage from such events includes cambial damage and bud death. Some mismanaged cultural practices can predispose cherry trees to such damage, such as irrigation or soil nitrogen applications or pruning too late during the growing season or into the fall, thereby promoting ongoing succulent growth that fails to cold-acclimatize.

Cherry buds begin cold-acclimatizing in fall concomitant with shorter photoperiods, and become most cold-hardy in response to decreasing temperatures in late fall and early winter (see Fig. 4.1).

During a cold period, cherry buds are capable of gaining ~1.5°C (~3°F) of hardiness per day, so long as buds are progressively exposed to sub-freezing temperatures below –1°C (30°F). If temperatures rise above freezing, buds can quickly lose their added hardiness at 1–2°C (2–4°F) per hour. The capacity to withstand sub-freezing temperatures during this

Fig. 4.2. Flower and spur damage on Cristallina sweet cherry, evidenced by tissue browning after temperatures fell to −20°C (−5°F) in mid-November, before trees obtained endodormancy (Oregon) (courtesy of L.E. Long).

period is largely due to the ability of plant tissue to supercool (cooling below the freezing point of a liquid without solidifying).

As noted above, during fall, buds that are exposed to rapidly declining temperatures are killed at relatively higher temperatures, irrespective of the previous temperature exposure. Buds that are subjected to extreme cold temperatures prior to developing full hardiness may show signs of tissue damage or death, as indicated by internal oxidative browning when cut open or abortion during winter (Fig. 4.2). Buds damaged in this manner, whether floral or vegetative, may open in the spring, but lack viable pistils or may collapse as temperatures begin to warm. In less extreme situations, where floral tissue is slightly damaged, fruitlets may even develop but drop later as water requirements increase.

Once fully dormant and cold-acclimatized, most healthy cherry trees can often withstand winter temperatures as low as −32°C (−25°F), as long as temperatures decline slowly over a period of several days or weeks, or have remained consistently cold following such acclimatization. Temperatures lower than this, however, may cause severe damage or tree death even in fully acclimatized trees during midwinter. Similarly, rapid declines in temperature can damage fruit or vegetative buds, kill branches or cause vascular damage to trunks or branches (Fig. 4.3.). A rapid increase in temperature for several days, followed by quickly falling temperatures, can be severely damaging. Thus, a location where temperatures fluctuate greatly in the winter may be more hazardous than a colder location where temperatures decline slowly in the fall and remain cold until the spring.

Once the chilling requirement has been met, buds become much more responsive to warm temperatures, with the ability to lose acclimatization

Fig. 4.3. Extreme winter temperatures can cause significant damage to cherry trees when not properly acclimatized (Oregon) (courtesy of C. Kaiser).

(de-acclimatize) rapidly and initiate growth processes such as remobilization of storage reserves to buds and final stages of bud development before opening. However, trees will remain cold-hardy as long as low temperatures persist, but exposure to warmer temperatures raises the threshold temperatures at which buds are damaged. These threshold temperatures for cold damage change dynamically with the stage of bud development in late winter/early spring (Fig. 4.4).(Table 4.1.)

Flower buds maintain the ability to regain hardiness until about the side green stage (stage 2 of 8), at which point they can no longer re-acclimatize

Fig. 4.4. Stages of bud development in the spring as trees move from dormancy through bloom to post-bloom (also see Table 4.1) (courtesy of Washington State University AgWeatherNet).

Table 4.1. Critical temperatures for flower buds

Bud stage	Phenology description	Critical temperatures °C (°F)		
		10% bud kill	50% bud kill	90% bud kill
0	Dormant	−16 (Flynn, 2009)	−20 (−4)	−29.5 (-21)
1	First swell	−9.5 (15)	−15.5 (4)	−21.5 (-7)
2	Side green	−4 (25)	−10.5 (13)	−16.5 (2)
3	Green tip	−1.5 (29)	−8 (18)	−14 (7)
4	Tight cluster	−1.5 (29)	−5.5 (22)	−9 (16)
5	Open cluster	−1.5 (29)	−4.5 (24)	−7 (19)
6	First white	−1.5 (29)	−4.5 (24)	−7 (19)
7	Balloon stage	−1.5 (29)	−4 (25)	−5.5 (22)
8	Full bloom	−1.5 (29)	−4 (25)	−5.5 (22)
9	Post-bloom	−1.5 (29)	−3.5 (26)	−5 (23)

From T. Einhorn, Michigan State University, and D. Gibeaut, Oregon State University, unpublished.

below the threshold temperatures for cold damage. At bloom, temperatures just below freezing can cause minor to significant damage, depending on the duration of the frost or freeze event.

In the PNW, chilling requirements generally are met sometime in late December to mid-January. In Michigan, chilling requirements may be met at about the same time, or if daily temperatures go below 0°C (32°F) in December and remain consistently below freezing through January and February, the chilling requirements may not be met until late February or even early March. In many northern hemisphere locations, the winters of 2014 and 2015 were among the warmest on record; thus, to evaluate potential orchard sites in low chilling regions, winter chilling data from these years may be valuable as 'worst case' scenarios to determine whether low chilling cultivars, dormancy-breaking chemicals or dormancy-enhancing orchard technologies may need to be considered at the site evaluation stage.

As the weather warms in spring, buds begin to swell, and the flower primordia within the buds become most susceptible to freezing; this susceptibility to frost damage persists through flowering and fruit set. Spring frosts may limit production by killing flower buds, flowers or young fruits, but mild frosts after fruit set can also affect fruit quality by causing superficial marking on the fruit surface. Low-lying areas in draws or valleys where cold air drainage is restricted, and high elevation sites, tend to be more prone to both spring frosts and extremes in winter temperatures. In temperate growing areas, these sites should be avoided.

Fig. 4.5. Cherries damaged by excessive heat (oregon). (courtesy of I.e. long).

During bloom, air temperatures of at least 13°C (55°F) are required for honeybee foraging, and maximum foraging occurs at temperatures above 18°C (65°F). Similar minimum temperatures are needed for pollen germination, pollen tube growth and ovule fertilization to take place in a reasonable amount of time to set the crop. Although temperatures near 24–27°C (75–80°F) significantly raise bee activity, high temperatures can result in poor fruit set due to rapid pollen desiccation, increased respiration of growing pollen tubes and accelerated ovule deterioration prior to fertilization. Some cultivars, such as Regina, are particularly sensitive to warm bloom temperatures, as ovule viability is significantly shortened at higher temperatures, substantially decreasing fruit set.

Following fruit set, hot weather above 35°C (95°F) during fruit development can cause tree stress, reducing photosynthesis, inhibiting fruit growth, and leading to softening of cherries during ripening. These same high temperatures also can scorch fruit, causing sunburn and fruit browning (Fig. 4.5). Certain cultivars, such as Lapins and Skeena, are particularly sensitive to high temperatures, and in regions subject to such temperatures, should be grown on vertical canopy training systems such as the upright fruiting offshoot (UFO) and Kym Green bush (KGB), in which branch aspect and/or leaf density help to protect the fruit from the heating effect

Fig. 4.6. As flower buds are forming after harvest, excessive heat can cause malformation of the developing ovary leading to conjoined 'double' fruit the following year (courtesy of L.E. Long).

of infrared radiation on hot days. Fruit grown on pendant wood seems to be most sensitive to this type of damage.

Approximately, one month after harvest, for a period of about two weeks, temperatures approaching 35–37°C (95°F–100°F) and above during the earliest stages of flower bud development for the next year's crop can cause malformation of the floral primordia. The floral malformation affects the developing ovary, often resulting in conjoined 'double' fruit or a single fruit with an attached shriveled second rudimentary ovary (Fig. 4.6.), both of which would be considered a cull (see Chapter 7, 'Sweet Cherry Training Systems' for control methods).

For some fruit crops, such as wine grapes, the number of heat units, or 'growing degree days' (GDD), that occur during the growing season determines whether certain varieties will fully ripen or not. Since cherries are an early- to midsummer ripening fruit in most typical growing regions, GDD generally are not a site consideration except for at extreme northern latitudes, such as Norway, Scotland or Alaska (there are no comparable land areas with extreme southern latitudes). Experience in New Zealand has found that midseason sweet cherries require a minimum of about 800 GDD (Celsius) or 1440 GDD (Fahrenheit) to ripen. The GDD for a particular region can be calculated by a mathematical equation using the seasonal sum of the average daily temperatures for each day subtracted from a threshold (or base) temperature of 4°C (39.2°F) using the equation:

$$(T_{max} - T_{min})/2$$

where T_{max} is the daily maximum and T_{min} is the daily minimum temperature. In the northern hemisphere, the typical season for calculations is from 1 March to 31 October; in the southern hemisphere, it is from 1 September to 30 April.

4.2.1.2   Precipitation

The ideal site for growing cherries would have minimal rain during the growing season, with precipitation largely occurring during dormancy in winter as rain or snow. This winter precipitation would recharge the soil moisture profile and replenish local ponds or regional reservoirs, providing irrigation water supplies when needed for crop growth. Rain during spring and summer can create an environment conducive to serious infectious diseases, including bacterial canker, brown rot and leaf spot, and rain during the final stage of fruit development (rapid fruit growth and ripening) often leads to fruit cracking. Even during late fall and winter, rain can disseminate bacteria and spores that lead to growing season infections. Site evaluation should include researching the average annual precipitation and when it typically occurs, particularly how much rain occurs during fruit maturation and ripening, and the duration of these events. Similarly, hail at any time during spring or summer can cause major damage to tree canopies and the fruit crop, sometimes resulting not only in crop loss but also bacterial (such as canker) or fungal (such as silver leaf) infections. Locations that are prone to frequent hail should be avoided or should include the cost of mitigation strategies such as hail netting (which could be defrayed somewhat if other benefits, such as protection of the crop from birds, could also be achieved).

Most cherries can survive some rain even at their most sensitive stage (as long as the fruit dries quickly), but extended rainfall can damage any cherry crop. As cherries begin to change color from green to yellow (straw color), they become susceptible to absorption of water on the fruit surface through the fruit cuticle, as well as uptake of water through the roots and into the vascular system of the trees that can cause internal pressure on the fruit. Both situations can lead to severe rain-induced fruit cracking (Fig. 4.7.). In some years, cracking can be a serious problem that destroys enough of the crop to make it economically unfavorable to harvest. The

Fig. 4.7. Rain-induced fruit cracking of sweet cherry fruits (courtesy of C. Kaiser).

potential for significant rain-induced crop loss can be mitigated somewhat by planting multiple cultivars that ripen at different times or that are less susceptible to cracking, such as Regina or Kordia (known in the USA as Attika). However, no cultivar is completely resistant to rain-induced fruit cracking.

To reduce the potential for rain and hail damage, some growers establish orchard structures that facilitate the covering of trees with solid plastic, or hail and/or bird netting, to protect the ripening fruit (see Chapter 9, 'Fruit Ripening and Harvest'). Although these structures can be the difference between a marketable crop and complete crop loss, they are also very expensive, and a careful analysis needs to be conducted in order to determine the value and typical frequency of the risks that such covers will mitigate, in order to consider the economic viability (return on investment) for such a strategy.

4.2.1.3   Wind

Winds of ≥15 km/h (≥10 mph) during bloom will negatively affect bee foraging, and bee flights may stop altogether if winds reach 25 km/h (15 mph). If winds continue through bloom, fruit set can be significantly reduced. High winds can also affect ripening fruit, causing scuff marks and bruising, increasing cullage and reducing packout.

Constant daytime winds during spring and summer can interfere with pesticide applications. In these situations, it is often necessary to spray at night. If orchards are near homes, however, night-time spraying may be disruptive to neighbors.

Winds can also make tree training difficult. In windy areas, growers often tie branches so that they face into the wind to prevent lopsided trees. It also may be necessary to support trees grown on weaker rootstocks or with poor anchorage (such as Gisela 6 with some cultivars) to prevent tree lean away from the direction of the prevailing wind.

Site evaluation should include determination of how often the wind blows at the potential site, the average wind speed during the growing season, and the direction of the prevailing wind. Planted windbreaks or some types of orchard covers may need to be considered to help mitigate wind problems.

4.2.2 Topography

A gentle slope facilitates good drainage of both rainfall and air. On heavier soils, a slope causes water to move away from root systems and provides better aeration. Cold air, being heavier than warm air, flows downhill like water, moving away from the trees. This helps reduce the potential for frost just before and during bloom. There should be a clear path for cold air to flow completely out of the orchard; woodlands and brush at the bottom of hills can create 'air dams' that cause cold air to 'back up' the slope. Sites at the bottoms of hills, on flat areas or in low spots may be at greater risk of spring frost damage. However, slopes that are too steep increase the potential for soil erosion and are unsafe for equipment operation. A slope of less than 10% (<6°) is ideal for cherry orchards as it will allow for drainage of both water and cold air away from the trees. Steeper slopes, up to 20% (~11°), are still quite manageable; however, slopes over 40% (~22°) become problematic and are not suitable for efficient orchard production. Where mechanization may be considered in the future, slopes in excess of 26% (~15°) should be avoided.

The orchard aspect (exposure) affects sunlight exposure patterns, air drainage, the potential for wind damage, time of bloom and harvest, incidence of winter freeze and spring frost injury, and disease incidence. In the northern hemisphere, north-facing sites do not warm as quickly during the day, retain moisture longer on trees, and delay bloom and ripening (which may or may not be a disadvantage, depending on markets, frost potential, etc.). Southern exposures are generally warmer and drier, and have more intense sunshine throughout the year, which leads to earlier bloom and harvest, as well as greater potential for sun-damaged fruit. The opposite north–south directional effects occur in the southern hemisphere. Across both hemispheres, western exposures warm up more slowly during the day than eastern exposures, thereby retaining moisture longer on trees and surrounding vegetation. This prolongs higher humidity within the canopy and may increase the potential for disease.

Orchard exposure in the direction of the prevailing winds increases the potential for wind damage to fruit and asymmetrical canopy development. Sites that are particularly windy also require more irrigation due to greater evapotranspiration. Living or artificial windbreaks, as well as plastic or netting orchard covers, can help mitigate some of the negative effects of wind during the growing season, but excessively windy sites are best avoided, as strong winds can pose dangers to covering structures themselves.

Air temperatures generally decline with higher latitude and elevation. According to the US National Aeronautics and Space Administration (NASA), air temperature declines 4°C (7°F) between sea level and 500 m (0 ft to 1640 ft) in elevation (NASA, 2020). However, the scale is not linear as elevation increases. Between 0 and 500 m, air temperature drops by 0.008°C for every meter of elevation gain; but between 500 and 1000 m (1640 to 3280 ft), air temperature drops by 0.006°C per meter (for a total of 3°C or 5.5°F). This may seem like a small difference, but when accumulated over the growing season, it can mean a delay in harvest of 14 or more days. It also can mean the orchard is more prone to frost and winter freeze damage.

Large bodies of water (e.g. lakes and reservoirs) can have a moderating influence on air temperature fluctuations in nearby orchards; for this reason, most cherry orchards in Michigan are planted within about 80 km (50 miles) of Lake Michigan, one of the Great Lakes. The larger the body of water, the greater the moderating effect, and the further the distance from the body of water, the less moderating the influence. Orchards within the zone of influence experience less extreme low air temperatures in winter, and remain cool for longer in the spring (particularly when there is significant ice cover on the lake), delaying budbreak while the risk of spring frost decreases.

4.2.3   Soil characteristics

Soils are important for providing anchorage, nutrients, water and oxygen needed for root growth. When choosing a site, a number of soil criteria need to be evaluated, including depth, presence of compacted layers, texture, slope, and drainage or moisture problems such as a high water table. Evaluating these soil conditions prior to planting is critical for determining site suitability and planning for a successful orchard operation. Once the site is planted, it becomes much more difficult to modify any physical limitations of the soil and correct any nutritional deficiencies, so the soil should be of at least moderate fertility levels, without extremes in pH or high electrical conductivity (EC).

Generally, soils are shallower at the tops of slopes as erosion carries soil downhill. For this reason, it is imperative to check the soil depth at the top, middle and bottom of the hill, which can be done with a mechanized soil auger or by digging a soil pit to examine the profile. Cherry trees need at least 90 cm (3 ft) of rooting depth and no lower impervious layers that would impede good drainage. Any impervious layers must be broken up and if the soil is over a layer of impervious bedrock, tile drainage should be considered to ensure excess water drains out of the soil. Cherry tree roots do not like 'wet feet'. Consequently, soils with a clay content of >30% should be avoided. Cherries thrive in sandy-loam and silt-loam soils.

The current crop or native vegetation growing on the proposed orchard site can reveal a lot about the potential of the site for growing high quality fruit. Walk the site and map areas of variability and poor growth. Use aerial photos (from piloted small planes, drones or Google Earth) to identify problematic areas of the field. If an orchard or vineyard is currently located on the site, multispectral imaging, which records an index of plant cell density, can be used to identify areas of high and low vigor within the field. If replanting on an existing orchard site, soil conditions in the former tree rows may be different from the alleys. These differences can particularly be seen in pH and compaction. If the site has a history of nitrogen (N) fertilizer applications to the tree row, the pH may be considerably lower than in the alley and should be treated differently. Similarly, the alley will generally have more soil compaction due to years of tractor traffic. If the new planting will not use the old spacing, then efforts should be made to homogenize soil conditions as much as possible to minimize the chance of a 'vigor wave' effect where a new row crosses a former row. If planting in the existing rows, then treatments can be targeted to this area. With this information, soil modification strategies can then be designed to address specific needs.

The USDA–NRCS has a 'Web Soil Survey' website (www.websoilsurvey. sc.egov.usda.gov/App/HomePage.htm, accessed 19 June 2020) that provides users with searchable, interactive maps of the USA, enabling them to examine soil characteristics for individual farms or parcels. By inputting the address or location of the parcel of interest, one can view the soil characteristics, slope, elevation and more. Printed soil maps may also be obtained from local NRCS offices.

Numerous soil/site classification applications for mobile phones, tablets or computers are also available. Dr Toby O'Geen, University of California at Davis (UCD), and Dr Dyan Beaudette, USDA, developed a computer-based application called 'SoilWeb' that uses aerial photos from Google Maps with soil survey map overlays that provide site-specific information, including the identity of the soil type along with soil texture, pH, organic matter (OM), EC, available water storage, drainage and much more. A second application, called 'SoilWeb Earth', is available at the same website (www.casoilresource.lawr.ucdavis.edu/soilweb-apps/, accessed 19 June 2020) and uses Google Earth to display SoilWeb data in a three-dimensional graphic display. A third application, the 'Soil Series Extent Explorer' (www.casoilresource.lawr.ucdavis.edu/see/, accessed 19 June 2020), allows one to search the USA for a particular type of soil. For example, if the soil type 'Cherryhill' is entered, a map reveals that this soil series is located only in and around The Dalles, Oregon. In contrast, a search for the soil type 'Walla Walla' reveals that this soil series is located in several locations in eastern Oregon and Washington. This application is particularly useful to identify sites with a particular soil type.

For California growers, there is another application available at the SoilWeb address to help select the best orchard sites. The 'Soil Properties' app allows users to search a map of California to find soils based on their chemical, physical or land use properties. For example, one can search for soils having a specific water holding capacity, silt content, EC and many other characteristics.

4.2.3.1   Soil texture

Soil texture is one of the more important factors determining the ability to grow cherry trees on a particular site. Cherries prefer light, well-drained soils. A silt-loam soil is best, but cherries can tolerate soils ranging from sandy-loam to clay-loam as long as there is good drainage. Cherries will die in waterlogged soils or where oxygen in the soil is limited.

4.2.3.2   Soil compaction

As previously stated, a soil depth of at least 90 cm (3 ft) is desired for adequate root growth and development. Locations with hardpans or other subsurface obstructions in this profile depth should be avoided. In the US, NRCS soil survey maps can be used to determine variability of soil types within the field. Soil profile pits should be dug with a backhoe, in each area with a different soil type, or in problem areas within the field (as identified by observing variations in the previous crop). These pits should be at least 90 cm (3 ft) deep so that soil texture, hardpans and soil layering can be determined (Fig. 4.8). Soil compaction can be a major impediment to cherry production. Ideally, soils should be free of compaction to a depth of 90 cm (3 ft); however, cherries can survive with a compacted layer as close as 60 cm (2 ft) from the surface as long as high water tables do not keep water in the rooting zone. Cherry trees will die quickly where drainage is impeded, whether caused by soil compaction or fine textured soils. In these cases, creating a raised bed or berm for planting can improve the effective root zone and alleviate some of the negative effects of heavy soils, hardpan layers and/or poor drainage (Fig. 4.9).

4.2.3.3   Soil pH

Cherries grow best in soils with a pH range of 6.0 to 6.8. They can grow and produce outside this range, although growth, yield and tree health may be affected. As soil pH extends beyond this narrow range, in either an acidic or alkaline direction, the tree roots lose the ability to take up key elements from the soil solution. For example, below pH 6.0, phosphorus

Fig. 4.8. A particularly deep soil pit dug in an existing orchard to ascertain root distribution and possible soil compaction problems (Chile) (courtesy of L.E. Long).

(P), molybdenum (Mo), calcium (Ca) and magnesium (Mg) become less available for uptake. As the pH becomes more alkaline (7.5–8.0), P, iron (Fe), manganese (Mn), boron (B), copper (Cu) and zinc (Zn) become limiting factors. Symptoms of nutrient deficiencies in cherry trees include interveinal chlorosis (Fe deficient), interveinal browning (Mg deficient), and small pointed leaves that form a rosette pattern at the shoot tip (Zn deficient). Evidence of any of these symptoms needs further evaluation by a simple pH analysis of the soil, which can be conducted by most commercial analytical laboratories. See Chapter 8, 'Managing the Orchard Environment' for more discussion on nutrient deficiencies.

In general, soils in climates with high rainfall tend to be highly weathered, which usually results in lower pH values. Different soil types and prior land use history also may impact soil pH, e.g. previous use as pasture or a pen for animals may result in elevated mineral nutrients and changed pH due to manure. Conversely, soils in low rainfall climates generally have higher pH values. In eastern Oregon and Washington and other desert areas around the world, soil pH can be as high as 8.5. At pH values less than 6.5 and greater than 7.5, mineral nutrient availability is adversely affected, making it necessary to adjust the soil pH. The best time to make this adjustment is prior to planting.

Fig. 4.9. Cherry trees are planted in berms in order to facilitate water drainage away from the roots (Chile) (courtesy of L.E. Long).

4.2.3.4 Soil salinity

In many parts of the arid western US, saline soils can be a serious impediment to growing cherries. A saline soil contains enough dissolved salts to detrimentally affect crop growth. Cherries have a low tolerance to salt and will be affected by soils that have an EC of 2.1–4 dS/m (Whiting *et al.*, 2015). According to the Food and Agriculture Organization (FAO), besides cherries, other stone fruits are also very sensitive to saline soils (UN FAO, 2020). For example, apricot begins to show crop yield reductions of about 10% at 2.2 dS/m and 50% at 4.1 dS/m. Trees grown in saline soils may appear wilted, even after irrigation. This is because the high salt content in the soil decreases the osmotic gradient between the roots and the soil, impeding the uptake of water by the roots. A white crust is sometimes visible on the surface of a saline soil (Fig. 4.10). Trees sprinkle-irrigated by saline water may show marginal leaf burn symptoms as the salty water moves to the leaf edge and evaporates, concentrating the salt at the margin (Waskom *et al.*, 2010).

Saline soils have several origins. In an agricultural situation, the most likely causes are: (i) irrigating with water high in salts; (ii) the overuse of inorganic fertilizers or organic fertilizers such as manure; (iii) soils that are inherently high in salts due to the breakdown of parent rock material; or

Fig. 4.10. A white powder, denoting a saline soil, is visible in this young orchard (Uzbekistan) (courtesy L.E. Long).

(iv) in arid areas, salinity builds up due to evaporation exceeding precipitation (Kratsch *et al.*, 2008).

4.2.3.5   Soil sodicity

Sodic soils have a high level of sodium (Na) as a percentage of the cation exchange capacity (CEC). The CEC measures the ability of the soil to hold cations such as Mg, Ca, and Na. When Na makes up 6% of the CEC, a soil is considered sodic. With a sodic soil, the sodium adsorption ratio (SAR) of the saturation extract (SAR_e) is at least 13 mmhos/l. Fine textured soils (i.e. clays, clay-loams, silts, silt-loams, etc.) may experience dispersion and crusting due to Na with SAR levels as low as 5 or 6 mmhos/l (Silvertooth, 2001). Sodic soils typically occur in semi-arid to arid areas and have poor structure which interferes with water infiltration, leading to poor water availability and plant growth. Table 4.2 classifies saline or sodic soils based on laboratory analyses for EC and SAR values.

Table 4.2. General classification of salt- and sodium-affected soils (from Silvertooth, 2001).

Criterion	Normal	Saline	Sodic	Saline–Sodic
Electrical conductivity (dS/m)	<4	>4	<4	>4
Saturation extract (SAR_e) (mmhos/l)	<13	<13	>13	>13

Soils that are high in sodicity swell excessively when they become wet. This separates, or disperses the clay particles and weakens the aggregates in the soil, causing the soil structure to collapse and closing off soil pores. In these soils, water tends to pool or run off rather than penetrate. Sodic soils will also sometimes exhibit a brownish-black crust on the soil surface due to dispersion of OM (Waskom *et al.*, 2010).

4.2.4   Soil analysis

Suitable orchard sites should have soil with at least moderate fertility levels and no extremes in pH or high EC. In many production regions, soil test element values developed for crops other than sweet cherry need to be interpreted carefully if being applied to cherry orchards. However, it is important to have as much information as possible prior to undertaking the expensive task of planting or replanting a cherry orchard. This includes chemical characterization of the major soil types in a proposed location. Soil pH is particularly important as it can identify acidic (low pH) soils which may require lime application that is more easily worked into the soil in the absence of planted trees and their roots. High pH soils (>7.5) are more susceptible to deficiencies of micronutrients such as Zn, Mn and Fe. Very high pH soil (>8.2) may have high accumulations of calcium carbonate or soluble and toxic accumulations of salts such as Na or chloride. Elevated soil salinity, which is much more common in semi-arid regions, can be determined by making measurements of soil solution EC. These measurements can be useful since cherry, as with other fruit trees, exhibits reduced growth when soil EC readings are high.

For most of the major plant fertility elements in soil, such as N, P, potassium (K), Mg and various micronutrients, soil test values can be useful for identifying orchards with extremely high or low availability of specific nutrients. Many of the nutrient inadequacies of orchard soils can be corrected by development of selective fertilization programs. However, many of the physical properties are not so readily changed, so it can be useful for growers to understand the limitations of their soil by making measurements of some of these key physical properties. Particularly important are measures of soil texture (percent sand, silt and clay-sized particles) and water holding capacity as a soil drains and dries. These properties are especially important if irrigation is being contemplated, as will be discussed further in Chapter 8.

To determine soil pH, nutrient levels, OM content, and CEC, soil samples should be taken for submission to an accredited laboratory for analysis. In the US, a list of accredited laboratories can be obtained from local university Extension Service offices. Soil analysis reports and formats may vary between laboratories. Samples should be collected from each soil type in the field, as indicated by soil survey maps or suggested by soil color and texture differences within the field. Separate samples should also be collected where cropping history, or fertilizer or other soil amendment treatments have varied. For organic growers, testing for soil OM prior to adding any amendments establishes a baseline from which 'soil improvement' can be measured in subsequent years. This will provide a record for the organic certification process.

Before planting an orchard, the OM content of the soil also should be assessed. The OM content of many native soils in the PNW region of the USA range between 1.5–2.5%. A target content should be ~5% OM. This will improve aeration, provide adequate drainage and increase the water holding capacity of the soil. Carbon is a source of nutrition to soil biota and these will be elevated in high OM soils. Soil biota are an important source of beneficial organisms which may help suppress root diseases. A standard soil test can provide total OM (or total carbon). Organic matter levels are easier to build and maintain in wetter, cooler environments compared to hot, dry climates. Soils with more clay tend to conserve more OM than coarse-textured soils. Organic matter influences the chemical, physical, and biological properties of a soil, and it is considered one of the main determinants of overall soil productivity. The target content of ~5% OM may be difficult to achieve in semi-arid orchard regions such as central Washington and Oregon. Increasing soil OM will help overcome limitations of soils at the extremes: increasing water holding capacity and cation exchange capacity of sandy soils; increasing aeration and drainage of clay soils.

Soils can also be sampled for nematodes, which are more easily controlled before planting than after. Nematodes can be either parasitic or predatory, so the types and their population densities are important data to know. Most nematodes are beneficial, contributing to the breakdown of OM in the soil, but parasitic nematodes can have significant detrimental effects on plant health, including feeding on roots and transmitting viruses. Parasitic types include root lesion, dagger, rasp leaf, and root knot nematodes. Control measures include fumigants (which are not effective if infected plant tissue, such as root knots, remain in the soil), chemical and biological nematicides (i.e. certain cover crops and green manures) or soil solarization.

4.2.4.1 Sampling for soil nutrient analysis

Remove any surface debris before collecting samples. Samples should be taken prior to adding soil amendments. Samples should be collected again after amendments have been added, but before planting. Use a soil probe, auger or shovel to collect soil samples; about 0.5 l (1 pint) of soil will be needed for each analysis. One sample should represent a maximum of 8 ha (20 acres) on a level, uniform field. Where hills predominate, the sample area should be no more than 2 ha (5 acres) in size. Within each area, collect 15 to 30 subsamples. Subsamples of about the same size should be collected in a zigzag pattern throughout the sampling area. Samples should be taken at two depths, 0–15 cm (0–6 in) and 15–30 cm (6–12 in). Subsamples should be combined in a plastic pail and mixed thoroughly. Problem areas should be omitted or sampled separately. If samples are collected wet, they should be dried before submitting for analysis.

Once the orchard is established, soils should be analyzed every 3 years. After establishment, soil samples should be taken from the tree row.

4.2.4.2   Sampling for pathogenic soil nematodes

Nematodes or eelworms are microscopic threadworms that occur ubiquitously in soils. They range in size from 0.5 to 1.0 mm (0.02 to 0.04 in) in length. There are many types of nematodes present in soils, most of which are beneficial or benign. Plant-parasitic nematodes, however, can attack cherry trees, damaging roots, creating wounds and generally weakening the trees by diverting normal metabolic activity into supporting nematode populations and repairing wounds. In addition, some nematode species, such as the dagger nematode, can serve to vector viruses. The cherry rasp leaf virus can be transmitted in this way from tree to tree within an orchard. Finally, nematode-weakened trees are known to be more susceptible to bacterial canker and *Verticillium* wilt diseases. Pathogenic nematodes can be very aggressive and when present in high numbers, they will need to be controlled.

Nematode pests are generally classified according to their feeding habits:

1. Migratory ectoparasites feed from outside roots, moving from cell to cell and piercing the root cells to feed, without entering root tissue. (Examples: dagger nematodes, *Xiphinema* spp.)
2. Sedentary ectoparasites tunnel partially into roots, their heads entering to establish permanent feeding sites while their bodies remain outside. The nematode does not move after this. (Examples: citrus nematode, *Tylenchulus semipenetrans.*)
3. Migratory endoparasites feed inside roots, tunneling inside and moving back into soil and to new roots at will. (Examples: root-lesion nematodes, *Pratylenchus* spp.)
4. Sedentary endoparasites tunnel into the roots, establishing permanent feeding sites from which they do not move. They may protrude from roots as they grow. (Examples: root-knot, *Meloidogyne* spp.; and cyst nematodes, *Heterodera* spp.)

It is wise to test for pathogenic nematodes at least 1 year before planting. Commercial laboratories may be able to test soil samples for nematodes. Sampling is best done at the end of the growing season in late summer or early fall, when populations are highest. Both fine feeder roots and soil should be sampled where feeder roots are prevalent, usually at depths between 15 to 90 cm (6 to 36 in) under the tree canopy dripline. Place the root and soil samples in separate containers. Take 10 to 20 subsamples from a given area and combine each type into a single composite root and single composite soil sample. Laboratory analysis should provide counts of all nematode types, including ring nematodes.

4.2.4.3 Herbicide residues in soil

Before purchasing land or planting an orchard, it is important to determine whether the site was previously farmed, and if so, what crops and herbicides were used to control weeds. Persistent herbicides used on agronomic crops (e.g. wheat, pea, chickpea, lima bean, soybean, alfalfa, lentil, corn, clover, hay grass, rangeland or pasture) may last 4 years or longer in the soil. Planting cherries in a soil with herbicide residues before they have broken down can result in stunted, debilitated plants that are unlikely to recover from the herbicide effects. Thus, the complete history of herbicide applications on the potential orchard site is critical. If possible, ask to see records of herbicide applications for the past 3 or 4 years. Lacking records, soil samples can be submitted to laboratories for herbicide residue analysis; however, this can be expensive as individual tests need to be conducted for each potential chemical. Alternatively, a simple bioassay may provide valuable information. Take surface samples of soil from several regions of the field and use this soil to grow leafy vegetable crops, such as lettuce and other plants that tend to be sensitive to herbicides (such as tomatoes). Look for leaf distortion, browning of the leaf, spotting or yellowing of the veins. If there is any sign of damage on the indicator plants, it is best to wait before planting the orchard.

4.2.5 Site water availability and quality

Many cherry growing regions of the world are winter rainfall areas, e.g. Mediterranean climates with hot dry summers and cool wet winters. Such climates are less prone to significant rain-induced fruit cracking and diseases during the growing season. This is not to say that summer rainfall climates are unsuitable for sweet cherry production; however, additional orchard inputs, such as protective covering structures and increased fungicide applications, may be critical requirements for such climates (e.g. the UK, Norway and Japan). Most commercial fresh market sweet cherry growing areas of the PNW and California are typically arid regions with primarily winter precipitation. In general, cherry trees require around 1000 mm (39 in) of water during the growing season. Furthermore, peak demand for water occurs during the summer months, so to establish and produce premium cherries, irrigation is required (particularly for trees on dwarfing rootstocks).

4.2.5.1   *Water pH*

The pH of irrigation water can directly affect, and eventually change, the soil pH to the point that plant uptake of nutrients will be hindered. Ideally, the pH of irrigation water should be between 6.0 and 7.0. A value outside of this range is a warning that the water needs further evaluation and possible remediation. Certain nutrients can become limited or even toxic at lower or higher pH levels. Extreme pH levels are also an indication that irrigation equipment must be chosen carefully to avoid failure of components due to the corrosive nature of such water.

Water alkalinity is a measure of the buffering capacity of water related to the quantity of bases in the water that will neutralize acids.. Potential problems due to water pH are directly proportional to its buffering ability. The more buffered the water and the more out of range from what is acceptable, the more severe the problem. Water of low salinity (EC_w <0.2 dS/m) will occasionally have a pH outside the normal range because of its low buffering ability. This is usually not a problem for crops or soil, but such water may be corrosive to orchard sprinklers and other equipment (Ayers and Westcot, 1994).

Due to the strong buffering ability of the soil, any change in soil pH due to irrigation water will be slow. In some cases, it may be possible to modify the irrigation water through additives, such as sulfur for alkaline soils, or gypsum if the alkalinity is due to high exchangeable Na (Ayers and Westcot, 1994).

4.2.5.2   *Water salinity*

Although water in the PNW region of the USA tends to be low in dissolved salts, saline water can be a significant impediment to cherry production in other parts of the world, especially in arid regions of the USA, Chile, Spain and elsewhere. High salt and Na content in irrigation water can ruin soils if used for long periods. Water analysis is extremely important for water that will be used to irrigate cherry orchards, as cherries have low tolerance to salt, similar to that of peaches and apricots, which are also considered sensitive to salt. Salt content of irrigation water is determined by measuring its EC. An EC of 130 mS/m will cause a 10% loss of yield in peaches and apricots and an EC 180 mS/m will cause a 25% loss in those same crops (Government of Western Australia, Department of Primary Industries and Regional Development, 2019). Salinity of soil extract may also be used to determine potential damage from salt (Table 4.3). The threshold salinity of soil extract for peaches is 1.7 mmhos/cm and for apricots 1.6 mmhos/cm. The reported decrease in yield when soil crosses the salinity threshold is 21% and 24% yield reduction, respectively (Spectrum Analytic, personal communication).

Table 4.3. Salinity hazard levels of irrigation water (mmhos/cm or dS/m) (adapted from Spectrum Analytic).

Salinity Hazard Levels						
Application	Units	None	Increasing	Significant	High	Severe
Field crops	Mmho/cm	<0.75	0.75–1.0	1.1–2.0	2.1–3.0	>3.0

4.2.5.3 Sodium adsorption ratio of water

As described under Section 4.2.3.5, 'Soil sodicity', irrigating with water high in Na can cause soil structural problems. The Na hazard of water is expressed in its SAR value. Sodium can also accumulate in soil to sufficiently large amounts such that Na uptake becomes toxic to the plant. Fine textured soils with poor drainage can be subject to degradation due to irrigation water that has moderate SAR values (3 to 6). From the perspective of inducing soil permeability problems, SAR and EC both need to be considered (Table 4.4). Low salinity water (usually low in Ca and Mg) increases the deleterious effect of Na in water (University of Missouri Extension, 2011).

4.2.5.4 Bicarbonates and carbonates in water

High levels of bicarbonate (HCO_3) and carbonate (CO_3) in water increase the concentration of Na. Bicarbonates are a concern in alkaline water (Table 4.5) where pH increases beyond 7.4. With high (>61 ppm) levels of HCO_3, Ca precipitates and is replaced with Na. In this situation, not only does Ca become limiting to the plant, but soil structure is detrimentally affected, leading to poor water penetration. Acidification of the water is the best way to manage HCO_3 issues. Water HCO_3 levels of < 1.5 mEq/l are safe; however, severe problems can develop with HCO_3 levels of > 2.5 mEq/l (Flynn, 2009).

Carbonates can become a problem if the water pH is above 8.5. They will not be a significant component of the water, or a problem when the water pH is below 8.0. Carbonates should be <100 ppm $CaCO_3$. High CO_3

Table 4.4. Guidelines regarding the hazard of irrigating with water of varying sodium adsorption ratio (SAR) and electrical conductivity (EC) values relative to the development of soil permeability problems (adapted from the University of Missouri Extension Service).

SAR	No restriction	Slight to moderate restriction	Severe restriction
0–3	EC_w <0.7	EC_w 0.7 to 0.2	EC_w >0.2
3–6	EC_w <1.2	EC_w 1.2 to 0.3	EC_w >0.3
6–12	EC_w <1.9	EC_w 1.9 to 0.5	EC_w >0.5

Table 4.5. Potential bicarbonate hazard showing threshold levels for damage on field crops. Bicarbonate levels above 3.3 mEq/l (200 ppm) will cause lime (calcium and magnesium carbonate) to be deposited on foliage when irrigated with overhead sprinklers (adapted from Spectrum Analytic).

Application	Units	None	Increasing	Significant	High	Severe
Field crops	mEq/l HCO_3	<1.0	1.0–2.0	2.0–3.0	3.0–4.0	>4.0
	ppm HCO_3	<61	61–122	122–183	183–244	>244

will cause Ca and Mg to form insoluble minerals, leaving Na. As described earlier in Section 4.2.3.5, this will lead to dispersion of the soil and poor water infiltration.

Water hardness is a measure of the total concentration of divalent cations, which should be <150 ppm. Looking at the individual components, both Ca (ppm × 2.5) and Mg (ppm × 4.1) should be <25 ppm and <20 ppm, respectively. Total dissolved solids (TDS) must be <1000 ppm and ideally <500 ppm. The EC should be <0.75 dS/m or mmho/cm. Chloride in irrigation water should be <70 ppm, whereas chloride in water used for foliar applications should be <40 ppm. Sodium concentrations should be <70 ppm and the SAR should be <2 mEq. If the water source exceeds these minimum values for any category above, action should be taken to lower the levels if an alternative source of purer water is not available.

4.2.5.5   Water sampling

In general, the water source that is intended for both irrigation and foliar sprays should be tested. When sampling from a new well or a well that has been used only sporadically, be sure to run the pump for 1 to 2 h before collecting the sample. For wells or water sources used more frequently, run the water long enough to bring fresh water through the piping system to the outlet. Since winter and spring precipitation can have an effect on water quality, samples should be taken during the irrigation season.

Use a clean plastic container for sample collection. Usually, 0.25 to 0.5 l (8 to 16 oz) is adequate for most quality assays. Accurate tests for certain components in water require the addition of preservatives during sampling. Contact the testing laboratory for special instructions. Deliver the samples to the laboratory within 24 h after collecting. Samples should be kept out of heat and direct sunlight and should be refrigerated until shipped.

4.2.6 Access to key auxiliary facilities

In addition to the orchard planting site that focuses on the development of rows of trees, tractor alleys, and support structures as needed, the orchard's proximity to various key auxiliary facilities is also a pertinent part of orchard planning. Storage facilities for orchard tractors, sprayers, mowers, pruning equipment, agrichemicals, harvest buckets, orchard ladders and associated equipment should be readily accessible. A facility for mixing agrichemicals and cleaning spray equipment, designed to include a water source and containment to prevent chemical contamination of soils and surface or groundwater during mixing and cleaning, should be accessible. Sanitary facilities for hand washing and toilet use is imperative for orchard workers.

Plans must also be made for the future handling of harvested fruit. A holding area for bins of fruit may need to be set aside, with shade, to collect the bins as they are filled and protect them from solar radiation until they are trucked to a packing facility (although reflective, insulated bin covers or tarps can also serve this purpose). Depending on the typical heat of the location during harvest and the distance to the packing house, an onsite hydrocooling operation and/or cold storage may be valuable. Hydrocoolers are more useful and effective in hot dry growing regions; in humid growing regions, Ca must be added to hydrocooler water to reduce the potential for fruit cracking during hydrocooling. At a minimum, an area to load trucks for transport to the packing facility should be included in the orchard design.

4.2.7 Environmental impact

Last, but not least, the impacts of developing the orchard site must be considered with regard to the wildlife and natural resources in the area. Consider utilizing an environmental consulting entity which can help plan the orchard to avoid damage to wildlife habitats and to prevent damage to aquatic life in waterways. Conversely, an assessment of wildlife in the area may identify possible problems for production, such as the potential for tree damage by such wildlife as deer, rabbits, mice, and even beavers, as well as damage to fruit by birds, raccoons, bears, etc. Such problems may require planning for fencing or other structures to specifically exclude certain types of wildlife. In some cases, identification of an endangered species or habitat may give the location a higher property value.

For More Information

Koumanov, K.S. and Long, L.E. (2017) Site preparation and orchard infrastructure. In: Quero-Garcia, A., Iezzoni, A., Pulawska, J. and Lang, G. (eds) *Cherries: Botany, Production and Uses.* CAB International, Wallingford, UK, pp. 223–243.

Historical climatic data (e.g. daily temperatures, precipitation, GDD) and some plant production decision models may be obtained from local university, state or government sources, such as:

- California

CIMIS, www.cimis.water.ca.gov

- Michigan

Enviroweather, www.enviroweather.msu.edu

- Oregon

Oregon Climate Summaries, www.wrcc.dri.edu/summary/climsmor.html

- PNW region of USA

Agrimet, www.usbr.gov/pn/agrimet/webarcread.html

- US Nationwide

NOAA, www.noaa.gov/weather

- Washington

AgWeatherNet, http://weather.wsu.edu

Acknowledgements

Portions of this chapter were adapted with permission from Oregon State University. *PNW 642*. Oregon State University Extension and Experiment Station Communications, Corvallis, Oregon.

References

Ayers, R.S. and Westcot, D.W. (1994) Water quality for agriculture. FAO Irrigation and Drainage Paper 29, Rev. 1.

Flynn, R. (2009) Bicarbonate. Irrigation water analysis and interpretation. New Mexico State University Extension Guide W-102.

Government of Western Australia, Department of Primary Industries and Regional Development (2019) Water salinity and plant irrigation. Available at: https://www.agric.wa.gov.au/water-management/water-salinity-and-plant-irrigation?page=0%2C0#smartpaging_toc_p0_s1_h2

Kratsch, H., Olsen, S., Rupp, L., Cardon, G. and Heflebower, R. (2008) Soil salinity and ornamental plant selection. HG/Landscaping/2008-02pr. Utah State University Extension Service.

NASA (2020) Available at: www.grc.nasa.gov/www/k-12/problems/Jim_Naus/TEMPandALTITUDE_ans.htm (accessed 19 June 2020).

Silvertooth, J. (2001) *Saline and sodic soil identification and management for cotton.* Publication az1199. University of Arizona Cooperative Extension, Arizona.

UN FAO (2020) Salinity problems. Available at: www.fao.org/docrep/003/T0234E/T0234E03.htm (accessed 19 June 2020).

University of Missouri Extension (2011) Interpretation guide for irrigation water. Available at: www.soilplantlab.missouri.edu/soil/waterirrigation.aspx (accessed 19 June 2020).

USDA-SCS (1977) Covar Sheep Fescue. Washington state university agricultural research center

Waskom, R.M., Bauder, T.A., Davis, J.G. and Cardon, G.E. (2010) Diagnosing saline and sodic soil problems. Colorado State University Extension Service. No. 0.521.

Whiting, D., Card, A., Wilson, C. and Reeder, J. (2015) Revised by Carter. S. Saline Soils. CMG GardenNotes #224. Colorado State University Extension Service.

Orchard Establishment and Production

5

5.1 Designing the Orchard

In the US, topographical maps of the property under consideration for orchards can be obtained from the Natural Resources Conservation Service (NRCS) of the US Department of Agriculture (USDA). These show soil types and potential problem areas. Growers should lay out the proposed orchard by populating the map with existing and intended access roads, buildings, windbreaks, chemical storage sites, water filling locations, the irrigation system, drive rows and loadout areas, etc. Wherever possible, tree rows should be oriented in a north–south direction to optimize uniform light interception. This may not be possible on moderate to steep slopes where it is important to follow the contour of the hill in order to reduce the potential for soil erosion. Also, in these cases, gaps should be left in the bottom rows for cold air to drain away from the orchard, reducing the potential for spring frost or winter cold damage.

Planting density and row spacing will depend on soil fertility, rootstock vigor and the canopy training system chosen. Spacing for modern canopy systems ranges from 0.5 m × 3 m (20 in × 10 ft) for the super slender axe (SSA) on dwarfing rootstocks to as much as 2.5 m × 5.0 m (8 ft × 16 ft) for the Kym Green bush (KGB) and 4.0 m × 5.0 m (13 ft × 16 ft) for the steep leader (SL) on semi-vigorous to vigorous rootstocks (Long *et al.*, 2015).

5.1.1 Surveying and staking the orchard site

There is a wide range of options available to growers for laying out and preparing a new orchard site for planting. Size, topography, surface variation, available labor and cost can help define the method chosen. On hills, or where the topography is uneven or the orchard is very large, growers will often hire a surveyor or geodetic engineer (also known as a 'geodesist') to provide measurement accuracy in laying out the orchard parcels. Orchard sites with variable, uneven surfaces should be graded and smoothed prior to the final layout of the orchard rows and tree locations.

On smaller plots, or on flat ground, it is possible for growers themselves to lay out and stake the orchard with some accuracy. In this case, growers measure distances by hand and stake each row and tree location. A rope with regular marks to signify exact tree spacing distances can be prepared and, starting from the outside row of a parcel, moved from row to row with wire flags or stakes placed at each mark to indicate a future tree. Holes can be dug with a tractor-mounted auger and planted by hand with shovels. This is a slow, labor intensive process, but one that is commonly used.

For greater accuracy and reduced labor inputs, surveyors or geodetic engineers use a global positioning satellite (GPS) unit to map and lay out the orchard, as well as to then guide a GPS-enabled tractor to provide row and tree placement accuracy of up to 2.5 cm (1 in). GPS-guided tractors can be equipped with a partially to fully automated tree planter that plants the trees at the designated spacing. Tree planters may be a good investment as they can be used to quickly plant large orchard areas in a very short period of time with only a few workers (Fig. 5.1). With tree planters, only the rows need to be staked. While the GPS unit provides extremely accurate row placement, between-tree spacing is not as accurate and planting super high density orchards where spacing is as dense as 50 cm (20 in) apart may be difficult.

Highly accurate rows established with GPS precision and orchard surface leveling is especially important where trellising and/or protective orchard covering structures for rain, hail or birds will be used, or where mechanization of tasks such as flower thinning or summer pruning by hedging is envisioned. Also, when establishing moderate to high density protected orchards (more than 1200 trees/ha or 485 trees/acre), it is most efficient to first determine the GPS placement of each structural support pole or post, then determine the placement of every tree. No matter which option is chosen to lay out the orchard, the plot plan should provide enough space at the end of each row to safely turn a tractor with an attached sprayer. To facilitate easy movement of fruit to the load out area, drive rows should be planned at about 6 to 7.3 m (20 to 24 ft), depending on canopy training system architecture. Drive row intervals should be planned at about every 45 to 75 m (150 to 240 ft).

5.2 Orchard Site Preparation

Once a suitable site is located, careful preparation needs to be made for its development. If there is an established orchard on the site, and renovation is determined to be the best option, then tree removal can begin immediately after harvest or any time the ground is firm enough to work. The best time to modify and improve soil conditions is prior to planting. Wherever

Fig. 5.1. Orchard workers use a tree planter to quickly plant this high density orchard (Oregon) (courtesy of L.E. Long).

possible, hardpans, nutrient deficiencies and pH levels need to be brought to within acceptable levels.

5.2.1   Previous vegetation removal and fumigation

If the future orchard site was previously under rangeland or agronomic crops, plowing the existing vegetation is a common way to begin conversion to orchard while incorporating organic matter (OM) into the soil. If the future orchard site had perennial bushes and small trees, or previously was an orchard, it is not uncommon to simply push bushes, trees and stumps into a pile with heavy equipment so that they can be burned. However, doing so wastes an important resource that can be used to improve the complex population of soil bacteria, fungi, nematodes, etc. (termed the 'rhizosphere'), in which the trees grow. In addition, due to air pollution issues, open burning is now restricted in many communities, forcing growers to look for more sustainable practices. Grinding trees and stumps and incorporating the wood chips into the soil, or using it as mulch around the newly planted trees, can enhance carbon (C) sequestration, and improve soil quality and tree yield. Growers are finding that water needs are reduced and soil biology is enhanced by utilizing this ready form of OM.

If the previous crop was an orchard, whether stone or pome fruit, newly planted trees may be subject to what is known as 'replant disease', in which the rhizosphere associated with the previous orchard trees tends to reduce growth of newly planted trees (Mai and Abawi, 1978; Mazzola *et al.*, 2005). This reduced growth and production can last the life of the orchard. If replant disease is suspected it must be treated before planting. Little can be done to improve conditions once new trees are in the ground. Soil fumigation can significantly reduce the impact of replant disease. However, the most common fumigant, methyl bromide, is no longer available, and some studies have shown that fumigation provides only a temporary boost in growth, with problematic soil microbial populations returning to the rhizosphere within a couple of years. Alternatives to methyl bromide are available.

The best time to fumigate the soil is in the fall, while soil temperatures are still warm and before irrigation water is shut off. This will give the fumigant time to work and totally dissipate by the time the new orchard is planted in the spring. Often, after trees are removed in the fall, little time is left for fumigation. In this situation, spring fumigation may be necessary. Fumigation cannot be done until soils are at the proper temperature for the fumigant to be active, but it is important to leave sufficient time before planting for the fumigant to be expelled. All of this may cause an unacceptable delay in planting.

Unfortunately, fumigation not only kills pathogenic microorganisms, but also beneficial organisms. For this reason, cultural methods to control replant disease may have some advantages over chemical fumigation. Some of these methods have shown promise, such as *Brassica* seed meal and/or green manure treatments which promote healthier soil microbial populations that can provide a longer lasting boost to growth of new trees on replant sites than fumigation (apple studies by Mazzola *et al.*, 2015). However, to date none of these cultural control methods have been adopted commercially. Researchers are also testing whether applying an organic amendment such as compost after fumigation can help reinoculate the soil with a more desirable microbial community. Initial results have been mixed. Combining fumigation and a soil amendment could give a short-term

Fig. 5.2. Cherry roots are sprayed with a mycorrhizae suspension prior to planting (Oregon) (courtesy of L.E. Long).

boost to newly planted trees along with a longer-term beneficial soil biological composition.

In order to re-establish beneficial microorganisms to the soil in lieu of, or after, fumigation, and enhance the overall health of the soil, compost from municipal waste and other sources can be mixed with high C mulch such as wood chips, corn cobs or other ready sources of C. Carbon is also a source of nutrition to soil biota and these will be elevated in higher OM soils. Both total OM as well as active OM will influence the biota; it is possible to stimulate the biota with inputs of more active C such as a green manure that may not increase the total soil C. Soil biota are an important source of beneficial organisms, which are critical to nutrient cycling and may help suppress root diseases. Among such soil biota are mycorrhizal fungi, which form a symbiosis with roots, increasing the surface area for metabolic exchanges with the host, as well as interfacing directly with the soil surrounding the roots, effectively acting like root hairs. This increases the potential of the root system to take up nutrients and water, especially phosphorus (P), a critical element needed for root growth and development. Mycorrhizal fungi also increase soil aeration and facilitate better water percolation through the soil. Before and at planting, nursery tree roots can be inoculated with mycorrhizal fungi, which can be purchased from many farm supply stores (Fig. 5.2).

In one Oregon orchard a benefit of growing Colt rootstock on a replant site was noted. Trees grew stronger, more uniformly and produced earlier crops than trees in the same field on Mazzard rootstocks (Fig. 5.3). Large differences in replant disease tolerance have been found among apple rootstocks, but this screening has not yet been done in cherry. Former apple orchard soil from central Washington was pasteurized and then

Fig. 5.3. In this replant site, where cherries followed cherries, trees on Mazzard rootstock (foreground) are stunted and growing poorly, whereas trees planted on Colt rootstock (background) are growing more vigorously (Oregon) (courtesy of L.E. Long).

cherry rootstock growth was evaluated in the pasteurized and untreated soils. The relative growth in untreated soil compared with the pasteurized was as follows: Gisela 49–69%; Mahaleb 53–74%; Mazzard 30–100%. Cherry rootstocks planted in apple replant orchard soils exhibited a significant increase in susceptibility to freeze damage relative to the same rootstock established in pasteurized replant soil (Mazzola *et al.*, 2004). Screening for genetic tolerance to replant disease is needed for cherry rootstocks.

5.2.2   Physical preparation of the soil

In preparation for planting, as many roots as possible should be removed from the previous orchard. Deep ripping, sometimes known as 'subsoiling', involves cutting through the lower soil levels, using a strong narrow tine implement, without inverting the soil. The soil should be ripped in at least two directions to break up compacted soils. A ripper that can cut through the soil within 60 cm (2 ft) of the surface, is most commonly used to break up hardpans. Although this works well to break up hardpans near the surface, layers tend to reform and subsurface hardpans remain undisturbed. Slip plows also can be used to break up hardpans. These tend to disturb soil to a depth of as much as 182 cm (6 ft), but can be expensive.

Although breaking up hardpans has been common practice in orchard establishment for years, research on California almonds suggests that under microsprinkler irrigation, the benefits of deep slip plowing may not provide an economical return. After nine growing seasons, there was no statistical difference between trees, yield or kernel size in trees planted in slip-plowed and non-slip-plowed soils (Edstrom and Cutter, 2008).

In order to increase rooting depth or reduce the effects of poor drainage or high water tables, berms (raised beds) can be formed by blading the area between rows to form an elevated planting row 20 cm (8 in) high and up to 1.5 m (5 ft) wide (please note that the initial height of these berms must be higher as they will settle over time). In severe cases of poor soil profiles, berms are sometimes made as much as 90 cm (3 ft) high and up to 3 m (10 ft) wide. Berms increase the rooting depth in shallow soils, help to drain water away from the crown, and keep roots above any high water tables that might exist. Such improved drainage can help reduce the risk of rain-induced fruit cracking resulting from uptake of excessive soil water following rain events. Where soils are heavy and winter temperatures are low, improved drainage can improve winter survival and help to warm soils more quickly in spring to promote root nutrient uptake activity. However, berms can hinder the movement of equipment, and make both weed control and harvest more difficult.

Subsoil tiling is another option where drainage is a problem; this can be quite effective and is well worth the effort and expense. Drainage tiles are installed in a grid pattern throughout the orchard to prevent excessive water accumulation in the root zone.

5.2.3   Compositional preparation of the soil

5.2.3.1   Remediation of acidic soils

The amount of lime needed to increase soil pH varies with cation exchange capacity (CEC), which is related to a combination of factors,

Table 5.1. The quantity of agricultural limestone needed to increase the pH of a soil to the optimal pH (target pH) for the crop or crop rotation using the Shoemaker–McLean–Pratt (SMP) value (Shoemaker *et al.*, 1961).

	Lime to apply[a] to attain desired soil pH (t/acre)				Lime to apply[a] to attain desired soil pH (t/acre)		
	Desired soil pH				Desired soil pH		
	5.6	6.0	6.4		5.6	6.0	6.4
SMP value				SMP value			
6.7	0	0	0	5.7	2.8	4.2	5.8
6.6	1	1	1.0	5.6	3.2	4.6	6.3
6.5	0	1.0	1.7	5.5	3.6	5.1	6.8
6.4	0	1.1	2.2	5.4	3.9	5.5	7.3
6.3	0	1.5	2.7	5.3	4.3	6.0	7.8
6.2	1.0	2.0	3.2	5.2	4.7	6.4	8.3
6.1	1.4	2.4	3.7	5.1	5.0	6.9	8.9
6.0	1.7	2.9	4.2	5.0	5.4	7.3	9.4
5.9	2.1	3.3	4.7	4.9	5.8	7.7	9.9
5.8	2.5	3.7	5.3	4.8	6.2	8.3	10.4

[a]Lime to apply values are based on application of 100-score lime and 6-inch soil sampling depth. For example, lime to apply = 1.7 t/acre when desired soil pH is 5.6 and the SMP lime requirement value is 6.0.

including soil texture, type of clay present, and soil OM content. As clay and OM content increase, CEC increases. In turn, the amount of hydrogen ions (H+) that need to be neutralized by lime also increases. Thus, variation in CEC provides the basis for variable lime application rates. The Shoemaker–McLean–Pratt (SMP) lime requirement test was developed to make lime recommendations for soils that differ substantially in CEC. Results may, however, necessitate variable-rate lime applications. Lime application rate is determined using the SMP lime requirement test (SMP buffer method).

Lime reacts with H+; therefore, the amount of both soluble and exchangeable H+ must be measured or estimated to determine lime rate. When soil is mixed with water to measure soil pH in a laboratory, very little H+ is present in soil solution. Most of the H+ is exchangeable, or electrostatically attracted to the soil particles. The SMP buffer measures H+ attracted to the soil particles as well as the soluble H+. The SMP test is not perfect, and because soil sampling techniques add variability, it is recommended that the test be used as a guide for lime requirement. Measuring soil pH after lime application verifies the adequacy of the lime application rate. 'Lime to apply' recommendations given in Table 5.1 are usually accurate to ±0.5 t lime.

Liming materials, including oxides, hydroxides, carbonates and silicates of calcium (Ca) and/or magnesium (Mg), vary in effectiveness. [Note: Calcium alone does not increase soil pH. For example, gypsum (calcium sulfate) and other additives contain Ca, but do not contain a basic anion (carbonate, hydroxide, oxide or silicate). Therefore, they do not neutralize soil acidity.] Liming material characteristics should be evaluated based on effectiveness (lime score) and cost per ton of 100-score lime. The carbonate in traditional agricultural lime (calcium or magnesium carbonate) reacts with soil acidity to neutralize it. By-product lime materials can be a cost-effective substitute for traditional agricultural lime. For certified organic crops, only use lime approved by your certification agency. Liming materials have very limited movement into the soil without incorporation. Thorough mixing (tillage) of the lime in the profile is required for maximum efficacy, increasing effectiveness by mixing lime materials into the root zone, and allowing them to make contact with acidic soil before the materials can react and change soil pH. The ideal time to fix low pH is before the orchard is established. After establishment, lime can only be placed on the soil surface since tilling it in would destroy roots and possibly cause severe tree damage.

Post-establishment frequency of lime application is determined primarily by CEC and crop management practices, especially nitrogen (N) fertilizer rate. Soil pH declines faster in sandy (low CEC) soils than in soil with moderate to high clay content. The typical rate of pH decline is approximately 0.1 pH units per year when 110 kg ammonium N/ha (100 lb/acre) is applied. It is advisable to test soil and leaves annually, to determine nutrient values available in the soil as well as whether those nutrients are being taken up by the plants. Poor uptake may be caused by nutrient imbalances in the soil, e.g. iron-induced manganese toxicity. For established orchards, a topdress lime application (e.g. 2.2–4.4 t/ha or 1–2 t/acre) may be beneficial.

5.2.3.2 Remediation of basic soils

Lowering the pH of alkaline soils is more difficult and expensive than raising the pH of acid soils with lime. For cherry, soils with pH greater than 8.4 require the addition of elemental sulfur (S) to lower pH and the addition of gypsum (calcium sulfate) to lower sodium (Na) content. Soils with pH less than 8.4 generally require only elemental S to lower pH. Soils with high pH and carbonates are extremely difficult to acidify and solve iron deficiency problems. Poorly drained soils or areas that are continually wet also are extremely difficult to amend. In these soils, high water tables limit root growth, reducing a plant's ability to take up iron. Artificial drainage should be considered before attempting soil acidification. Without good soil drainage, lowering pH is not economical. The primary material used to acidify soil is elemental S. An important point to remember is that S in the form of sulfate (SO_4^{2-}), including gypsum, is not an acidifying material. Sulfur is oxidized by bacteria to form sulfuric acid:

$$SO + O_2 + H_2O \Rightarrow H_2SO_4$$

(elemental S + oxygen + water + *Thiobacillus* + time $\Rightarrow$ sulfuric acid)

This reaction is temperature dependent, and the process requires several years to complete. Generally, a minimum broadcast rate is 0.5 t/ha (500 lb/acre), and rates can easily exceed several tons per hectare. Elemental S and gypsum rates depend on pH, soil texture, crop and the grower's desire to correct pH. As an alternative to elemental S, the application of acid to the soil (for example, sulfuric acid) can quickly correct alkaline soil. Unfortunately, acids can be very dangerous to use and must be handled with special equipment. Growers commonly inject or apply small quantities of acid, but these rates rarely have a significant effect on soil pH. Most soils require large quantities of acidifying material per hectare in a single application to effectively lower pH for only a growing season. Some fertilizers also act as acidifying materials, such as those containing ammonium. This process is slow compared to acidification by elemental S, but one benefit is that the rhizosphere pH is lowered, making mineral nutrients in close proximity to roots more available.

For established orchards, if the soil pH is >6.8 and Na, carbonates and electrical conductivity (EC) are high, elemental S applications in four to eight holes under the dripline around each tree should be considered. Apply 235 ml (8 oz or 1 cup) of elemental S in each hole.

5.2.3.3 Remediation of soil salinity and sodicity

Where soil drainage is good, saline soils can be reclaimed by the application of a sufficient amount of pure water with low salt, to flush the salts from the root zone profile of the soil. This water can come from rainfall or from an irrigation source that is lower in salts, such as may exist in the early spring when rainfall has diluted the salt content of surface water. The amount of salt removed depends on the quality and quantity of water. The following levels of water applied in a single, continuous irrigation will dissolve and decrease soil salts by these fractional amounts (Whiting *et al.*, 2015):

- 15 cm (6 in) of water will leach about half the salt;
- 30 cm (1 ft) of water will leach about four-fifths of the salt; and
- 60 cm (2 ft) of water will leach about nine-tenths of the salt.

Where soil pH is below about 6.5 (measured in calcium chloride solution), the use of lime or a gypsum/lime blend may be the most effective and profitable way of dealing with a sodicity problem (McKenzie, 2003).

Gypsum improves soil structure because it contains Ca, which displaces Na and Mg from the spaces between the clay particles. The Na should then be leached out of the soil profile by a single continuous irrigation as described above. Once a soil has been stabilized by Ca ions, additional stability can be achieved by adding OM.

5.2.3.4   Remediation of mineral nutrients

If soil analysis indicates a need for specific nutrients, these should be incorporated into the soil before planting. In particular, since P does not move quickly through the soil when applied to the surface, it needs to be incorporated into the soil profile. Phosphorus can be added to just the future tree row area or carefully added to the backfill as the trees are planted. To incorporate at time of planting, add 100g (3.5 oz) of mono-ammonium phosphate fertilizer to the backfill before shoveling the soil into the hole. To prevent phytotoxicity, be sure that the fertilizer is well distributed in the backfill; fertilizer near the roots will promote strong root growth.

5.2.3.5   Remediation of organic matter

Prior to planting trees is the ideal time to make major additions of OM. Cover crops can be grown and incorporated if time allows. Green manure crops can provide significant benefits to soils by improving fertility, structure and increasing OM content. In deep soils, alfalfa can grow roots to 18 m (60 ft) below the surface bringing up nutrients that can add N, P, potassium (K) and micronutrients to the orchard soil. Legumes, such as clovers and peas, fix N and add it to the ecosystem. Others create an extensive vegetative network of roots and leaves that prevent wind and water erosion during the winter. If possible, plant a green manure crop such as sorghum, Sudan grass, hybrid kale, sugar beets, winter rye, clovers, field peas or hairy vetch, at least one year before planting the orchard (Fig. 5.4). Multiple crops should be grown and plowed under just before they bloom, until the orchard is finally planted. Incorporate OM amendments throughout the rooting zone if possible, working the soil in several directions to evenly distribute the material.

High C, low nutrient amendments such as a compost can be added at rates as high as 22–44 t/ha (10–20 t/acre). Ask the vendor of the amendment for a recent or representative laboratory analysis. This will help identify what levels of nutrients may be added, and what nutrients may set the upper limit for application rate (e.g. P or K). Ask about potential contaminants, such as heavy metals or herbicide residues. Make sure the EC is not so high that it will raise the bulk soil EC higher than desirable for cherry trees. Check the pH of the compost since it may be in conflict with the need to alter soil pH; many composts are quite alkaline.

Fig. 5.4. A green manure crop of multiple species is grown on this site to improve soil fertility and structure prior to planting the orchard (Oregon) (courtesy of L.E. Long).

For maximum cost effectiveness, once the new tree rows are marked out, apply the amendments to only the tree row to concentrate the benefit where most of the future roots will be growing. It is important to take the carbon-to-nitrogen (C:N) ratio into account. An ideal microbial diet has a C:N ratio of 24:1. Mature alfalfa hay has a C:N ratio of 25:1 which is very close to this ideal. Other components, however, such as poultry manure is richer in N than the ideal, while wheat straw has a much higher C component than ideal (see Table 5.2). Incorporating a high C material will cause a shortage of N in the orchard system, whereas poultry manure will add additional N to the system. To obtain a balanced mix of components use the compost mix calculator developed by Klickitat County, Washington: www.klickitatcounty.org/1030/Compost-Mix-Calculator (accessed 19 June 2020).

Specialized spreaders are available to help with an application to tree rows only. Tillage that incorporates the amendment will be needed, such as a rotovator or disc plow. Good mixing throughout the tillage depth is desirable. Up to 4 t/ha (2 t/acre) of OM should be added before plowing and 0.5 t/ha (500 lb/acre) per year as a topdressing should also be considered, to continue increasing the OM content of the soils.

Opportunities to add carbon during the life of the orchard are discussed in Chapter 8, 'Managing the Orchard Environment'.

Table 5.2. Soil organic matter amendments and associated carbon-to-nitrogen ratios (adapted from (USDA NRCS, 2011).

Organic substance	C:N ratio
Wood chips	600:1
Sawdust	400:1
Rye straw	82:1
Wheat straw	80:1
Pea straw	29:1
Rye cover crop (vegetative)	26:1
Mature alfalfa hay	25:1
Compost	15–25:1
Ideal microbial diet	24:1
Barnyard manure	20:1
Legume hay	17:1
Sweet clover (green)	12:1
Hairy vetch	11:1
Grass clippings	<10:1
Poultry manure	<10:1

5.2.4　Irrigation planning

The orchard irrigation system should be designed before planting trees, often with the expertise of a consultant. It can also be helpful to talk with other growers in the area, particularly if the irrigation supply is part of a regional irrigation district. A sand filtration and UV sterilization system may be necessary if the water source is an open waterway (such as an irrigation ditch or river) that may be contaminated with water molds, e.g. *Pythium.* Acidification may be necessary if carbonates and bicarbonate concentrations are too high. There are numerous design and delivery options for cherry orchard irrigation. These include drip (trickle) systems, microsprinklers, impact sprinklers, furrow and flood irrigation. The most advanced and water efficient systems include drip and microsprinklers and will be the only two systems discussed here.

5.2.4.1　Drip irrigation

Drip irrigation enables maximum conservation of water, giving it a significant advantage over other systems. In a study conducted in The Dalles, Oregon, use of a double-line drip system reduced irrigation water use by 58% compared with the use of microsprinklers (Yin *et al.*, 2012). For optimal watering efficiency that allows for the maintenance of a between-row cover crop, many growers install a drip system for the trees and use microsprinklers to irrigate the cover crop with short irrigation sets. These water savings are due to the fact that a much smaller area is wetted with drip irrigation than with any other system and the water de-livered in this way is less prone to evaporation. Drip irrigation delivers the water directly to the rooting zone in small amounts in frequent intervals, maintaining relatively uniform moisture in this area. This reduces the stress on trees and de-veloping fruit that can be common with other systems, which may allow soils to dry between irrigations. In this way, the growth of trees and fruit are optimized, and fruit avoid some of the shrinkage and expansion that may be common with other irrigation systems. Furthermore, as smaller trees and dwarfing rootstocks become more common, the effective root zone within the tree row is more lim-ited and so directed delivery of frequent small amounts of water optimizes both tree growth and water use efficiency.

In most cases, irrigation distribution lines (usually polyethylene tubes) are positioned on the ground in the tree row, or are suspended just above the ground, supported by the tree crotch or a trellis wire. A single- or double-line tube system may be used, depending on tree size and emitter output. Emitter sizes range from 2–8 l/h (0.5–2 gal/h); application rates should be slow enough to prevent surface runoff. Emitters are either located directly in the row's drip tubes in an inline system or on smaller 'spaghetti' tubes that originate from the row's 'feeder' tube (Fig. 5.5). Each row's feeder tube is connected to a buried main line that is equipped with a filter, pressure gauge, injector, check valve and vacuum breaker near the water source. The filter should be located after the injector to filter out any possible precipitates. An anti-siphon device should be located between the injector and the water source to prevent backflow into the water source. Other pressure gauges are located in the system as needed, depending on the extent and distance from the pump.

Main and manifold pipelines are usually buried for increased durabil-ity and convenience. Row tubes and spaghetti lines are normally located above ground. This helps to detect and more easily correct clogged emit-ters before damage to tree growth occurs. It is imperative that irrigation water is clean and free of silt, sand or debris to prevent such clogging. For this reason, a high quality filtration system is imperative. In most cases, water from rivers, canals and other surface sources will need more careful filtration than water from a well.

5.2.4.2 *Microsprinkler irrigation*

Microsprinklers are relatively small emitters that throw water anywhere from 3–9 m (10–30 ft), delivering 30 l/h up to 340 l/h (8–90 gal/h), de-pending on orifice size and pressure. The discharge rate should closely match the soil infiltration rate. Due to larger orifice sizes, microsprinklers are less likely to clog than drip emitters. In addition, compared to impact sprinklers or furrow and flood irrigation, microsprinklers use up to 50% less water while increasing yields and decreasing water use, fertilizer and labor requirements (Reich *et al.*, 2009).

Fig. 5.5. A double-line, inline drip system with four emitters surrounding the trunk in this newly planted apple orchard (courtesy of G.A. Lang).

Microsprinklers apply water relatively evenly to the soil surface, allowing the water to slowly penetrate into the soil over a relatively wide area. In most situations, microsprinklers are arranged in a pattern that will ensure total coverage of the orchard floor. Besides providing irrigation to the trees, this allows for the maintenance of a cover crop in the tractor alley as well as a potential tool for mitigating frost damage. However, a significant disadvantage to microsprinklers (compared to drip irrigation) is increased germination and growth of weeds in the larger wetted area.

5.3   Orchard Structures

5.3.1   Tree support systems

For establishment of high density spindle or upright fruiting offshoot (UFO) orchards, particularly if partial mechanization of thinning or hedging is envisioned, it is recommended to utilize a trellis system for precise orientation and development of the tree canopy. Unlike apples, the trellis

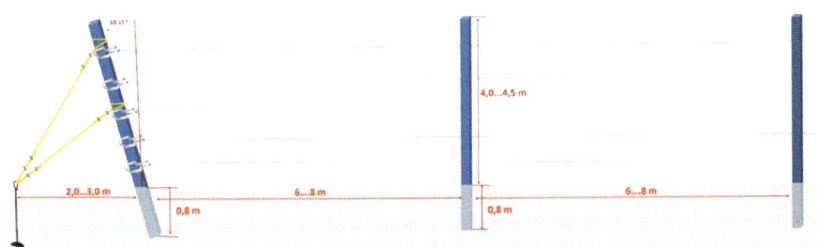

Fig. 5.6. A typical trellis system that provides support for trees, tiedown wires for branches and the possibility of covers for light, rain and hail (courtesy of E. Gudumac).

system is generally not needed to support the future crop load, but may be needed to prevent trees from leaning during the first few years of tree establishment, or to precisely orient fruiting branches or leaders allowing the most efficient interception of light for optimal cropping and harvest access. Furthermore, if protective covers are planned for rain/hail/bird protection, the support structure must be integrated into the orchard tree rows. For new orchards, it is advantageous to design and stake out the trellis system prior to planting, but usually it is not installed until just after the trees are planted since posts and wire would interfere with the use of tree planters or the augering of planting holes.

The most widely used orchard support system is the simple monoplane post trellis. This trellis consists of a network of angled posts (wood, steel or cast from concrete cement reinforced with pre-stressed steel rods) at the beginning and end of each row, with intermediary vertical posts, three to five wires for tree orientation, and subsoil anchors that are installed at the ends of the rows (Fig. 5.6).

Important trellis considerations include the installation of the end posts at a declining angle (18–30°) towards the row end, and the subsoil anchors placed at 1.5–2.0 m (5–6 ft) from the end posts, depending on the length of the rows to be supported. End posts should generally be of a larger diameter than intermediary posts. The intermediary posts are generally placed 8–9.5 m (25–30 ft) apart in the row; close spacing of the intermediary posts is not so much for support of the crop load weight (as would be the case with high density apples), but rather to maintain the trellis wire height and tension for precise orientation of the tree canopy. A special case may be required for trellising of UFO canopies, where an additional mini-post or anchor that extends only to the bottom wire may be situated halfway between intermediary posts, to provide additional stability to the bottom wire, since the horizontal orientation of the UFO cordons exerts a significant upward force on the lowest trellis wire. All posts should be installed, preferably by pounding, to a minimum depth of at least 0.75 m (2.5 ft).

The height of the first trellis wire layout above the soil depends on many factors, including the canopy training system, and potential use to support drip irrigation tubing, etc. Usually, the first wire is 50–60 cm (20–24 in) from the ground level, and the remaining wires are installed at intervals of about 70–80 cm (28–32 in) above the bottom wire. Trees and branches should be attached to the trellis wires with vinyl tubing or plastic or rubber ties, generally on the leeward side of the trellis, so that prevailing winds blow the plant away from the wire to reduce bark rubbing and the potential for bacterial canker infection. Similarly, on angled V- or Y-trellises, the tree leaders or branches should generally be tied to the interior of the angled planes, such that they pull upward and away from the wires. Such ties can be affixed to the wire, twisted several times, and then affixed to the tree so that they provide a barrier to reduce rubbing between the bark and the bare wire. Alternatively, old drip irrigation tubing can be repurposed by slitting along one side and slipping over the wire where trees are to be tied, preventing contact between the tree and the bare wire. Finally, high tensile plastic trellis wire is now available, which can reduce the potential for bacterial canker infection compared to standard galvanized steel wire. When using high tensile plastic wire, it is often advantageous to use white instead of black or gray, since white is easier to see and therefore less likely to be accidentally cut during pruning. Tensioning and retightening of both steel and plastic trellis wires can be achieved by many methods, including gripple locks or racheting gears, and tension should be checked and adjusted accordingly at the beginning of every new growing season.

V-trellises may be developed as individual rows with stand-alone trellises or as trellis rows conjoined with cross-bracing wires or cables (Fig. 5.7). Cross-arms or partial hoops may be used at the apex of adjacent angled supports to impart structural strength perpendicular to the trellised tree rows (Fig. 5.8).

Fig. 5.7. Free-standing (left) and cross-braced (right) V-trellis fruiting wall sweet cherry canopies (courtesy of G.A. Lang).

Fig. 5.8. Extra cross-arm bracing for V- or Y-trellises to counter the strong tendency of sweet cherry trees to pull angled trellis wires toward vertical orientation (courtesy of G.A. Lang).

5.3.2 Orchard covering systems

Modification of the cherry production microclimate with various types of orchard covering systems has become easier with higher density, intensive plantings of smaller trees and greater attention paid to optimization of light distribution throughout the canopy. There are particular synergies possible with training systems that require trellis structures, since these offer the potential to integrate the support structure for the covers with the support structure for the trees. Orchard covers are not only suitable for modifying the microclimate (e.g. to protect from rain, frost, hail, as well as to alter ripening times by changing growing degree day/heat unit accumulation), but also to protect from birds with nets (or a combination of nets and covers), protect from rain-disseminated diseases such as cherry leaf spot (*Blumeriella jaapii*) and bacterial canker (*Pseudomonas syringae*), and possibly from some insects such as Japanese beetle (*Popilla japonica*)

Fig. 5.9. Insect exclusion drapenets with rain-impermeable top panel for protection against cracking (Italy) (courtesy of G.A. Lang).

and spotted wing *Drosophila* (SWD, *Drosophila suzukii*) (Fig. 5.9). On the other hand, some covers can create conditions that increase the incidence of certain diseases (such as powdery mildew, *Podosphaeria clandestina*) and insect pests (such as mites).

Orchard covering technologies can be grouped into three types, although there are gradations of each:

1. Pole-and-cable tent-like structures that support long narrow plastic sheets and/or continuous joined net panels, usually installed or opened seasonally, particularly during bloom and ripening;
2. Steel hoop-like structures (high tunnels) that support large, long sheets of plastic that are usually deployed seasonally from bloom through ripening or beyond; and

3. Greenhouse-like structures with the potential for full enclosure throughout the year.

The structural integrity of all types of cover is closely aligned with how well they are anchored to the earth, including poles and anchors driven or screwed into the ground, multiple anchoring points on sides and especially strong anchors on corners, and even concrete footers for the structural supports.

There are many formulations of polyethylene for row covers and high tunnels, which can vary in plastic thickness and well as light transmissive properties (e.g. see www.tunnelberries.org/uploads/5/3/8/2/53821521/high-tunnel-brambles2017.pdf, accessed 19 June 2020). The typical thickness for orchard covers is 6 mm, which has a predicted lifespan of 3 to 7 years, depending on how long it is deployed in the orchard each season and annual wear and tear. Typical polyethylene formulations transmit relatively high levels of photosynthetically active radiation (though transmission tends to decrease over time, particularly with dust, dirt and spray residues) and are stabilized for ultraviolet (UV) light to increase their lifespan. Polyethylene can be formulated with a variety of additives that alter transmission of UV and/or infrared (IR) light, as well as increasing the diffusion of light as it passes through the cover into the orchard. Plastics that reduce IR light transmission can reduce the daily heat gain within the tunnel by at least a few degrees. This can be a significant advantage in summer, but a disadvantage in spring if earlier heat unit accumulation is desired to advance bloom and ripening.

Pole-and-cable structures are the least expensive orchard covering technology, and can be established with poles made of wood, steel or cement; the latter two materials usually being most durable but also most expensive (Fig. 5.10). Trellis wire can be used, but multi-wire cable is stronger and more durable. The simplest structure consists of a cable or wire running the length of each tree row over the peak of the tree canopy, supported periodically (perhaps every 10–12 m or 30–40 ft) by poles in the tree row. Long narrow sheets of plastic are supported by the wire, draping tent-like on both sides of the peak cable and tied off to maintain their clearance above the tree canopy. In the case of individual tree row nets used solely to protect ripening fruit from birds or insects, the sheet of netting should extend to the ground or be tied or clipped off to the other side underneath the canopy to prevent entry by birds or insects into the canopy from below as well as from above.

More extensive pole-and-cable row covers often include either: (i) a lower cable over the middle of each tractor alley; or (ii) two lower cables located parallel to the tree row and at each edge of the tree canopy. With either of these two systems, the lower edge of the plastic tent is periodically clipped, tied, or affixed to a low cable on each side of the tree row, to create a cover that provides dry air clearance above the canopy and a gentle slope

Fig. 5.10. Robust pole-and-wire plastic covers with cross-arm wires and clips for retraction (cherries trained in upright fruiting offshoot, UFO, system in New Zealand) (courtesy of G.A. Lang).

to shed rainwater into the tractor alley (Fig. 5.11). A relatively flat angle to the tent and an open gap between the dual low canopy edge cables can minimize the retention of heated air under the tent during fruit ripening (Fig. 5.12). The gap can be created by a rope or elastic 'bungee' cord tied from the plastic cover edge to the single low cable over the tractor alley. Such gaps also improve honeybee pollination under the covers, due to access to open sky which is necessary for good honeybee navigation. These gaps and the use of bungee-type elastic cords for tying to the low cables can reduce wind stress on the plastic covers as well. Tying together end poles with cables perpendicular to the tree rows also improves torsional strength of the covers where wind speeds can be significant.

Cherry growers and orchard-related businesses have developed many modifications of pole-and-cable covering systems. Some have used metal rings or clips (such as carabiners from mountain climbing) attached to the low support cable, allowing the covers to be quickly retracted or re-covered, either by hand or mechanically. This can facilitate daily opening during bloom for pollination (with covering at night for minor frost protection) or for exposure to full sunlight during ripening, except when rain

Fig. 5.11. Voen hybrid plastic/net covers with plastic clips to join over the tractor alleys (spindle cherries in the Netherlands) (courtesy of G.A. Lang).

is forecast. Another innovation is the use of a combination of hail netting and overlapping panels of plastic sheets that form a 'shingles'-type dual material orchard cover (e.g. see the Voen company in Germany) that minimizes rain penetration, but since the plastic panels are bonded only along their upper edge to the hail netting, they become flaps that can shed wind and passively vent rising hot air to some extent. With all types of pole-and-cable row covers, and nearly all bird nets, the covers are either removed entirely before the winter and stored in barns or other outbuildings, or are unclipped from the low support cables and rolled up and tied off around the top wire for winter 'hibernation' (Fig. 5.13). Such rolled up covers should be covered with black plastic to block exposure to UV light and to keep excess moisture (and ice) out during winter; this will extend the life of the cover.

Most pole-and-cable row covers are adequate to excellent, depending on cover design and canopy architecture, for keeping sweet cherry fruit dry during rain events. However, they all shed rainwater into the tractor alley, which then has the potential to saturate the root zone and promote cracking of fruit due to root-based water uptake and excessive internal water pressure. Maintaining dry fruit only solves half of the rain-induced fruit cracking problem. Therefore, it is often wise to combine row covers with improved tractor alley drainage (such as subsurface drainage) and/or

Fig. 5.12. Bungee-corded rain covers under bird/hail netting, with cable cross-bracing of cement posts (Italy) (courtesy of G.A. Lang).

Fig. 5.13. Hibernated Voen row covers (Michigan) (courtesy of G.A. Lang).

Fig. 5.14. Cherries grown in high tunnels, such as this one in the UK, are protected from rain as fruit ripens (courtesy of G.A. Lang).

planting on raised beds (which can be especially effective with the smaller root systems of dwarfing rootstocks) to maintain a significant portion of the root system above the potential saturation zone.

The use of high tunnels in fruit production has been pioneered primarily by strawberry, raspberry and sweet cherry growers in the UK, where prevalent rainy weather that creates ideal conditions for fruit diseases and fruit cracking are a serious threat to sustainable production and retention of markets (Fig. 5.14). The use of high tunnels for sweet cherries has subsequently expanded around the world, probably nowhere more than in China. Chinese agriculture has a long history of using passively heated 'half-tunnels', that is, tunnels formed by south-facing partial hoops (usually steel, but historically made of bamboo) with the north side being a wall of earth (when located on south-facing hillside slopes) or brick (Fig. 5.15). The partial hoops are covered with plastic and the earth or brick wall absorbs heat during the day and radiates heat during the night. The hoops are covered with plastic to create the tunnel, and an insulative material layer may be rolled up from the bottom of the hoop to the top of the wall during the day for solar heating, then rolled down at night to trap the heat and protect from frosts. Thousands of acres of these half-tunnels have been planted in recent years for sweet cherries in China, to promote early bloom and ripening during advantageous marketing windows.

Fig. 5.15. Inside south-facing half-tunnels with cherries (China) (courtesy of G.A. Lang).

The more typical type of high tunnel in use around the rest of the world is based on steel hoops that connect to steel legs with a 'Y' fork at the top, so that multi-bay tunnels of adjacent hoops can be created over contiguous cherry orchards (Fig. 5.16). The most basic of these have steel bracing usually just below the 'Y' to connect the legs and provide some level of rigidity to the hoops. Structural integrity increases with closer spacing of legs, added cross-bracing of hoops (either by steel or by cross-cables), and how peaked or flat the top of the hoop is. The greater the amount of steel that is used, the greater the strength of the structure, but the higher the cost. Most multi-bay high tunnels are considered to be three season structures, with flatly rounded hoops that cannot support any significant snow loads in winter. Peaked (or 'gothic') tunnels, with adequate cross-bracing, can support most winter snow loads in typical cherry producing areas. However, since sweet cherries require exposure to winter chilling temperatures to break endodormancy, and may lose cold hardiness in winter if exposed to periodic warm temperatures, most high tunnels are not covered until enough cool weather for the alleviation of endodormancy

Fig. 5.16. Haygrove multi-bay high tunnels (Michigan) (courtesy of G.A. Lang).

has occurred. Furthermore, fully covered multi-bay tunnels would collect snow in the 'Y' gutters where the tunnels meet.

The high tunnel steel hoops are usually covered with a long sheet of polyethylene plastic, one per tunnel, with the edge of each sheet of plastic meeting in the 'Y' gutter between tunnels. Some designs include the installation of a separate true plastic gutter in the 'Y', so that rain is shed from the plastic into the gutter and is carried to external drains on the perimeter of the tunnels rather than simply running off the plastic sheets into the tunnel leg row (and possibly seeping back into the cherry tree root zone in the tunnel). Alternatively, subsurface drainage pipes can be installed along the base of each tunnel leg row, with the shallow trench backfilled with gravel, so that water that does run off the tunnel plastic is efficiently drained away at ground level. These steps are important, since cherry fruit can crack from excess soil moisture, separately from exposure to rainwater in extended contact with the fruit.

One of the key challenges of high tunnel management for sweet cherries is to reduce the potential build-up of heat in the orchard during ripening. If the plastic covers extend from gutter to gutter, air heated by solar radiation is trapped and tunnel temperatures can easily exceed 35–40°C (94–104°F) or more during a late spring or summer day. Orchards on sloping land will create a natural chimney effect, allowing the hot air to escape from the higher end, but those on relatively flat land will trap the hot air. Typically, the sides of the tunnel cover plastic are pushed upwards ('venting') by 1–2 m (3–6 ft) to create air gaps that facilitate some escape of hot air and the potential for exposure to ventilating cross-breezes, but the majority of the hot air remains trapped from the point of venting to the top of the hoop. This venting also means potential exposure of the lower portion of the tree canopy to rainfall, and the runoff from vented plastic no longer flows into the 'Y' gutters or the leg row subsurface drains, but rather closer to the tree row root zones. If this is the case, rows may need to be planted on raised beds to reduce the potential for root zone saturation during rain events, but this in turn reduces the height of the fruit-bearing canopy since the top of the tree is limited by the height of the tunnel.

Venting has traditionally been done manually, which is why the level of venting during the growing season is rarely changed to coincide with weather events. Some of the most recent innovations in high tunnel design have incorporated motorized venting capabilities, which conceivably could be tied to thermostats for automated venting, perhaps even with changing temperature thresholds for different stages of reproductive development. However, even automated venting that occurs by raising the sides of the plastic covers will still trap hot air in the peak of the tunnel, unless the sides are raised almost to the top. It is also conceivable that a programmable automated, motorized tunnel venting mechanism could be tied to a rain gauge rather than, or in addition to, temperature sensors, to close just moments after rain begins and then open for full exposure to sunlight. Such a system may be close to development, but at this time is not commercially available. Perhaps the most recent commercial innovation to address improved labor efficient venting of heat is the use of polyethylene with a netting panel bonded into the plastic at the peak of the tunnel, so that rising hot air does not become trapped but rather escapes through the netting. The width of the netting panel is directly proportional to how well it vents heat, but of course it is also proportional to how much incidental rain may come in through the netting. Consequently, a combination of plastic and netting tunnel covers may vent peak heat, but growers may need to orient the tree rows to minimize incidental rain entry, and such covers may be problematic for capturing early season heat to advance bloom and ripening.

Sweet cherries have been grown in greenhouses to a very limited extent, since the costs of greenhouse production are extremely high. Consequently, greenhouses are economically viable only for off-season production for high value markets. Greenhouses generally offer the greatest control over the production environment, even including the potential for CO_2 enrichment and absolute protection from damaging low temperature events due to heating efficiency (Fig. 5.17). Perhaps the most economical use of expensive greenhouse space is with potted trees that can be moved in and out of greenhouses, cold storage, and outdoor environments to cycle multiple harvest seasons during the calendar year, but potted trees also require a higher level of intensive management, usually with high frequency, short duration irrigation and fertigation to provide the necessary water and nutrients to such limited root systems. Recent advances in such microclimate modification include automated, programmable retractable roof

Fig. 5.17. Modifying the climate with the use of propane heaters in a retractable roof greenhouse is possible, but expensive (courtesy of G.A. Lang).

greenhouses for in-field or in-pot production (Fig. 5.18). Both high tunnel and greenhouse sweet cherry production often utilizes bumblebees for pollination, rather than honeybees, since bumblebees are highly efficient, work at lower temperatures, and have no trouble navigating under plastic covers that disrupt the polarized light that honeybees use for navigation.

5.4 Selecting a Nursery Tree Source

Nurseries should have a proven track record of delivering quality stock that is true-to-type, pest and disease free, and the chosen nursery should provide excellent after sales service. In the USA, it is advisable to deal with nurseries that subscribe to the American National Standards Institute (ANSI) for Nursery Stocks. Reputable nursery sources should be identified at least 2 years before planting, as rootstocks will need to be ordered by the nursery, budded with the desired cultivars, and grown for one season before digging, grading and delivery for spring orchard planting. Some

Fig. 5.18. Cravo brand greenhouse retractable roof partially open to balance heat retention and facilitate honeybee navigation for pollination (courtesy of G.A. Lang).

nurseries now offer greenhouse production of potted trees to provide a faster order-to-delivery timetable (Fig. 5.19). It is often wise to check with several nurseries to determine the availability of the cultivars, rootstocks and nursery tree type desired. However, if the cultivar is new or unusual, even more time may be necessary so that the nursery can locate a reputable source of budwood.

If at all possible, nursery budwood should be certified to be free of known viruses. There are no measures for controlling viruses in the orchard other than starting with virus free rootstocks and scion wood, or removing existing trees once a pathogenic virus has been identified in the orchard. Currently, the 'little cherry' virus and the 'western X-disease' phytoplasma have negatively affected many orchards in the PNW region of the USA, and sometimes individual viruses such as prune dwarf virus (PDV) and Prunus necrotic ringspot virus (PNRSV) may cause trees to exhibit only mild symptoms, but in combination with other viruses or in certain scion–rootstock combinations, the symptoms become limiting to crop value or even lethal.

Phone and e-mail contacts with trusted nursery staff or brokers can help to determine whether current or anticipated plant availability will be sufficient or whether a custom order must be arranged several years in advance of planting. In any case, a deposit will be expected to guarantee orders. Occasionally, nursery grown trees become available when another customer cancels an order or production in a nursery grades out better

Fig. 5.19. These bagged trees were grown for 1 year in a greenhouse before being ready for planting in field (Chile) (courtesy of G.A. Lang).

than expected, but the availability of such unexpected cultivar–rootstock combinations should not tempt one to deviate significantly from the orchard business plan.

Unlike apple nursery trees, sweet cherry nursery trees often have only a few or no branches, and cherry training systems usually call for the removal of any nursery branches unless the tree is well feathered and the training system is a variation of the spindle tree form. Well feathered nursery cherry trees are more common in Europe where two-year-old knip-boom tree production occurs. The important characteristics for good cherry nursery trees include: (i) a good root system; (ii) a moderate to somewhat vigorous caliper; and (iii) many strong, intact buds along the portion of the leader where future shoots are needed to develop the desired tree structure. Short, weak nursery trees of small caliper are difficult to handle (especially with tree planters), difficult to establish in competition with weeds, and generally will take at least an extra year to develop adequate structure for fruiting. Tall, overly vigorous nursery trees of very large caliper often increase the difficulty for promoting lateral shoots or achieving balanced new shoot development along its length; if headed to reduce leader length, the resulting new shoots below the heading cut are often too vigorous for many training systems except where multiple strong leaders are desired. Achieving adequate and balanced shoot development on overly vigorous nursery trees can also delay development of adequate structure for fruiting, though in climates with warm spring seasons where plant growth regulators like Promalin® are consistently effective, a more vigorous nursery tree has the potential to achieve significant early production. In general, moderate size nursery trees of 1.6–2.2 cm (5/8–7/8 in) caliper and 1.5–2.0 m (4–5.5 ft) in height, with a high proportion of strong buds intact above the graft union, provide excellent options for orchard establishment.

In the newly planted orchard, trees tend to develop new roots and shoots to establish a balanced 1:1 root-to-shoot ratio. Consequently, if the scion portion of the nursery tree is greater than the rootstock portion, new shoot growth may be minimal during the first year as the root system reactivates after the transplant shock and new roots grow to re-establish balance with the canopy. The implications of this include: (i) promoting new shoots on the nursery tree leader may be difficult due to transplant shock and root system re-establishment; and (ii) larger nursery trees that have insufficient roots at planting may benefit from heading back the leader to achieve a more balanced root-to-shoot ratio during recovery from transplant shock and tree establishment. This is particularly feasible for multiple leader training systems, but should be avoided for some training systems that utilize a single leader, such as the angled version of the UFO system and some spindle canopy architectures. Heading at planting of training systems that utilize dual leaders, such as the bi-cordon UFO and bi-leader SSA canopies, allow the root system to become better established before significant new canopy leader growth and leaf area develops.

5.4.1 Transport and storage of nursery trees

Unless the nursery trees are shipped directly to the orchard site by the nursery for immediate planting, they should be collected from the delivery point and transported in a closed refrigerated truck that is sanitized (see Table 5.3) before and after collection of the trees. Typically, nursery trees

Table 5.3. Some common chemical sanitizers for shipping containers and cold storage units (adapted from Schmidt, 1997).

Sanitizer	Product examples	Key aspects	Disadvantages
Acid-anionic	Include an inorganic acid plus a surfactant	For acid rinse and sanitation	Low activity on molds and yeasts
Chlorine	Sodium and calcium hypochlorite	Effective for 1 min at 50 ppm at 24°C (75°F)	May raise water pH by forming sodium or calcium salts
Chlorine dioxide	Stabilized ClO_2	Effective from 1–10 ppm; can be used as a foam	More expensive than hypochlorites and must be generated on site
Fatty acid or carboxylic acid	Phosphoric acids, organic acids	For rinse and sanitation with low foaming potential	Low activity against yeast and molds
Iodine	Iodophors	12.5–25 ppm for 1 min; used in hand sanitizing solutions	May stain porous surfaces and some plastics
Peroxides	Hydrogen peroxide	Limited application in the food industry	High concentrations of 5% and above can irritate eyes and skin
Peroxides	Peroxyacetic acid	Possible chlorine replacement	Affected by pH; any pH above 7–8 reduces activity
Quaternary ammonium compounds (QACs)	Environmental fogs and room deodorizers	Good against molds, ineffective with some bacteria	Inactivated in hard water

are shipped dormant and bareroot, often with damp sawdust or other such organic media packed around the roots to maintain them in a moist condition. Roots should never be allowed to dry out. Increasingly, some nurseries are now producing trees in tall narrow pots, which can even be shipped when fully leafed out. The potted root system, which should also never be allowed to dry out (especially if leafed out), tends to transplant better than bareroot trees and can facilitate planting later in the spring with less of the transplant shock associated with planting bareroot trees in the warmer conditions of late spring or even early summer. If the number of trees being picked up does not warrant using a climate-controlled truck and an open-bed truck is used instead, if the trees are not boxed, they should be protected from drying out (caused by exposure to the sun and/or wind) by the use of a tarp tied down securely over the trees, especially over the root systems for bareroot trees and over the leaf canopies if present on potted trees. The roots should be doused with water before leaving the nursery.

In anticipation of notification from the nursery that the trees are ready for collection, refrigerated storage should be identified and/or prepared for holding the trees until they are ready to be planted. Upon arrival, the trees should be transferred to the cold storage room immediately and the roots should be inspected for signs of dehydration. Bareroot nursery trees should be stored between 2 and 9°C (35 and 48°F) at >95% relative humidity (RH) until the field is ready to plant. The cold room (and storage bins, if used) should be sanitized at room temperature (23°C/75°F) prior to tree storage, then the temperature lowered to the target range. Wall and floor surfaces should be clean before sanitization, as any residual debris is a potential host to diseases. Common sanitizers are listed in Table 5.3. Additional sanitizers include ozone and acetic acid; however, generators and vaporizers are required in order to utilize these technologies. Sanitization at the recommended concentration is critical, since too little will be ineffective and too much could be corrosive. If a chloride-based sanitizer is used, the pH must be in the correct range since high pH (>7.5) will result in volatilization of chlorine gas, which has no effect as a sanitizer.

High humidity (≥95%) in cold storage, such as by fogging, is necessary to prevent root dehydration and death. However, free-standing water must be avoided, as this could result in the spread of any diseases present on the plant material when it arrived. If fogging is used, consider including a quaternary ammonium compound (QAC) sanitizer in the fogging solution. Some operations use burlap bags in which to store the bundles of rooted trees, with periodic irrigation of the bags to prevent the roots from drying out. However, this can lead to free-standing water and potential diseases. Similarly, storage of the bareroot system bundles in moist sawdust in bins can result in the rapid spread (between trees) of any root diseases present and should be avoided if possible. The longer trees are held in storage, the more buds will continue to slowly swell and become more susceptible to potential diseases or inadvertent bud loss from subsequent tree handling.

5.5 Successful Pollination

5.5.1 Pollinizers

As discussed in Chapter 2, 'Cherry Flowering, Fruiting and Cultivars', unless the sweet cherry cultivars to be grown are self-compatible, one or more additional cultivars may be needed, depending on the orchard design, to serve as pollinizers that provide compatible pollen for cross-pollination. Suitable pollinizer cultivars must have at least one S allele that is cross-compatible with the primary fruiting cultivar and the bloom periods must overlap. It is usually recommended to have more than one pollinizer cultivar in the orchard, particularly in growing regions where winter and/or

Fig. 5.20. A typical pattern is to plant pollinizers every third tree in every third row (courtesy of E. Gudumac).

spring weather can be quite variable year to year. Relative time of bloom can vary between cultivars from one year to the next, depending on each cultivar's unique genetic responses to variations in both winter chilling units to break endodormancy and spring heat units to alleviate ecodormancy. When growing a cultivar of low productivity, it is best to choose a pollinizer with two compatible alleles so that all of the pollen is active.

In moderate to low density orchards, it is common for pollinizers to be planted every third tree in every third row, alternating the pollinizer cultivars within each row to distribute diverse pollen sources evenly throughout the orchard (Fig. 5.20). In this configuration, there are no pollinizer trees in rows 1, 3, 4, 6, 7, 9, etc., resulting in each solid cultivar row being adjacent to a row with pollinizers. This is a normal practice when the pollinizer cultivar is of lesser value than the primary cultivar.

A higher proportion of pollinizers to the primary fruiting cultivar should be used if the primary cultivar tends towards moderate to low productivity (such as Regina). In this case, a highly productive dwarfing rootstock is used and pollinizers are planted in a reduced space about every fifth tree, with this fifth tree designation offset in adjacent rows. These low value pollinizer trees can then be pruned to very narrow spindle canopy architectures to maximize the orchard area devoted to the primary (high value) cultivar, considering the pollinizer as simply a pollen source with minimal market value.

It is important to bear in mind that honeybees are known to forage from only two to three trees per visit and they tend to work down tree rows, rather than across, especially in orchards with hedgerow canopy orientations, such as UFO. Thus, modern high density orchards with smaller trees may benefit from including pollinizers in every row, but at wider spacings

Fig. 5.21. Alternating solid rows of the primary and pollinizer cultivars can facilitate management and harvest ease where both cultivars have high market value (courtesy of E. Gudumac).

(farther apart than every third tree) and at staggered tree positions from row to row.

However, alternating solid rows of a primary fruiting cultivar with a different, but fully pollen-compatible second high value fruiting cultivar also has been shown to provide good yields in some high density orchards (Fig. 5.21). This type of orchard design with two high value, cross-compatible cultivars can also be accomplished with two or three rows of one, then two or three rows of the other cultivar. With a 2 + 2 configuration, every row is still adjacent to a compatible pollen source. This also makes for a less complicated harvest, as pickers can work down full rows without encountering pollinizer trees along the way.

It is convenient when possible for pollinizers and the primary fruiting cultivar to be similar in terms of bloom time, as well as ripening time and disease susceptibility, from not only a harvest point of view but also regarding pest spray management concerns. Pollinizers that are more suscepti-ble to key diseases than the primary cultivar should be avoided if possible. Pollinizer trees can be marked with color-coded paint on the trunks to help workers prune and harvest them differently from primary cultivars, as appropriate.

Many newer cherry cultivars are self-compatible, in which case pollin-izers do not need to be incorporated into the planting plan. The possible exception to this is when less productive self-compatible cultivars, such as Benton, are to be planted as the primary cultivar, and productivity may be improved with additional pollen sources.

5.5.2 Pollinators

Regardless of whether the cultivars being grown are self-incompatible or self-compatible, insects are needed to move pollen from the anthers to the pistil for pollination. A common misconception among some growers is that self-compatible cherry cultivars don't need pollinators – after all, some self-compatible fruit plants like grapes have flower structures that facilitate pollen transfer from anther to pistil without any insect intervention. This is not the case with sweet cherry, which evolved with attractive floral nectaries and pollen that is relatively heavy and sticky, to facilitate cross-pollination by insects that visit the flowers, unlike some forest trees such as oaks and pines that have light, wind-blown pollen. Many insects are attracted to sweet cherry flowers, including honeybees, bumblebees, mason and other solitary bees, syrphid flies, and other flies. Typically, honeybees are the preferred pollinator for orchards due to their commercial availability and the ease by which beekeepers can move large populations in hives, in and out of the orchard to coincide with annually variable bloom times. Ideally, honeybees should be made available when the first bloom reaches about 10%, to assure that the early flowers (which usually yield the largest fruit) are pollinated and to attract the bees to the cherry flowers rather than have them establish alternative foraging habits if placed in the orchard before cherry blossoms are present.

Sweet cherry blossom nectar tends to be low in sugar content, so honeybees often will seek out competitor flowers that have a higher sugar content. Consequently, weeds that flower at the same time as cherries, such as dandelions, mustard, and wild radish, should be eradicated since they attract bees away from cherry blossoms. Eucalyptus trees also will attract bees away from a cherry orchard. European honeybees are not active at temperatures below 13°C (55°F) and activity increases markedly with temperature. Ideal pollination temperature is about 20°C (68°F), which promotes good honeybee activity and good flower longevity. Higher temperatures, particularly above 25°C (77°F), lead to shorter effective pollination periods (EPP) since flower ovules degenerate before pollen tubes may grow for successful fertilization at higher temperatures. When temperatures are low during flowering, providing bees with supplemental sugar helps maintain hive strength for better bee activity in the orchard. Research has shown that a solution of 50% glucose is preferable to 50% sucrose (C. Kaiser, personal communication). Feeders with footholds containing the glucose solution should be placed equidistant from hives within the orchard, provided hives are strategically placed in pairs on either side of the orchard perimeter (Fig. 5.22). To attract the bees to the glucose, it may be enriched with 1% honey or Coca-Cola®.

Sweet cherry orchards typically need four to six honeybee hives per hectare (2 to 2.5 per acre) to ensure sufficient pollinator activity. Hive orientation is also a factor as bee activity is encouraged by direct sunlight. Consequently, orienting the hive entrance to face the rising sun (east) is recommended. Elevating the hives 1 m (3 ft) off the ground helps to ensure that they stay warmer and bees will start working the orchards sooner than those at ground level. If hives are to be used for trees under tunnels or shade nets, they must be placed in the open, 3 m (9 ft) back from the perimeter of the structure as bees will exit the hive and go up, looking for the sun for orientation. If the hives are under shade nets, the bees will climb through the nets and then be unable to find their way back to the

Fig. 5.22. Chicken feeders containing 50% glucose solution helps to maintain hive strength for better bee activity during times of low bloom temperatures (courtesy of C. Kaiser).

hive. This can result in a significant loss of the foraging bees from the hive. When plastic covers are in use during flowering, commercial bumblebee hives can be placed within the covered orchard since bumblebees do not need direct light for orientation and navigation. Bumblebees also work at lower temperatures than honeybees.

5.6   Planting the Orchard

Orchard site soil preparation must be thorough to ensure that settling of the soil does not result in newly planted trees sinking in the tree rows. This is especially pertinent if mulch has been applied before planting in the tree rows. Consider planting the trees on raised beds or mounds if this is the case, to avoid eventually having sunken areas around the tree trunks once they are established. Such sunken areas can lead to pooling of water during heavy rainfall, which can increase disease susceptibility as well as become potential reservoirs for cold air during a freeze in late fall, winter, or early spring.

   Time of planting also can be a factor in orchard establishment success. In areas with extremely severe winter freezes, waiting until spring to plant may be worth considering. In areas with milder winters, if nursery trees are available, fall planting often results in better spring growth since some es-tablishment root growth will occur during the fall, winter and early spring, minimizing transplant shock. In any event, if trees are planted in fall when typically dry locations and cold temperatures are expected that could nega-tively affect the roots, it is important to irrigate the soil to keep it close to field capacity during the winter months, since moist soils have been shown to result in less winter root damage than dry soils. On the other hand, in cold wet environments, like that of the north-eastern USA, if the orchard site is characterized by heavier soils, tree rows should be planted on raised beds or berms for adequate winter drainage, as saturated winter soils also have been associated with greater susceptibility to winter damage.

If the roots of any stored trees appear dehydrated the day before plant-ing, consider immersing the root systems in sanitized water for up to 12 h to rehydrate them. On the day of planting, the tree bundles should be sepa-rated carefully so that each tree and root system can be inspected. Trees should be handled with care to avoid any injury to bark, branches, etc., and to minimize the loss of buds, especially if the buds have begun swelling. A loss of buds prior to or during planting equates to a loss of future branch structure and cropping potential. Any marginal plants should be set aside, such as those with crown gall (enlarged globular growths on the roots or crown caused by the bacterium *Agrobacterium tumefaciens*), any with gum-ming on the leader (possible evidence of bacterial canker), and those with discolored, rotted or decayed portions. Any broken, damaged or torn roots should be removed with pruners (these could provide easy entry sites for pathogens). If any marginal trees must be planted, they should be grouped together in the same row or edge of the orchard for ready evaluation and possibly greater care or removal if they fail to recover adequately.

Trees should be laid out for planting in moderate numbers to prevent desiccation of roots. Dipping or spraying roots with an inoculant suspen-sion of mycorrhizae before planting can improve the establishment, growth and early yield of trees. Mycorrhizae can be purchased at agricultural sup-ply stores. Consider rebalancing the root-to-shoot ratio at planting, as it should be about 1:1 to enable the trees to grow without putting undue stress on the tree in spring and early summer. Planting holes should be large enough to encompass the roots without having to bend them. When planting in augered holes in heavier soils with clay content, ensure that the sides of the planting holes are free of glazing (as this will restrict new root growth) by roughing up the soil on the walls of the hole, and the hole diameter should not force roots to bend up or down. Spread the roots as evenly as possible, with the tips pointing slightly down. When placing soil around the roots, ensure that there are no air pockets or rocks directly in contact with the roots. Be sure to press the soil around the trees without compressing the soil too much, as this can also restrict root growth. Trees should be planted at about the same depth as in the nursery. Unlike ap-ples, cherries rarely scion root, so it is not critical to keep the bud or graft union above the soil line, though typically the graft union is planted several inches above the soil level to ensure no possibility of scion rooting.

Ensure that the newly planted trees are irrigated immediately after planting to bring soil nutrients and water in contact with the root system, and maintain good soil moisture to hasten budbreak and active root extension and uptake with concomitant canopy photosynthesis and growth. Newly developing fine roots and root hairs should not be placed under any stress associated with insufficient (or excess) water, or weed competition, during the first few months of orchard establishment. After planting, it may be beneficial to sow a cover crop or orchard grass between the tree rows (i.e. in the tractor alley). These can reduce erosion and compaction, aerate the soil and keep the orchard cooler in the summer. Agricultural suppliers often sell seed mixes acclimatized to local conditions. In silt-loam or sandy-loam soils where trees are irrigated by drip emitters, Covar sheeps fescue can serve as a non-irrigated ground cover. Covar is a hardy sheeps fescue selection that will survive without supplemental irrigation in areas that get 250–430 mm (10–17 in) of annual precipitation.

5.7 Use of Plant Growth Regulators in Cherry Production

A number of plant growth regulators that supplement, or affect the synthesis of, natural plant hormones can be important management tools for cherry tree development as well as fruit production and quality. In general, plant growth regulators are not translocated significantly within the tree, so they are usually applied as dilute solutions in enough spray volume to achieve full, uniform coverage of tree canopies for uptake and localized responses. Depending on tree age, size and density, such volumes would range from ~935 l/ha (~100 gal/acre) to ~3750 l/ha (~400 gal/acre). This trait of limited translocation also means that application effects can be localized to certain portions of the canopy by directed sprays if desirable, such as to apply growth inhibitors or promoters to reduce or increase vigor, respectively, in specific sections of the canopy.

5.7.1 Aminovinylglycine

Aminovinylglycine (AVG), marketed in the US as Retain®, is an inhibitor of the naturally occurring plant hormone ethylene. It can be applied at the beginning of flowering to extend ovule viability and increase the EPP, which can be particularly advantageous when temperatures during pollination are higher or with cultivars that tend to have short EPPs, such as Regina. As cherries flower, ethylene is produced naturally, which eventually promotes floral senescence. With most cultivars, the ovule remains viable long enough for pollen to germinate on the flower stigma and pollen tubes to grow down the length of the style, ending with fertilization of the ovules before the flower senesces. AVG application can extend the viability of the ovule, providing a longer EPP and potentially a higher fruit set (Rothwell and Pochubay, 2016). AVG also may be useful to increase fruit set when there is a low population of viable flowers due to winter freeze or spring frost damage (Schwallier, 2012). The currently recommended application rate is a single application at 823 g/ha in 934 l water (11.7 oz/acre in 100 gal). The addition of an organosilicone adjuvant of 370 ml (12.5 fl. oz.) is recommended. Although some studies have shown AVG application timing from 10–80% full bloom can be equally effective (Castagnoli *et al.*, 2016), this may depend on weather, so it may be best to apply at the earlier stages of bloom to assure the prolonging of the EPP for a greater proportion of flowers in anticipation of unknown subsequent weather conditions.

5.7.2 Gibberellic acid

Gibberellic acid (GA) is a naturally occurring plant hormone that is involved in numerous plant processes, including the promotion of cell enlargement, which are very important for flower bud formation, fruit set, fruit development and tree growth. More than 130 different chemical formulations of GA have been identified from plants, fungi and bacteria. Commercial plant growth regulator products utilize several forms of laboratory-derived GA, including GA_3 (e.g. ProGibb®, Falgro®), GA_4 (e.g. Novagib®), and GA_{4+7} (e.g. ProVide®). Commercial production of GAs is accomplished through fermentation, and some formulations of GA are certified for organic production.

5.7.2.1   *Canopy development*

Branching of young sweet cherry trees can be promoted with commercial formulations of GA_{4+7} that include the addition of the cytokinin 6-benzyladenine (6-BA), such as Promalin®. This can be accomplished in early spring at the green tip stage of budswell (see Fig. 4.3) by painting the growth regulator (usually one part Promalin® to four parts water or latex paint) directly on the buds or on an incision made immediately above the bud. Incisions should be just deep enough to reach the cambium layer, below the bark. One-year-old wood responds best, but success has been obtained on branches up to 4 years old. Individual buds can be selected, or if branches are desired along the entire trunk, such as for SSA training (see Chapter 7, 'Sweet Cherry Training Systems'), incisions that girdle the trunk can be made every 30 cm (12 in) (Figs. 5.23 and 5.24). Experimental research also has shown some promise for post-budbreak treatment with spray applications at 600 ppm as leaves are emerging. Generally, greater success in promoting new shoots is achieved on trees with well established

Fig. 5.23. Incisions are made and painted with Promalin® every 30 cm (12 in) to completely girdle the trunk (Oregon) (courtesy of L.E. Long).

root systems (e.g. the year after planting) and when temperatures following application are relatively warm (greater than 15 C [59 F]).

Alternatives to the use of plant growth regulators for inducing outgrowth of vegetative buds to form lateral shoots on young tree leaders, include heading (described in Section 7.2.1: 'General central leader canopy training'), scoring/notching, or bud selection/removal. All of these techniques alter the hormonal status of the buds targeted for growth. The application of GA_{4+7} + 6-BA overcomes the inhibition of growth by natural auxins. Scoring or notching the cambium just above buds targeted for growth, which can be done as buds swell, interrupts the flow of auxin to the inhibited bud that originates from upper growing points (apical dominance). If done too early or if the score is too superficial, the cambium may heal and restore the flow of auxin before growth can occur.

Bud selection/removal is usually accomplished by removing all of the buds 10–15 cm (4–6 inches) below the leader terminal, which are the buds that would naturally initiate growth in spring. Following this, a pattern of

Fig. 5.24. Induced branching as a result of scoring and Promalin® application (Chile) (courtesy of L.E. Long).

selecting a bud where a branch is desired and removing the next three to four buds, selecting another bud in a desirable position and removing the next three to four buds, etc., is imposed to create a continuous whorl of potential branches down the leader (Fig. 5.25). Alternatively, three consecutive buds can be retained to create a tier of potential branches, then all buds can be removed below the tier until the next tier of three branches is desired. In general, either form of bud selection/removal results in the targeting of individual buds for growth in desirable locations on the leader, coupled with the removal of about 65 to 75% of the remaining buds.

5.7.2.2   *Fruit quality*

Several fruit quality parameters can be enhanced significantly with treatments of GA$_3$ (40–120 g active ingredient/ha; 16–48g active ingredient/acre), applied at 'straw color' (as fruit is changing from green to yellow; Stage III of fruit growth, see Chapter 9, 'Fruit Ripening and Harvest'),

Fig. 5.25. Bud selection and removal to promote a whorl of lateral branching on one-year-old shoot growth: blue arrows indicate retention of the terminal bud and one out of every four to five buds selected in a whorled pattern down the leader; red arrows indicate removal of intervening buds (courtesy of G.A. Lang).

including improved firmness and fruit size, increased fruit sugar levels and delayed harvest. In Europe, fruit firmness is often tested in the field and sometimes postharvest in the laboratory using a Durofel® penetrometer. Applications to improve fruit quality are made as early as 2 weeks before the change in fruit color from green to straw, and as late as 10 days before expected harvest, with reduced effects from later timings. The most typical

Fig. 5.26. Gibberellic acid is typically applied at straw color, as cherries change color from green to yellow (courtesy of C. Kaiser).

treatment is 20 ppm applied at straw color (Fig. 5.26). A second application, about 1 week later, may be useful if the trees are carrying a heavy crop, since GA_3 is most effective under moderate to light crop load conditions. Maintaining good tree vigor with proper pruning, nutrition and irrigation, as well as avoiding heavy crop loads, are important for achieving optimal GA_3 responses. Although research is lacking, grower experience suggests that GA_3 can be mixed in a tank with a wide range of pesticides and some nutrients. The mixture pH should be neutral or only slightly acidic. If the natural water pH is outside of this range, it should be modified to near 7.0 before adding the GA_3. Understanding the various fruit responses and the factors that influence the action of GA_3 will help growers achieve more consistent results:

- *Firmness* – Application of GA_3 has become an important and consistent standard practice in many sweet cherry production regions worldwide to optimize fruit firmness and texture, which is both highly desired by consumers as well as important for long distance shipping, storage and export. Fruit firmness levels can increase by 20% or more.
- *Fruit size* – GA_3 application also consistently improves fruit size by 1–2 mm in diameter, increasing the proportion of the fruit in the next higher size class for which buyers will pay a premium. Thus, GA_3 can be

an important tool for growers to improve returns, especially for cultivars with moderate genetic fruit size or for years when cropping is heavy enough to negatively impact fruit size.

- *Soluble solids* – Treatment of fruit with GA_3 can increase fruit sugar levels by as much as 1–2°Brix. However, this effect is not as consistent as increased firmness and fruit size.
- *Harvest timing* – GA_3 treatment will usually delay harvest by 3 to 5 days, which can be useful for extending the harvest window for individual cultivars or managing harvest labor across large blocks of a specific cultivar, by treating adjacent blocks selectively with a full, reduced or no rate. For late ripening cultivars that command higher prices as market supplies dwindle, GA_3 can help achieve this economic advantage. Conversely, a delay in harvest for early ripening cultivars may shift the fruit into a less valuable market window, reducing returns. For this reason, many growers apply a half rate of 10 ppm GA_3 to early ripening cultivars to increase fruit firmness and size without significantly delaying harvest.
- *Fruit color* – Since GA_3 can delay harvest and fruit size, fruit skin coloration also can be delayed or reduced. Generally, this is not noticeable for dark red-fleshed cultivars, but with yellow-fleshed cultivars that develop a red blush on the skin, such as Rainier, GA_3 treatment can prevent the blush color from appearing, considerably reducing fresh market value. To avoid this, growers of blush varieties will reduce GA_3 rates to 10 ppm or avoid use altogether.
- *Fruit pitting* – Fruit treated with GA_3 are less prone to pitting, a mechanical injury that occurs during picking or packing and appears as small depressions in the fruit skin during cold storage and sale.
- *Rain-induced fruit cracking* – Many growers believe that cherries are more susceptible to fruit cracking immediately after GA_3 application. Research suggests that GA_3 application can reduce transpiration for up to 48 h after treatment, so applications should be avoided if the forecast is for extensive rain over the coming couple of days, when the major danger of fruit cracking will be from water taken up through the roots and transported to the fruit internally, causing pressure to build within the fruit flesh until the skin cracks. Research does not simply determine that fruit treated with GA_3 are more susceptible to cracking during light to moderate rains, when the major danger of cracking is from water absorbed through the fruit epidermis. Cherries grown in soils allowed to dry out, in arid regions or under rain exclusion high tunnels, and then exposed to significant soil water due to rainfall or irrigation, may exhibit greater fruit cracking following GA_3 treatment. GA_3 has not led to higher cracking when the soil moisture profile has been maintained close to field capacity during fruit development. GA_3 treated fruit may exhibit larger cracks (perhaps due to the larger fruit size and firmer flesh texture) than untreated fruit, but research has not always shown that GA_3 treatment leads to a higher incidence of cracking.

5.7.2.3 Flower spur bud formation

Low rates of GA_3 (10–45 g active ingredient/ha; 4–18 g active ingredient/ acre), usually applied about 3–4 weeks after budbreak when three to five terminal leaves have expanded or 2.5–7.5 cm (1–3 in) of shoot extension has occurred, can inhibit the formation of non-spur basal flowers on new shoot growth. This maintains the meristem in the axil of the basal leaves as vegetative, rather than converting to reproductive development, which will allow the node to develop into a vegetative spur the subsequent year, with the potential to become a flowering spur the following season (2 years after application). This is a common practice in sour cherry production to prevent young trees from fruiting until the trees are large enough to be harvested mechanically with trunk shakers. There has been some research in this area with sweet cherries to balance crop loads in highly productive cultivars, but since excessive crop loads usually form primarily on spurs, and this treatment timing mainly inhibits the flower buds that form at the previous season shoots, it generally has only a minor effect on crop load. However, for naturally low yielding cultivars that form a higher proportion of solitary basal buds compared to spurs, this GA_3 treatment may have some promise for increasing yields and reducing the formation of blind wood.

5.7.3 Prohexadione-calcium

Prohexadione-calcium (P-Ca), marketed as Apogee®, Regalis®, or Kudos®, interferes with gibberellin synthesis and reduces vegetative growth in cherry trees. The main purpose for using P-Ca is to improve light penetration and distribution throughout the tree canopy. This can have the benefit of developing more uniform fruit color and sugar levels in cherries located throughout the canopy, as well as improve flower bud formation for the subsequent year. Improved fruit coloration is particularly important when extended bloom has caused fruit maturity to vary within the tree canopy. It is also important for yellow-fleshed blush cultivars since a partial red blush on the skin is needed to attract premium fresh market returns. With Lapins and Sweetheart trees grafted to CAB-6P rootstocks in Chile, P-Ca reduced vegetative growth and increased the number of flower buds and fruit firmness, without affecting other fruit quality factors such as fruit size and soluble solids (Cares *et al.*, 2014).

The first P-Ca application should be made when initial growth has not exceeded 5 cm (2 in). Later applications are less effective, as endogenous GA levels become too high to control. The most effective rates are between 125 and 250 ppm. Tree growth should be evaluated after 2 or 3 weeks to determine if a second application is necessary. Trees on vigorous rootstocks or with a history of high vigor may need a second application at half rate to prevent excessive compensatory late season growth. Some research has shown that P-Ca applications during flower bud initiation (early to mid-May in the PNW) can result in greater flower bud density and greater yields the subsequent year (Castagnoli *et al.*, 2016). Depending on the cultivar, this may or may not be advantageous.

5.7.4   Hydrogen cyanamide

Hydrogen cyanamide (Dormex®) is used to break bud endodormancy in cherries grown in areas with mild winter climates, by inhibiting the enzyme catalase, which plays an important role in plant metabolism. Dormex® tends to advance and increase both flower and foliar budbreak percentage, improves bloom synchronization and concentration, and can advance harvest by up to 2 to 3 weeks. This can increase yield, especially in years when chilling is insufficient for fully breaking endodormancy, and will shift yield to earlier, more valuable market windows in warm climates where the risk of spring frost is low. The budbreaking effect of Dormex® is best achieved in combination with 2–4% winter oil, applied once the cultivar has received at least 70% of its chilling requirement. Therefore, it is important to monitor chilling models for the local cherry producing area, especially with the changing climate of warmer winters being experienced in many northern latitudes.

Typically, Dormex® concentrations are based on each individual cultivar (some cultivars may be sensitive to Dormex® and exhibit some phytotoxicity) and orchard situation. The orchard needs to be evaluated in terms of cultivar chilling requirement, tree vigor (vigorous trees may have a higher chilling requirement to break endodormancy), history of delayed foliation and the amount of chilling experienced as the winter season progresses. The best results occur from full tree coverage, the inclusion of non-ionic wetting agents, and spray droplet sizes not smaller than 150 µm. Dormancy-breaking effects or phytotoxicity may be attenuated by stress conditions like 'wet feet', nematode injury and stem cankers. Treatment also can negatively affect return bloom the following year.

5.7.5   Nutrient-based compounds for breaking dormancy

Similar to the use of hydrogen cyanamide, several N-based fertilizers have shown plant growth regulator-like activity for breaking dormancy in mild climates. These include Erger® (a solution of calcium, ammonium, nitrate, urea and polysaccharides), CAN-17 (a calcium–ammonium nitrate formulation), and potassium nitrate (KNO_3). Generally, these are considered to be somewhat less potent or consistent than Dormex®, but they have been shown to have similar effects on flower and foliar budbreak, improved bloom, advancement of harvest and increased yields, as well as being safer for applicators. As with Dormex®, these should be applied once the cultivar has received at least 70% of its chilling requirement.

References

Cares, J.K.X., Sagredo, T. and Retamales, J. (2014) Effect of prohexadione calcium on vegetative and reproductive development in sweet cherry trees. *Proceedings International Symposium Integrating Canopy, Rootstock and Environmental Physiology in Orchard Systems. Acta Horticulturae 1058*. DOI: DOI:%2010.17660/ActaHortic.2014.1058.9.

Castagnoli, S., Long, L.E., Shearer, P., Einhorn, T., Pscheidt, J.W. *et al.* (2016) 2016 Pest management guide for tree fruits in the mid-Columbia area, EM 8203-E. Extension and Experiment Station Communications, Oregon State University.

Edstrom, J.P. and Cutter, S. (2008) Slip plow tillage effect in almonds. University of California at Davis. Available at: www.cecolusa.ucdavis.edu/files/65221.pdf (accessed 19 June 2020).

Long, L., Lang, G., Musacchi, S. and Whiting, M. (2015) Cherry training systems. Pacific Northwest extension (PNW) publication 667. Available at: www.catalog.extension.oregonstate.edu/pnw667 (accessed 19 June 2020).

Mai, W.F. and Abawi, G.S. (1978) Determining the cause and extent of apple, cherry, and pear replant diseases under controlled conditions. *Phytopathology* 68(11), 1540–1544. DOI: 10.1094/Phyto-68-1540.

Mazzola, M., Granatstein, D. and Mullinix, K. (2004) Employing biological elements of orchard ecosystems for enhance tree health. final report AH-01-71, wash. Tree Fruit Research Commission, Wenatchee, WA. Available at: www.treefruitresearch.org/report/employing-biological-elements-of-orchard-eco-systems-for-enhanced-tree-health/ (accessed 19 June 2020).

Mazzola, M., Granatstein, D. and Mullinix, K.M. (2005) Employing biological elements of orchard ecosystems for enhanced tree health. Final report AH-01-71, wash. *Tree Fruit Research Comm.*

Mazzola, M., Hewavitharana, S.S. and Strauss, S.L. (2015) Brassica seed meal soil amendments transform the rhizosphere microbiome and improve apple production through resistance to pathogen reinfestation. *Phytopathology®* 105(4), 460–469.

McKenzie, D. (2003) Salinity and sodicity – what's the difference? *The Australian Cotton Grower* 24(1 February to March), 28.

Reich, D., Godin, R. and Broner, I. (2009) Micro-sprinkler irrigation for orchards. 4.703. Colorado State University Extension.

Rothwell, N. and Pochubay, E. (2016) Retain use to increase sweet cherry yields. Michigan State University Extension. Available at: www.canr.msu.edu/news/retain_use_to_increase_sweet_cherry_yields (accessed 19 June 2020).

Schmidt, R.H. (1997) Basic elements of equipment cleaning and sanitizing in food processing and handling operations. University of Florida IFAS Extension Publication FS14.

Schwallier, P. (2012) Using retain to set shy bearing or frosted sweet cherries. Michigan State University Extension. Available at: www.canr.msu.edu/up-loads/files/2014_NW_orchard_show/Using_ReTain_to_Set_Shy_Bearing_or_Frosted_Sweet_Cherries_Schwallier.pdf (accessed 19 June 2020).

Shoemaker, H.E., McLean, E.O. and Pratt, P.F. (1961) Buffer methods for determining lime requirement of soils with appreciable amounts of extractable aluminum. *Soil Science Society of America Journal* 25(4), 274–277.

USDA NRCS (2011) Carbon to nitrogen ratios in cropping systems. Available at: www.nrcs.usda.gov/Internet/FSE_DOCUMENTS/nrcseprd331820.pdf (accessed 19 June 2020).

Whiting, D., Card, A., Wilson, C. and Reeder, J. (2015) Saline soils. Revised by Carter, S. CMG GardenNotes #224. Colorado State University Extension Service.

Yin, X., Huang, X.-L., Huang, X.-L., Jaja, N., Bai, J. *et al.* (2012) Transitional effects of double-lateral drip irrigation and straw mulch on irrigation water consumption, mineral nutrition, yield, and storability of sweet cherry. *HortTechnology* 22(4), 484–492.

Sweet Cherry Pruning Fundamentals

6

6.1 Fundamentals of Pruning

Until the commercial advent of dwarfing, precocious rootstocks in the late 1990s and early 2000s, traditional cherry orchards planted on primarily vigorous seedling rootstocks were comprised of large, complex canopies that took years to fully develop. This made it difficult to clearly and systematically explain pruning strategies to field labor, and made canopy development and management relatively imprecise. The fundamentals of training and pruning contemporary sweet cherry orchards usually incorporate plant materials or techniques that promote: early fruiting for a more rapid return on investment; optimization of light interception and distribution, with minimal intra-canopy shading; greater precision in, and/or simplification of, canopy development for balancing yields with fruit size and quality, as well as ease of teaching to less experienced labor forces; and systematic processes for annual renewal of fruiting wood to maintain consistent yields of high fruit quality as trees age.

Cherry training systems around the world traditionally have been free-standing, but recent years have seen a significant increase in trellised systems as tree spacing has become more dense and tree canopies have become more two-dimensional in the form of narrow fruiting walls. Compact trees are easier to manage since they are easier to prune, spray and pick. First and foremost, pruning allows one to shape the tree and train the whole tree to a preferred canopy structural model that enables certain levels of simplification and efficiency in managing a crop load and leaf area balance and/or ease of picking. Pruning also reduces the risk of disease infection by removing dead and infected materials (pruning tools must be sterilized between trees where infections are known to occur), and improving the uniformity of protective spray penetration into all parts of the canopy. Pruning also ensures the plants stay healthy and vigorous. In the case of really old trees, pruning provides an opportunity to rejuvenate senescing wood. If bark inclusion is an issue in older trees, pruning of one of the limbs will be necessary to prevent splitting of the trees under heavy crop loads. Timely pruning in young trees allows one to select branches with wider crotch angles. Pruning can ensure adequate sunlight interception and distribution throughout the tree canopy for both photosynthesis and induction of the flower buds. This also increases airflow in the canopy which helps to reduce disease incidence (e.g. brown rot and powdery mildew). Additionally, pruning increases secondary thickening of shoots, which results in stronger, more compact trees.

6.2   Managing the Natural Growth Habit of Sweet Cherry

To most effectively manage sweet cherries in an orchard environment with the goal of efficient fruit production in mind, one should first consider the natural growth and fruiting tendencies that are to be manipulated horticulturally. Sweet cherry trees evolved over the centuries to compete and thrive in forest ecosystems, ultimately attaining the greatest height of any temperate zone fruit tree. The tallest documented sweet cherry trees have reached heights of 38.5–41.5 m (125–135 ft), clearly a disadvantage for managing efficient (and safe) production and harvesting of fruit! This is nearly twice the natural maximum height of the next tallest documented tree fruit species, pears (20 m or 65 ft), nearly four times the maximum height of the tallest apple trees (10–12 m; 32–39 ft), and five times the maximum height of the tallest sour cherry, plum, apricot, or peach trees (6–8 m, 20–26 ft). This forest survival growth habit has significant implications for sweet cherry orchard management, even with the advent of dwarfing rootstocks. Some of the components of this growth habit, and their implications, include:

1. Apical dominance and vigorous acrotonic growth, which means that the uppermost growing point (the apex) of any shoot (especially vertically-oriented shoots or leaders) strongly suppresses the growth of lower growing points (buds, shoots, and branches) and thereby promotes the greatest vigor in the uppermost portions of the tree canopy (Fig. 6.1). This also means there are fewer potential shoots for developing fruiting zones lower in the canopy. Obviously, this is exactly the opposite of desirable orchard development, for which the majority of the fruiting canopy should be easily accessible from the ground, or from small orchard ladders or motorized platforms.

   This apical dominance can be temporarily alleviated by removing (pruning) the uppermost growing point(s), which interrupts the flow of inhibitory plant hormones being produced in the terminal apices and allows lower buds to break free of their inhibition and begin elongation into new shoots. Even then, however, the buds that are primarily activated to grow are those very near the point of the pruning cut (due to the acrotonic growth habit), rather than further away, so the vigorous new growth is essentially regrowth that is still located primarily in the uppermost areas of the canopy. More apical growing points generally mean a greater inhibition of lower growing points, although more growing points also tend to diffuse the carbon (C) and nitrogen (N) resources available for growth, resulting in proportionally fewer individual lengths of new extension shoots compared to that resulting from a fewer number of growing points that would each receive proportionally more resources.

2. Delayed reproductive maturity, so that growth resources are directed to vegetative growth for many years, promoting successful tree establishment by maximum canopy extension growth (primarily vertical, but also laterally from branch growth) to improve the potential for competitive light interception in the forest. Not only is reproduction delayed, but since the majority of light intercepted is in the upper canopy, the lower portions of the canopy can become shaded to the point that flower buds fail to develop or abort after initiation, and shoots die for lack of light. Once again, this is counter to desirable fruitwood characteristics for efficient orchards.

3. Large shoot leaves (that is, when N for growth is sufficient and not limited by uptake from the soil or translocation competition with heavy crop loads), which improve light interception by vigorously-growing new shoots, but conversely decrease light distribution to the interior and lower portions of the canopy. Photosynthesis by these large shoot leaves primarily provides carbohydrates that are used in the growth of the new shoot, with a minor allocation to any fruit that are present in reproductively mature sections of the canopy. This contrasts with the usually smaller spur leaves that develop on all portions of the canopy, that are older than current season shoots; photosynthesis by the spur leaves provides the majority of the carbohydrates used for development and ripening of the crop load. Consequently, excessive shoot numbers with large shoot leaves can create a significant amount of shading of interior, spur-fruiting sites by midsummer when the flower buds are beginning to differentiate.

The advent of vigor-limiting rootstocks (like the Gisela series and others) have improved the situation for maintaining smaller trees. Closer planting may also increase the potential for root competition, which can reduce tree vigor. However, some annual shoot growth is critical for keeping the tree in a good balance between fruiting and vegetative growth, as will be discussed in the next section. Young trees in a forest situation tend towards excessive vegetative extension growth and minimal branching, while old trees in the forest tend towards minimal extension shoot growth as well as minimal new vegetative growth within the canopy. Therefore, one challenge for sweet cherry growers is to promote the annual development of adequate new growth to continually create young fruiting wood, even as the primary structure of the tree ages, to maintain a long (25–30 years or more) productive lifetime.

6.3   Managing the Natural Fruiting Habit of Sweet Cherry

There are two fruit population types that develop in sweet cherry canopies: spur fruit and non-spur fruit. Different training systems, and even different pruning methods, can focus on or impact one or the other type. There are also three leaf population types: shoot leaves (always on new shoot growth), non-fruiting spur leaves (usually on the section of shoot that grew the previous season), and fruiting spur leaves (usually on two-year and older shoot/branch sections).

Spur fruit generally constitute the majority of the yield potential of a sweet cherry tree. Fruiting spurs require two growing seasons to develop, yielding flowers and fruit in the third year (and beyond) after that portion of the shoot has formed. That is, a new shoot forms a single (usually large) 'shoot leaf' at every node during its initial season of growth. By the fall of that season, a single vegetative bud has formed in the axil of each leaf. The exception to this rule is when the shoot is growing rapidly, the bud may elongate into a new shoot in the same season, forming a 'sylleptic' shoot. The following (second) growing season, if that axillary vegetative bud did not form a sylleptic shoot the previous season or does not elongate in spring into a new lateral shoot, it will generate a whorl of usually five to eight new leaves (non-fruiting spur leaves) and terminate its growth again as a vegetative bud. These spur leaves are usually smaller than the individual shoot leaf that was present the previous season. During late spring and early summer, hormonal signals within the plant act upon the new buds in the axil of each of these spur leaves, inducing them to either remain dormant or to begin developing as a flower bud. By the fall of the second season, the spur may be comprised of none to as many as eight flower buds, plus the terminal vegetative bud. In spring of the third growing season, these buds open with generally one to four flowers per bud, and consequently a single spur may have from a few to as many as 24 flowers or more, plus another whorl of five to eight fruiting spur leaves. This spur fruit and leaf pattern may repeat for many years, or the spur may eventually elongate into a shoot or it may die.

Non-spur fruit only require one growing season to develop, yielding flowers and fruit in the second year (and only then) after that portion of the shoot is formed. That is, non-spur fruit buds may form in the axils of the single leaves at the most basal nodes of the new shoot during its initial season of growth. By the fall of that season, the single bud that has formed in the axil of each basal leaf is entirely reproductive, with no vegetative meristem (future growing point). Consequently, in the following (second) growing season, these buds open with generally one to three flowers per bud, with no spur leaves, and after flowering (and fruiting if pollinated successfully), that node will become 'blind' wood. The proportion of non-spur and spur flowers, then, varies based on the proportion of the canopy that is comprised of one-year-old shoots from the previous season and older sections of growth that have retained spurs.

Due to these locational and age differences, non-spur flowers/fruit are close to both the non-fruiting spur leaves and the shoot leaves, which means there is a lot of leaf area for a small population of fruit (a high leaf area-to-fruit ratio, LA:F). Accordingly, non-spur fruit tend to be the largest, highest quality fruit in the canopy. The generally more dense spur fruit clusters are associated with the more limited leaf area of that same spur, in competition with other fruiting spurs for any extra carbohydrates available from the more distant non-fruiting spur and shoot leaves. Consequently, spur fruit tend to contribute to higher yields, but are more limited in fruit size due to having lower LA:F ratios. Besides training and pruning, cultivar and rootstock can also influence the proportions of spur and non-spur fruit. Precocious rootstocks tend to increase the number of basal nodes that may form non-spur flower buds, and some cultivars are more prone to flower on spurs (e.g. Lapins) or on basal non-spur nodes (e.g. Regina).

The advent of precocity-inducing rootstocks (like the Gisela series and others) and some precocious new varieties (like Sweetheart) have improved the situation for developing orchards that reach fruitfulness at an earlier stage. Earlier, precocious reproductive maturity not only has the advantage of earlier cropping in the life of the orchard, but it also helps divert growth resources to fruit while the tree structure is still developing, thereby reducing their availability for vegetative growth and moderating the vigor of young trees. When annual average shoot growth falls below about 50–60 cm (20–24 in), flower bud formation may become excessive, especially towards the terminal portions of the shoot, creating heavy crops of small fruit. Conversely, annual average shoot growth that exceeds about 125–150 cm (4–5 ft) may produce relatively few flower buds, especially on the lower portions of the shoot. Young trees in a forest situation tend towards minimal fruit, while old trees in the forest tend towards excessive quantities of small fruit. The challenges and goals of the sweet cherry grower are to horticulturally manage the genetics of the forest tree in the confined space of an orchard to achieve an ideal balance of vegetative growth and fruiting, with a tree structure that has a major proportion of fruit in the lower and middle part of the canopy.

So, these natural leaf and fruit population relationships can affect sweet cherry yield and fruit quality (size, sugar and firmness) potential. In general, higher yields may result in moderate to lower fruit quality, and moderate to lower yields tend to result in higher fruit quality. Certainly, there are other natural factors involved as well, such as the amount of daily and seasonal sunlight versus cloudiness (higher sunlight levels generally result in higher yields and/or higher fruit quality), and daily temperatures and diurnal variation between daytime highs and night-time lows (cooler nights mean trees and fruit respire less, preserving more of the previous day's photosynthetically acquired carbohydrates for growth and sugar accumulation). However, growers have many orchard tools with which to influence these LA:F and yield-to-quality ratios, including pruning, training, irrigation and fertilization.

6.4 Canopy Pruning Decisions and Tools

Tree fruit training systems provide a structured canopy framework for production of fruit for specific target market criteria (and hence, fruit unit value per kilogram or pound) and specific target crop yields (and hence, crop value per hectare or acre). Profitability is a function of achieving these target values above the cost of the various inputs (labor, machinery, chemicals, capitalization, etc.). Specific business goals can vary from grower to grower, depending on climatic limitations, the cost and availability of essentials like land and appropriately skilled labor, or their proximity to population centers or transportation options to target markets. The foundation of developing and maintaining canopy training systems is the horticultural tool of pruning, although there are potential supplemental training tools to modulate growth as well, such as trellises, limb spreaders and tiedown strategies for re-orienting tree growth and canopy architectures, and plant growth regulators (PGRs) for stimulating or inhibiting growth.

Pruning is both an art and a science, best implemented by those who can both envision the fundamental structure of the canopy, and anticipate the natural growth and fruiting responses to such interventions. Before pruning, the pruner should assess the whole tree and think about what goals are to be achieved. Remember that every pruning cut removes some leaf area, some potential fruiting wood, and some growth resources, and is possibly stimulating new growing points and altering the responses of nearby growing points to the localized change in the environment. There are different pruning techniques for removal of the targeted plant parts, which are mainly determined by where in the plant canopy the pruning cut is made, i.e. tissue age, structural location relative to growing points or points of tissue age divergence, and ultimate orchard management purpose. The timing of pruning also is critical for achieving specific goals.

6.4.1 Timing of pruning

Dormant pruning in the winter is typically used for significant modification of the tree structure as well as for detailed manipulation of cropping potential, since the leafless canopy provides the clearest evaluation of canopy structure and flower and leaf bud balance. Dormant pruning removes both growing points and some storage reserves, but since the majority of the storage reserves are in the trunk and roots, winter pruning can invigorate the growth response of the remaining growing points at budbreak. Hence, dormant pruning can promote more rapid establishment of young orchards. However, in climates with extreme cold, it can be best to delay pruning until after the risk of extreme temperatures has passed. Typically, in the cherry production areas of the Pacific Northwest (PNW) region of the USA, this is after 15 February, unless the operation is constrained by size of the orchard and labor available to service it. The goals of dormant pruning (structural modification and crop load manipulation) can be achieved until budbreak occurs.

Spring pruning, generally 3–5 weeks after budbreak, is usually conducted as a selective follow-up to dormant pruning, to modify canopy structure with less of a strong regrowth response. At this timing, the majority of the stored growth resources have been remobilized to active growth and the tree has largely transitioned to current season sources of C (through photosynthesis by new spur and extension shoot leaves) and N (through uptake by new root growth in the soil). The vascular pipelines supplying this C and N to the active growing points have been established. Consequently, pruning of larger diameter branches, especially in the tops of trees where vigorous regrowth is not wanted, can be delayed until spring rather than imposed during dormancy. This has the added advantage of improving light penetration into the canopy where the pruning cut was made.

Summer pruning can be done prior to harvest or, more commonly, soon after harvest. Summer pruning before harvest is usually to remove watersprouts and/or excessive new extension shoot growth to improve light penetration and distribution into the canopy, improving fruit ripening and flower bud formation. Heading cuts (such as by hedging or topping) at this time will usually result in some regrowth. When summer pruning to reduce extension shoot growth is imposed in late spring or early summer (during pit hardening through early resumption of fruit expansion, Stage II to early Stage III of fruit growth; see Chapter 9, 'Fruit Ripening and Harvest'), some detrimental effects on fruit quality have been reported, such as reduced soluble solids and increased fruit cracking following rain events. This may be due to stimulation of extension shoot regrowth in competition with the developing fruit for allocation of new photosynthates from the spur leaves.

However, when summer pruning to reduce extension shoot growth is imposed 1–2 weeks before harvest, as rapid fruit expansion is slowing down and sugars are accumulating rapidly (late Stage III of fruit growth, see Chapter 9), beneficial effects on fruit quality have been reported, such as increased soluble solids, better coloring of blush cultivars, and reduced rain-induced cracking. These effects may be due to the improved light environment of the fruiting spur leaves, increasing their photosynthetic efficiency and ability to compensate for the loss of the extension shoot leaves that are primarily providing carbohydrates to extension shoot growth (and eliminating the extending shoot terminals during the final stage of fruit ripening).

Summer pruning heading cuts or hedging/topping soon after harvest improve canopy light distribution at the earliest stages of flower bud differentiation and may result in less regrowth than earlier hedging. (When regrowth occurs after harvest, it may be somewhat susceptible to superficial winter damage.) Heading cuts or hedging/topping later in the summer (after harvest) may not induce regrowth, but may allow detrimental levels of shade to persist long enough to be less favorable for flower bud development. Summer pruning also removes leaf area that is actively contributing to the building of storage reserves for the next spring, so pruning a bit later after harvest can not only minimize the regrowth response, but may also reduce the growth response during budbreak the next spring. Hence, increasing the level of summer pruning can be used to promote the least vigorous growth response of mature trees.

6.4.2 Types of pruning

Beyond timing of pruning, the types of pruning cuts can be used to elicit different responses in developing or maintaining the tree canopy structure and fruiting wood. Types of pruning cuts include:

- *tipping*, which removes only a small percentage of apical buds;
- *heading*, which removes up to 70% of a shoot or branch;
- *stubbing*, which removes 80% or 90% of the shoot or branch;
- *thinning*, which removes the entire shoot or limb;
- *topping*, which removes all of the canopy beyond the point of its desired maximum height; and
- *staghorning/stumping/rejuvenating*, which removes most of the canopy back to the main framework branches or even a single stump.

6.4.2.1   Tipping

Tipping is the removal of the very terminal point of the shoot, which thereby removes the inhibitory effect of apical dominance (associated with the shoot apical meristem's production of the plant hormone, auxin). This will promote the elongation of usually one to three lateral buds near the tipping cut (Fig. 6.1), depending on time of the growing season. Activation of the buds is greater following dormant pruning or pruning close to spring budbreak, when the maximum level of storage reserves is being remobilized; later in the season, the level of activation declines as growth resources (C and N) divert away from growth to build storage reserves for the next year.

Fig. 6.1. The natural growth habit of sweet cherry on a vigorous rootstock: arrows demonstrate strong vigor (acrotonic growth of new shoots) and apical dominance even after 35 years and diffusion of vigor into multiple leaders (courtesy G.A. Lang).

6.4.2.2   Heading

Heading is the removal of up to 70% of a leader, shoot or branch. Heading the leader is the most common and effective technique used for releasing the lateral buds just below the pruning cut from apical dominance, to form multiple leaders or lateral branches during tree structural development (Fig. 6.2). Consequently, heading of the leader just above the level where a tier of branches or scaffolds is desired is used in the development of many training systems.

When heading removes 15–30% of the length of a shoot that grew the previous season, it can reduce the future cropping potential by a higher percentage because the most terminal section of shoot growth has the highest density of spurs, and those tend to have the highest number of flower buds per spur. Therefore, this type of heading can remove a greater amount of future crop compared to the amount of future leaves removed (in fact, such heading usually stimulates the formation of two to three new shoots, replacing the leaf area that was removed). The highest quality cherries are usually produced at the base of previous season shoots, so

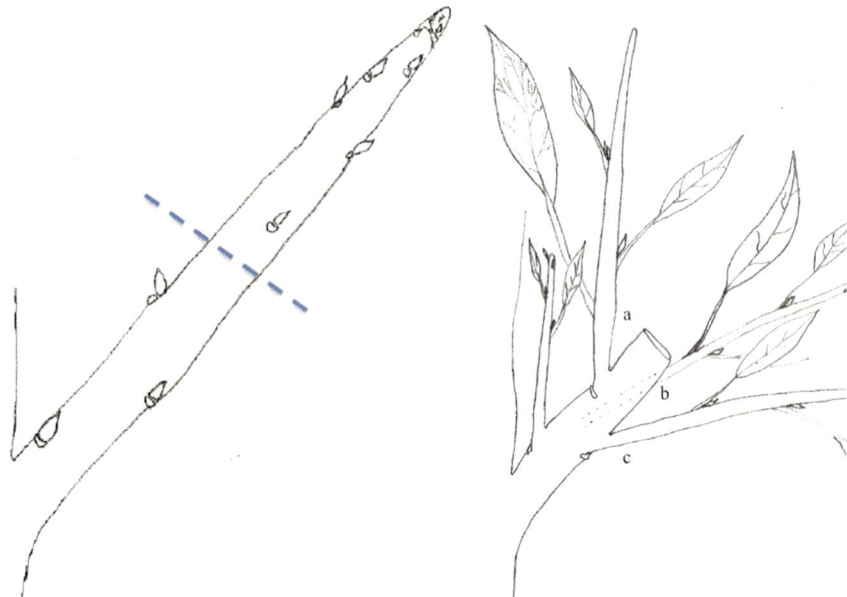

Fig. 6.2. The effect of heading a sweet cherry shoot or branch (courtesy of C. Kaiser).

the formation of new shoots from heading also increases the proportion of these high quality basal fruit relative to spur-borne fruit. Consequently, heading is an important and valuable pruning tool for managing crop load and fruit quality, especially on highly productive varieties and rootstocks.

When a significant, but incomplete, portion of the shoot or branch is removed by heading, the remaining buds have strong vascular connections and, given the removal of the inhibitory effect of apical dominance, therefore may result in more (three to five) lateral buds that elongate into new shoots (Fig. 6.3). The most terminal (distal) new shoot is the most vigorous and tends to exert apical dominance over the other new, subtending lateral branches. This trait can be used to orient and direct growth through a type of heading cut strategy known as 'sectorial double pruning'. Where shoot orientation is tending towards upright vertical growth, but a more horizontal lateral orientation is desired, three potential horticultural tools can be considered for achieving this outcome. One is to use a limb spreader to change the angle of the branch or shoot (Fig. 6.4), and another is to bend the branch or shoot to the desired angle and tie it to a trellis, post, another part of the tree, or to a stake or clip anchored in the soil (Fig. 6.5). The third method is to head prune, during dormancy or in early spring, the branch or shoot immediately in front of a vegetative bud on the upper side of the shoot, whereas the next two buds behind the upper bud are on the sides or bottom of the shoot (Fig. 6.2). This stimulates the upper bud to

Fig. 6.3. A heading cut made to a dormant one-year-old shoot stimulated the subsequent elongation of several lateral shoots. The leaf area of the new shoots will be greater than the leaf area removed, providing more resources for fruit that develops from the flower buds at the base of the new shoots (courtesy of L.E. Long).

grow vigorously upward (a), while exerting a modulated apical dominance over the next two buds which grow less vigorously and more horizontally. Sectorial double pruning means that a second pruning is then imposed later in summer or during the next dormant season to remove the terminal vertical shoot (a), leaving the horizontal shoots that are growing at the desired angles (b and c).

6.4.2.3   Stubbing and thinning

Stubbing, which removes 80–90% of a shoot or branch is used for training and rejuvenating structural fruiting wood (Fig. 6.6). Stubbing also may be performed for bacterial canker control to promote formation of a natural zone of suberization. Thinning differs from stubbing by removing the entire shoot (Fig. 6.7a) or branch/watersprout (Fig. 6.7b) where it intersects the larger, older branch or leader from which it arose, with the removal of all visible lateral buds and little or no stub remaining. The goal of this type of pruning is to open up a channel for light to penetrate into the canopy without the occurrence of regrowth. This is also the primary strategy for

Fig. 6.4. Limb spreaders were used to increase the branch angle, which opens up the tree and helps to control branch vigor (courtesy L.E. Long).

removal of vertical shoots (e.g. watersprouts) and weak or pendent lateral shoots (which typically may have smaller fruit). Removing these branches prior to bloom can eliminate a significant amount of small cherries before they develop.

In some cases, making thinning cuts back to significantly older wood may promote the eruption of dormant, epicormic buds from the collar of the stub or lower on the older branch or leader that can develop into a new shoot (Fig. 6.3). However, this process of epicormic shoot formation currently is poorly understood, generally does not occur if the area is shaded, and can be highly variable.

6.4.2.4   Topping

Topping refers to the removal of the upper part of the canopy beyond its desired maximum height (Fig. 6.1). This can be done mechanically and

Fig. 6.5. Hop clips are used to tie down branches to help form these steep leader (SL) trees (courtesy of G.A. Lang).

indiscriminately by hedging with a sickle bar or using circular saws in series, or selectively by a manual thinning cut to remove the main leader back to a weak side branch or shoot in the hope of minimizing regrowth by channeling growth resources into the intact side shoot, rather than initiating many new growing points with a heading cut. Mechanical topping is usually done in late summer after harvest when regrowth is less likely. Many traditional growers simply top their standard trees at a set height, typically

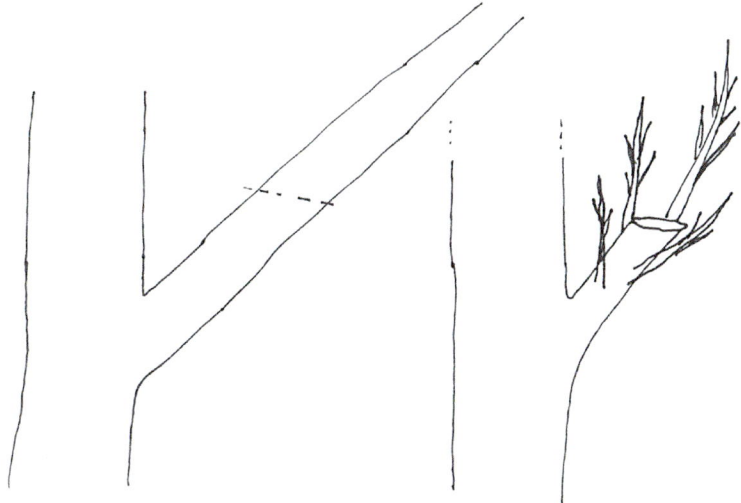

Fig. 6.6. The effect of stubbing a sweet cherry branch (courtesy of C. Kaiser).

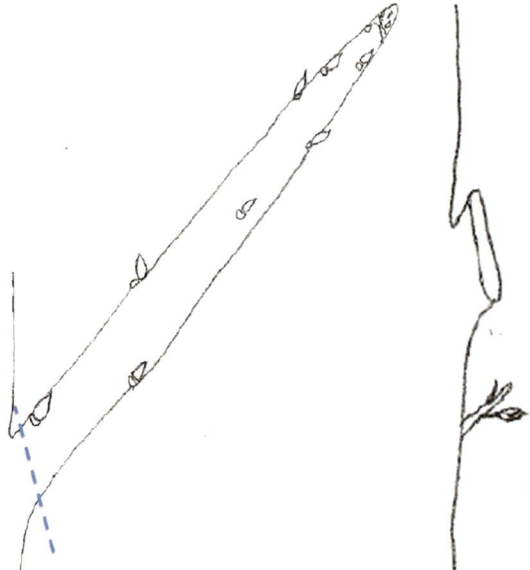

Fig. 6.7a. The effect of thinning a sweet cherry shoot (courtesy of C. Kaiser).

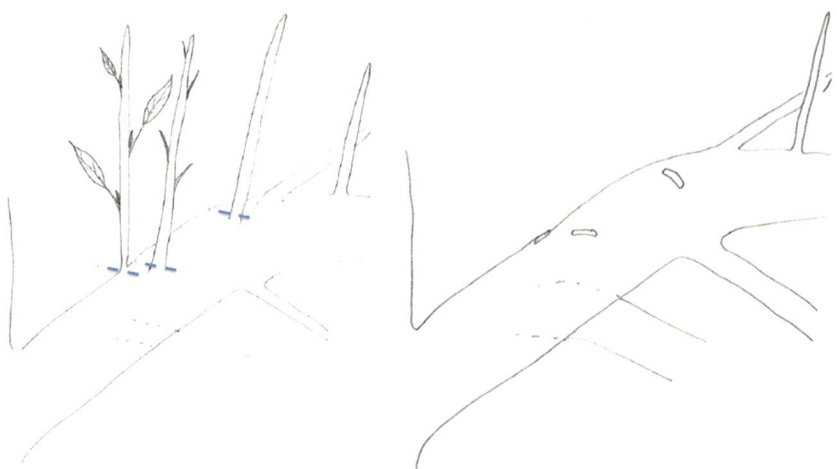

Fig. 6.7b. The effect of thinning watersprouts on a branch of a sweet cherry tree (courtesy of C. Kaiser).

4.3 m (14 ft) in the PNW, based on the traditional use of 3.7 m (12 ft) ladders. If done during dormant pruning, this unfortunately stimulates a massive vegetative flush at each pruning cut and creates an inverted pyramid in terms of canopy density. This creates numerous potential problems, including increased shading, difficulty with spray coverage, and invigoration of upper canopy fruiting while reducing lower canopy fruiting, with its associated harvest inefficiencies. Manual topping is usually done in late summer as well, but alternatively may be done in spring after the initial flush of new growth has occurred.

6.4.2.5   Staghorning/stumping/rejuvenating old trees (and whole tree renewal)

A critical factor for maintaining high quality fruit production over many years is the renewal of young fruiting wood on the primary structure of the tree. When trees have been allowed to become too large or the majority of the fruiting sites too old due to insufficient annual renewal, it may be worthwhile to rejuvenate the entire tree rather than remove the orchard and start over. One advantage of this strategy is that a rejuvenated orchard will usually come back into full production more quickly than a newly planted orchard. A second advantage is that by rejuvenating an existing orchard (presuming the trees are still healthy and the existing cultivar is still marketable), the grower can avoid potential replant problems that may occur with replanting the same site.

In recent years, a corollary to this entire orchard rejuvenation has been called 'whole tree renewal' (Figs 6.8, 7.13 and 7.27), in which whole trees are rejuvenated annually for a set percentage of the orchard. For example, the plan may be for 12.5% of the tree rows to be rejuvenated annually so that no fruit-bearing wood is older than 8 years. Two key factors in considering whole tree renewal are: (i) the target age for renewal should be when productivity or quality is expected to begin to decline (e.g. fruit quality may decline on spurs as they reach 5–8 years or older); and (ii) once the fundamental tree structure has been established, the target annual percentage for whole tree renewal should commence even though the first few years will be wood that is renewing and has not yet reached its full productivity. This latter point instills discipline and commitment to the annual renewal process and assures that, at the end of the first full renewal cycle (e.g. 8 years later), no proportion of the orchard has exceeded the target age for optimum yield and quality. In theory, then, an annual schedule of whole tree renewal should keep an orchard producing consistent yields of consistent quality fruit through many cycles over several decades. One challenge of imposing whole tree renewal is that the vigorous rate of regrowth that often occurs can lead to sylleptic (secondary) branches even as the renewal (primary) shoot is growing. This can be advantageous for central leader trees in the development of branched new scaffolds, but is detrimental to the development of spur-bearing vertical fruiting wood as comprise the canopies of Kym Green bush (KGB) and upright fruiting offshoot (UFO) trees. In these situations, the most vigorous regrowth may need to be headed again at an early stage, to diffuse its inherent vigor into two to three less vigorously growing replacement shoots.

Fig. 6.8. This central leader orchard has been renewed by cutting all lateral branches back to the trunk (Washington) (courtesy of G.A. Lang).

There are a number of different pruning techniques for rejuvenation or whole tree renewal of an orchard. The cutting of all major fruiting wood during dormancy back to the primary tree structure is known as staghorning (Fig. 6.9). Such major cuts tend to stimulate a lot of epicormic buds to erupt into new shoots from seemingly random places on the remaining stubs and trunk; when cuts can be made back to where visible buds exist, the resulting regrowth may be more predictable. The advantage of staghorning is that the multiple stubs increase the potential for more regrowth sites and more rapid complete canopy replacement.

A related rejuvenation technique, often called 'window pruning', is to rotate, perhaps by east–west–north–south canopy section, the annual removal of the largest main limb in the upper canopy. As a result, light

Fig. 6.9. The effect of staghorning a sweet cherry tree to rejuvenate an old orchard or for selective whole tree renewal on an annual schedule (courtesy of C. Kaiser).

penetrates into the canopy window and latent or epicormic buds on the remaining stub may erupt into new shoots. However, given the competition for light and resources with the remaining major limbs, these new shoots can vary in vigor and orientation (and therefore usefulness), and sometimes may fail to grow due to apical dominance or other unknown factors. Shoots that become pendent may set small soft cherries and after a couple of years become shaded out and die. Shoots that grow vigorously upward (watersprouts) will become apically dominant and must not be pruned, so that they form spurs and produce large firm fruit 2 years after they elongate, at which time they can be pruned to develop moderate lateral shoots. If pruned before fruiting, more vigorous watersprout growth will likely occur. Continuous annual window pruning, beginning with the south section in the northern hemisphere or north section in the southern hemisphere, will allow the most light into the canopy. Follow-up pruning in other sections of the tree in subsequent years will rejuvenate the canopy over time, rather than all at once as with staghorning. However, the responses to window pruning may be more variable and require more diverse follow-up pruning decisions compared to staghorning, and it is less effective for rapidly and uniformly reducing the height of the tree.

A third rejuvenation technique, also suitable for whole tree renewal, is stumping, the dormant removal of the entire tree canopy back to a stump

Fig. 6.10. The stumping of sweet cherry trees and their resulting regrowth (drawing courtesy of C. Kaiser, photographs courtesy of C. Kaiser).

or relatively low primary trunk or leader structure (Fig. 6.10). The result-ant flush of growth from almost entirely epicormic buds the following sum-mer can be trained into multiple leader canopy architectures, such as the Spanish bush (SB), KGB or even UFO (discussed in more detail later in this section). When stumping the trees, it has been found that a 'nurse' branch (one with lateral buds developed the previous summer) helps stim-ulate regrowth from the stump in the spring. When trees were stumped at a height of 1.5 m (5 ft), those with no nurse branch took 22 days longer to develop shoots in spring than trees that were left with a nurse branch (C. Kaiser, personal communication). Trees that were stumped at 1 m (3 ft) with no nurse branch failed to develop any shoots after 6 months, possibly due to the lack of evapotranspirational water flow for moving cytokinins (which stimulate cell division) from the roots to transpiring leaves on the nurse branch.

Key to controlling the vegetative flush from stumping is to allow it to grow with minimal pruning during the first growing season. The resultant growth will be mostly upright and, as long as that wood is not pruned, it will form spurs the following year for fruiting 2 years hence. The caveat to this is that the strongest, most vigorous new upright shoots may remain overly vegetative, so heading back this minor, most vigorous population of upright shoots when they are about 45 cm (18 in) long will split each into two to three similarly vertical resulting shoots, but at about 33–50% of the vigor of comparable unpruned shoots.

6.4.3 General considerations for pruning non-precocious trees on vigorous rootstocks

Cherry trees grown on full-size rootstocks, such as Mazzard, Mahaleb and Colt, are generally more vigorous and less precocious than trees grown on highly productive rootstocks like the Gisela series. For this reason, the pruning strategy needs to be different. When pruning trees on productive rootstocks, growers must focus on reducing crop load and maintaining or increasing tree vigor. On full-size rootstocks, the focus should be to encourage precocity and productivity in the tree while managing tree vigor. This can not only change the number and kinds of pruning cuts, but also the timing of those cuts, since summer and early fall pruning is a devigorating strategy that may reduce future growth since photosynthetic capacity for building storage reserves is being reduced with the removal of active leaves. In general, thinning cuts are used more frequently than heading cuts.

To increase the precocity of such rootstock–scion combinations, pruning of young trees should be kept to a minimum. Since cherry trees tend to be highly apically dominant, some kind of intervention must occur in order to obtain adequate branching. Heading the trunk or other upright leaders is the easiest and most common method of creating lateral branches. However, heading cuts stimulate vegetative growth and delay production. Other methods can be used to stimulate branching that do not invigorate the tree. These are described in Section 5.7.2.1: 'Gibberellic acid / Canopy development'.

When cherry trees on full-size rootstocks are planted at traditional low orchard density spacing, the philosophy of many growers was to fertilize the tree heavily in the first few years after planting so that the orchard space was filled prior to the tree coming into production. For moderate to moderately high density plantings with vigorous and semi-vigorous rootstocks, the emphasis is not so much to fill the space quickly, but to promote early production and keep the tree from overgrowing its more restricted space. To achieve these goals, growers withhold fertilizer, minimize pruning and often manipulate branches by tying them below horizontal orientation, thereby slowing their growth and promoting precocity.

The pruning steps below are followed for crop load management and vigor control on vigorous rootstocks:

6.4.3.1   Step 1. Thinning cuts

Timing: late summer, fall or dormant season, annually
Target: remove pendent and weak shoots that are less than pencil-size in diameter
Purpose: remove wood that can overset and produce small fruit

6.4.3.2   Step 2. Thinning cuts

Timing: late summer, fall or dormant season, annually
Target: thin out new shoots to a single branch, especially at the top of the tree. Once the tree reaches its desired maximum height, the leader should be removed with a thinning cut, leaving a weaker lateral shoot to become the new leader. The length of lateral branches in the top of the tree should be reduced by a thinning cut back to a shorter and weaker lateral branch to improve light distribution to lower branches. Upright and vigorous branches should be removed; horizontal branches and those with slight upright orientation should be left. The ultimate shape of the tree leader or leaders should be pyramidal, with the longest branches being at the base of the canopy
Purpose: allow light to penetrate to the center and bottom of the tree, maximizing photosynthesis and producing fruit throughout the canopy. Helps to manage tree vigor

6.4.3.3   Step 3. Heading cuts

Timing: late summer, fall or dormant season
Target: on mature trees, it is often desirable to tip or head branches in the bottom of the canopy to keep them invigorated. Heading/tipping any new growth that can be reached from the ground is acceptable. Branches in the top of the tree should not be headed or tipped. An exception can be made for horizontal wood of Sweetheart, Lapins or other highly productive varieties, for which heading removal of one-quarter to one-third of new shoot length can help reduce future crop loads and maintain a favorable LA:F ratio. On young trees, heading cuts should be minimized to avoid undesirable invigoration
Purpose: stimulate new fruitwood formation and invigorate the canopy base, reducing excess production on highly productive cultivars such as Sweetheart and Lapins

6.4.3.4   Step 4. Stubbing cuts

Timing: late dormant season through bloom, annually, beginning with the third year of significant fruiting

Target: one to three of the largest branches should be renewed. To keep trees from becoming overly invigorated, be careful not to take out too many branches in any given year. If a tree has 20 to 25 lateral branches, approximately three branches per year should be renewed. Head branches back to leave 7.5–13 cm (3–5 in) stubs at the top of the tree and longer stubs at the bottom or where light intensity is less. Regrowth will occur where light and tree vigor are sufficient for replacement branches

Purpose: renew fruiting spurs, improve light distribution in the canopy

6.4.4 General considerations for pruning precocious trees on vigor-limiting rootstocks

Pruning and training cherry trees on any of the productive rootstocks, including Gisela 3, 5, 6 or 12, Krymsk 5 or 6, Edabriz or W172, requires techniques that are generally counter to pruning trees on vigorous standard rootstocks. For moderately high to high density plantings with precocious, highly productive, and sometimes vigor-limiting rootstocks, the emphasis is on reducing crop load and maintaining or increasing tree vigor. To achieve these goals, growers provide more frequent irrigation and fertilizer, and prune to eliminate excess flowers and stimulate vegetative growth. In general, heading cuts are used more frequently than thinning cuts.

Maintaining vigor even in small tree canopies is important because more and larger leaves mean more carbohydrate production to support larger cherries and higher yields. The production of high quality cherries requires a gross canopy LA:F ratio of at least 200–250 cm^2 of leaf area per fruit, which roughly translates to five spur leaves per fruit (Whiting and Lang, 2004; Neilsen *et al.*, 2016). Trees with a lower LA:F ratio are unable to produce enough carbohydrates to produce premium cherries.

The pruning steps below are followed for crop load management and vigor on highly productive and sometimes vigor-limiting rootstocks:

6.4.4.1   Step 1. Thinning cuts

Timing: dormant season, annually

Target: remove pendent and weak shoots less than pencil-size in diameter

Purpose: remove wood that can overset and produce small fruit

6.4.4.2   Step 2. Thinning cuts

Timing: dormant season, annually

Target: thin out new shoots to a single branch, especially at the top of the tree. Once the tree reaches its desired maximum height, the leader should be removed with a thinning cut, leaving a weaker lateral shoot to become the new leader. The length of lateral branches in the top of the tree should be reduced by a thinning cut back to a shorter and weaker lateral branch, to improve light distribution to lower branches. Upright branches should be removed; horizontal branches and those with slight upright orientation should be left. The ultimate shape of the tree leader or leaders should be pyramidal, with the longest branches being at the base of the canopy

Purpose: allow light to penetrate to the center and bottom of the tree, maximizing photosynthesis and producing fruit throughout the canopy

6.4.4.3   Step 3. Heading/tipping cuts

Timing: dormant season, annually, beginning the year after planting

Target: tip or head all new growth, removing ~15–30% of the previous season's growth; this represents the terminal section that will have a relatively dense formation of vegetative spur buds, which would become the site of dense flower clusters one year later

Purpose: promote branching and creation of new leaf area. Reduce future crop load by eliminating the area of dense spur buds (which also tend to have the most flower buds per spur). This reduces the dense fruit clusters in cultivars that are prone to oversetting

6.4.4.4   Step 4. Stubbing cuts

Timing: dormant season, annually, beginning the year after significant fruiting begins

Target: one to two major branches should be headed back for renewal, leaving 7.5–13 cm (3–5 in) stubs at the top of the tree and longer stubs at the bottom or where light intensity is less

Purpose: reduce current season crop, renew fruiting spurs, improve light distribution in the canopy

6.4.5   Mechanical pruning/hedging

Next to harvest costs, pruning is the most labor intensive and expensive orchard operation for producing fresh market sweet cherries, often requiring periodic activity during dormancy, bloom, early spring and in midsummer or even early fall. Reducing the time and cost of pruning can significantly improve the profitability of an orchard while improving worker safety.

Scientists worldwide are evaluating the potential for mechanical pruning of sweet cherries, including in France, Italy and the USA. While some orchards have long been topped mechanically, the focus has now turned to fruiting wall canopy architectures such as the super slender axe (SSA) and UFO. In most cases, a combination of mechanical and hand pruning appears to be most successful, particularly with mechanical summer pruning for light and winter hand pruning (possibly preceded by mechanical dormant pruning) for detailed crop load management. Commercial growers in Italy have indicated that a combination of mechanical pruning and hand pruning reduced labor needs by ~50% when compared to hand pruning alone. A study conducted at Washington State University (Courtney and Mullinax, 2016) showed that mechanical pruning was 29 times faster than hand pruning in a UFO block and 17 times faster than a combination of the two. Consequently, for growers planning orchards anywhere that labor costs are anticipated to be significant in the future, careful consideration should be given to sweet cherry training systems that best accommodate the potential for mechanized pruning.

6.5 Root Pruning

Prior to the development of dwarfing rootstocks, the primary means of controlling sweet cherry tree vigor (other than pruning of the canopy) was through selectively reducing irrigation (this is a useful tool only in arid production regions) or root pruning. Root pruning is done with a tractor-mounted steel blade (also called a subsoil knife) that is lowered into the soil and then pulled along the tree row to sever the root system at a specified distance from the trunk (Fig. 6.11). This limits water and nutrient uptake, thereby reducing canopy growth. Root pruning can reduce canopy vigor by as much as 30%. The depth of the blade and the distance from the tree, as well as timing and whether the pruning is done to one or both sides of the root zone, are the factors that affect

Fig. 6.11. Root pruning one side of a young sour cherry orchard (courtesy of R. Perry).

the severity of the pruning and consequently the extent of vigor control. Root pruning tends to be more common in Europe than it is in North America.

Root pruning is of particular interest to growers with high density orchards in which the scion–rootstock combination has turned out to be a bit too vigorous for the tree spacing, necessitating more extensive canopy pruning than is desirable (since pruning is usually an invigorating process, even as it temporarily reduces the size of the canopy). Root pruning is most readily accomplished in soils that have few significant rocks in the root zone. The optimal timing to impose root pruning (based on apple research) is from bloom to fruit set, when the tree is translocating growth resources from storage reserves to the spring growing points (leaves, fruit and extending shoots) in competition with roots. A greater response results from the earlier timing, with the response diminishing the longer that root pruning is delayed. Cherry growers need to test different timings for their soil/rootstock/scion/climate combinations. Typical pruning depth is 35–45 cm (14–18 in) and typical distance from the trunk is 30–45 cm (12–18 in) in high density orchards.

The distance from the trunk also can be determined in proportion to the trunk diameter of the trees, e.g. root pruning at a distance that is three times the trunk diameter will likely cause a severe reduction in growth, while pruning at four to five times the trunk diameter will likely cause a moderate reduction, and six to seven times the diameter may only slightly reduce growth. Trees with heavy crops will exhibit a greater reduction in vigor than trees with light crops.

The potential risks associated with root pruning include potential infection sites for root diseases like crown gall and potential reductions in fruit set and size. Tree anchorage can be affected by root pruning, particularly when pruning is on the windward side of the tree. This is less of a concern for trellised cherry trees than those that are free-standing. Fertility and irrigation should be reduced in proportion to the anticipated canopy response, but if the climate becomes significantly stressful (such as during a heatwave), more frequent, low dose irrigation and nutrient management may be required to compensate for the significantly reduced root system.

For More Information

Long, L., Lang, G., Musacchi, S. and Whiting, M. (2015) Cherry training systems. Pacific Northwest Extension Publication, Oregon State University, 667, 63 pp. www.catalog.extension.oregonstate.edu/pnw667 (accessed 19 June 2020).

Acknowledgements

Portions of this chapter were adapted with permission from Oregon State University. *PNW 592 Four simple steps to pruning cherry trees on Gisela and other productive rootstocks*, copyright 2007. Oregon State University Extension and Experiment Station Communications, Corvallis, Oregon. www.catalog.extension.oregonstate.edu/pnw592 (accessed 19 June 2020).

References

Courtney, R. and Mullinax, T.J. (2016) Keeping limbs in line with mechanical pruning. Good Fruit Grower. Available at: https://www.goodfruit.com/keeping-limbs-in-line-with-mechanical-pruning-video (accessed 1 April, 2016).

Neilsen, D., Neilsen, G.H., Forge, T. and Lang, G.A. (2016) Dwarfing rootstocks and training systems affect initial growth, cropping and nutrition in 'Skeena' sweet cherry. *Acta Horticulturae* 1130, 199–206. DOI: 10.17660/ActaHortic.2016.1130.29.

Whiting, M.D. and Lang, G.A. (2004) `Bing' Sweet Cherry on the Dwarfing Rootstock `Gisela 5': Thinning Affects Fruit Quality and Vegetative Growth but not Net CO_2 Exchange. *Journal of the American Society for Horticultural Science* 129(3), 407–415. DOI: 10.21273/JASHS.129.3.0407.

Sweet Cherry Training Systems

7

7.1   General Principles of Orchard Training Systems

Canopy architectures and training systems for fresh market sweet cherry production have been evolving rapidly over the past two decades, generally accelerating with the advent of vigor-controlling and precocious rootstocks. Training systems should be considered to be dynamic and continuously developing, as every grower and orchard site is different, with inherent traits that lead to subtle modifications of initial ideas and training concepts that can significantly affect their ultimate degree of success. Sweet cherry cultivars vary in both growth habits (from extremely upright, such as Lapins, to extremely spreading, such as Sweetheart, to even somewhat pendent, such as Anderson) and fruiting habits (from predominately spur-bearing, such as Lapins, to predominately non-spur-bearing, such as Kordia and Regina). Adoption and modification of canopy training system concepts and guidelines certainly must consider and adapt to these inherent growth and fruiting habits.

Similarly, canopy training decisions must take into account the various factors that comprise overall orchard vigor, such as soil fertility, depth, and texture (e.g. water holding and cation exchange capacities); climate (growing season length, cloudiness/humidity and daily solar radiation levels, diurnal temperatures, windiness, rainfall); and rootstock vigor. As well as having effects on vigor, rootstocks can also affect anchorage and scion growth habit: e.g. Mazzard, Mahaleb, Colt, Krymsk 6 and Gisela 12 tend to confer more upright growth to scions, whereas Gisela 3, 5 and 6 tend to confer more spreading habits. Finally, each orchard operation may place different priorities on management factors that can affect training system outcomes, such as labor skill level, availability of workers, and prevailing hourly wages; scale of the overall orchard operation; and capacity for attention to detail.

The goal for imposing an orchard training system is to establish a structural tree canopy framework that best provides the potential to produce profitable quantities of marketable fruit. Large trees take many years to fill their allotted orchard space and generally result in complex canopies that have non-uniform light interception and distribution, simply due to their size and difficulties for labor access. These larger trees tend to have low input costs for tree training and high input costs for harvest, as well as mixed fruit quality. Modern training systems have focused on higher densities of smaller trees that have higher orchard establishment costs, but allow earlier yields and therefore earlier returns on investment; facilitate more efficient labor inputs and more systematic canopy development and renewal of fruiting wood; and more uniform light distribution for more uniform fruit quality. This chapter will address more than a dozen potential sweet cherry canopy training systems, which certainly is not a comprehensive list of all of the ways growers can and have pruned cherry trees over the years. What is intended is a discussion of some of the key cherry training techniques and concepts, and particular benefits and limitations, to help growers determine how they might adopt or mix-and-match training systems for their goals, cultivars, rootstocks, orchard sites and labor situations.

A recent award-winning sweet cherry training guide is available for free as an electronic publication (Long *et al.*, 2015; www.catalog.extension. oregonstate.edu/pnw667 as well as www.canr.msu.edu/uploads/files/ PNW_667_Cherry_Training_Guide.pdf, both accessed 19 June 2020). This guide, which can also be downloaded as a smartphone or notebook computer application, describes the establishment and maintenance of eight commercially successful sweet cherry training systems in step-by-step format: Kym Green bush (KGB), Spanish bush (SB), steep leader (SL), super slender axe (SSA), tall spindle axe (TSA), upright fruiting offshoots (UFO), upright fruiting offshoots 'Y' Trellis (UFO-Y) and Vogel central leader (VCL). The following discussion draws extensively from this guide. All of these training systems, as well as others by growers and scientists around the world, have been shown to be profitable in the right circumstances, yet the caveat remains that probably no combination of scion cultivar, rootstock, training system and climate is optimal for all situations. Therefore, it is important for growers to understand their orchard vigor factors, target markets, the fundamental aspects of sweet cherry growth and fruiting, and how the techniques used in different training systems affect those fundamentals.

Some key questions to ask with respect to training system decisions include:

- How quickly is the fruiting structure developed and when are economic yields (and therefore returns on investment) expected to begin?
- How easy is it to teach the orchard labor team to train and prune the orchard? How many and what type of decisions need to be made to achieve or maintain the desired canopy architecture?

- How is light interception and distribution within the canopy optimized, and where is the fruit borne relative to the light being harvested by the leaves?

7.1.1   Training system canopy architectural traits

The discussion of training systems in this chapter is divided into four general types of canopy architectures, based on single or multiple leaders, and three-dimensional or 'two-dimensional' (also known as 'planar', 'fruiting wall', or 'hedgerow') row volumes. Sweet cherry is a naturally vigorous tree that evolved in forests, and all this vigor tends to be concentrated over time at the top of the leader. This tendency is reduced in modern training systems, where single leader trees benefit significantly from vigor-controlling rootstocks. Multiple leader trees distribute the vigor among the multiple leaders, so even though vigor also becomes concentrated over time at the top of each of the multiple leaders (compared to lower along each leader), overall vigor is relatively lower than for single leader trees. Hence, multiple leader training systems are generally paired with semi-vigorous to vigorous rootstocks, while single leader systems generally are paired with semi-dwarfing to dwarfing rootstocks.

In addition to leader number, the dimensionality of tree row volumes also is useful as a major factor for classifying training systems. Three-dimensional tree canopies tend to be symmetrical; hence, orchard row spacing will vary somewhat proportionately with tree spacing. For example, symmetrical trees planted 2.5 m (8.1 ft) apart in the tree row are maintained at a mature width no greater than 2.5 m (8 ft) from tree to tree, and extend generally no further than 1.25 m (4 ft) on either side into the tractor alley. So, the minimum row spacing should be 2.5 m (8 ft) plus the width of the tractor (for example, 1.5 m or 5 ft), and this minimum row spacing of 4.0 m (13 ft) may need to be wider, in proportion to planned mature tree height, to minimize the potential shading of the bottom of one row's canopy by the top of the trees in the next row. Traditionally, in high density apple orchards, rows were spaced 1.3 times the height of the tree, so in the scenario above, the minimum row spacing of 4.0 m (13 ft) would equate to trees that are maintained to be no taller than 3.1 m (10 ft). If taller trees are desired, the row spacing would be widened. If the spacing of three-dimensional symmetrical trees is more or less than illustrated above, for example 3.0 m (9.8 ft) or 2.0 m (6.5 ft), the minimum row spacing (with a 1.5 m wide tractor) would change to 4.5 m (14.6 ft) or 3.5 m (11.4 ft), respectively.

However, in the case of two-dimensional (narrow, planar) tree canopies, trees are rarely symmetrical, because planar training systems maintain a target narrow canopy width into the tractor alley, regardless of how tree spacing down the row differs. Furthermore, as the canopy becomes extremely and uniformly narrow from top to bottom, the potential for tree height to create too much shade at the bottom of the canopy in the next row also decreases, since some sunlight may pass through a very narrow uniform canopy. Consequently, in very narrow planar orchards, it has been feasible to space rows at 1.0–0.8× the height of the tree, or perhaps even tighter. So, in the case of a narrow UFO canopy only 0.75 m wide, a minimum row spacing of 2.25 m (7.3 ft, tree width plus tractor width) is feasible, and the maximum tree height could range from 2.25 m (7.3 ft) to 2.8 m (9 ft). With the potential for over-the-row sprayers, labor platforms and/or more narrow tractors, some experiments with even closer row spacing have been initiated in New Zealand and the USA.

A third component of row spacing and canopy volume dimensionality that may be advantageous in certain circumstances is variable spacing of double tree rows. In the past, high density double and triple tree rows of three-dimensional trees, alternating with tractor alleys, have been attempted to increase productive fruiting area per orchard and minimize orchard space 'wasted' for tractor access. Such strategies have been developed in both open field and expensive protective covering technologies like high tunnels. However, these have generally performed poorly as the orchards start to mature due to more difficult labor access for pruning (and brush removal) and picking (especially when ladders are needed), increased shading, and poor or non-uniform canopy spray coverage to control insects/diseases/weeds in the interiors of the double/triple rows that lack tractor access. With two-dimensional tree canopies, however, specialized double rows of narrow vertical or angled planar trees, alternating with narrow tractor alleys, can provide increased fruiting area per orchard while maintaining easy access for labor (and mechanical platforms and hedgers), reduce shade potential in the double row interior, and achieve suitably uniform spray coverage.

Finally, with the recent innovations in sweet cherry canopy architectures that develop markedly different fruitwood orientations, such as primarily horizontal (e.g. central leader/spindle, tabletop [TTP] and espalier [ESP] trees), vertical (e.g. KGB and UFO training), pendent (e.g. solaxe trees, SLX), or diagonally-angled upward (e.g. palmette [PLM] and some central leader/spindle), the research verdict is still out on whether particular fruitwood orientations alone can promote significant advantages for fruit size or quality that are independent from leaf area-to-fruit (LA:F) ratio or light exposure effects. Clearly, certain shoot growth orientations affect branch and tree vigor, which can influence flower bud formation. Orientation can also influence ease of tree management decisions.

7.1.2 Nursery tree considerations for training systems

At present, sweet cherry nursery trees are produced primarily as single leader trees, although in rare instances, some nurseries may offer custom or specialty trees such as dual leader (also known as 'bi-axe') trees that are advantageous for certain training systems like bi-cordon UFO or Y-SSA – if the two leaders are uniform in vigor. Some nurseries, particularly in Europe, offer 'feathered' (also known as 'knip') single leader trees that have a significant whorl of relatively uniform lateral branches. These can be advantageous for central leader or spindle tree training systems – if the lateral branches have good crotch angles, are at the desired height in the canopy, are uniformly distributed around the leader and are relatively uniform in vigor. Specialty nursery trees such as knip or bi-axe trees can advance the development of the fruiting canopy in selected training systems by 1 year. However, in many cases, it is more important to establish an optimal tree structure in the first 2–3 years (which will optimize fruitfulness over many years) than to harvest a small initial crop in the second or third year that may slow the tree from achieving its optimal mature canopy structure.

Knip nursery trees can be disadvantageous for multiple leader or planar training systems, as the lateral branches are not likely to be of much use in developing the initial tree structure. Most cherry trees produced in US nurseries are 'whips' with no or only a couple of lateral shoots. More often than not, the few lateral branches on whip nursery trees must be removed at planting since they are insufficient to provide a full complement of desirable primary canopy structure, or have poor crotch angles, or are not uniform in vigor, or are not at the desired height. What is critical for good nursery tree quality, then, is a good complement of vegetative buds on the leader, particularly at the height(s) where lateral shoots are desired for future structural development, and a strong root system. With nursery trees that are too vigorous and of large caliper with long internode distances, it may be difficult to initiate the lower branches on a central leader tree without heading it; therefore, for such training systems, nursery trees of moderate vigor and moderate internode distances may be best. Potted nursery trees are a recent advance, with research ongoing to examine their potential advantages and disadvantages with respect to transplant shock, rapid establishment and growth, often smaller size than bareroot nursery trees, ease or challenge of handling and planting, and ability to plant later in the spring or even summer.

7.2 Free-Standing or Trellised (3-Dimensional) Single Leader Trees

Free-standing trees are more common for sweet cherry orchards than for apple orchards, as most cherry rootstocks are relatively well anchored (Gisela 6 with some cultivars is a notable exception) and the smaller fruit size and earlier harvest of cherries generally requires less structural support than the longer fruit development period and larger fruit of apples. Free-standing cherry trees may be self-supporting or each tree may be supported by a post or stake. Support may only be needed in the early years when the upper canopy can develop too quickly before the caliper of the trunk becomes strong enough to support it.

Conversely, single leader tree support may be required for much of the life of the orchard if the rootstock does not provide strong anchorage (e.g. some cultivars on Gisela 6). Also, some training systems may benefit from a single wire trellis for precise orientation of the central leaders, particularly if any type of mechanized pruning will be utilized. Finally, the TTP training system for single leader trees requires a trellis with two cross-arms and wires that support a lower and upper 'table' of continually recycled fruit-bearing branches, as will be described below. Obviously, establishment costs for systems that utilize posts, wires, trellis cross-arms, etc., are higher than for truly free-standing systems, so the potential benefits and durability of the trellis must be evaluated with respect to anticipated improvements in precision, yield, fruit quality (such as reduced bruising), labor efficiencies, etc.

7.2.1　General central leader canopy training

At planting, remove any branches on the nursery tree, unless there are three to four of moderate vigor and with good crotch angles at the desired height for the lowest portion of the fruiting canopy. If branches are retained, they can be headed to remove about 50% of their length to promote further lateral branching and, on precocious rootstocks, reduce the future second year yield in favor of further canopy development. Alternatively, particularly on non-precocious rootstocks, the retained branches can be bent slightly below horizontal to slow their growth and promote earlier fruiting.

If nursery tree branches are removed, lateral branches must be promoted. This can be done by selecting buds where branches are desired and removing the intervening buds between those selected after budswell, in the year of planting. Alternatively, scoring or notching just above the buds where branches are desired, or by using plant growth regulators (PGRs) can also induce branching (see Section 5.7.2.1: 'Gibberellic acid / Canopy development'). Scoring, notching and the use of PGRs to induce branching is most successful when conducted in the late dormant season of the year after planting. These techniques have the advantage of retaining the full height of the nursery tree and can help fill canopy space and reach production more quickly. However, if stimulation of lateral branches is not successful, the filling of canopy space is not only delayed, but it generally becomes more difficult to initiate new lateral branches in subsequent seasons. As each section of the central leader ages beyond the spring following its initial formation, the buds become more inhibited from growing out into new shoots due to increased apical dominance exerted by the combination of all shoots that have formed above it. This can result in sections of essentially 'blind' space on the leader that lack fruiting branches, thus reducing yield potential, a situation that is extremely difficult to rectify.

Fig. 7.1. Clothes pins are used to increase the branch angle in young developing shoots. This clothes pin was attached when the shoot was about 5 cm (2 in) long (Germany) (courtesy of L.E. Long).

Alternatively, the most tried and true method for lateral branch formation is to head the leader about 30 cm (12 in) above the point where the lowest canopy branches are desired, usually about 75 cm (30 in) above the ground. Of the buds that form new shoots, the strongest will become the new leader and the subtending shoots will need to be managed to develop wide crotch angles, as described below. Any lateral branches below about 60 cm (24 in) should be removed, and trunk tree guards (such as cardboard milk cartons or plastic spiral trunk wraps) should be installed to protect the young trunk from herbicides, string weed trimmers and gnawing animals like rabbits and mice. This annual heading of the leader to create tiers of lateral branches can be done until the leader reaches its mature height. While it is highly effective in creating the preferred branches at the desired heights, such annual heading slows the filling of canopy space and tends to create stronger growing lateral shoots than when PGRs, scoring or bud selection are used on an unheaded leader.

When new elongating lateral shoots are about 5–12 cm (2–5 in) long, if their angles of emergence are upwards, clothes pins can be attached to the leader just above the shoots to reset the angles to a more desirable horizontal orientation (Fig. 7.1). Eventually, the growing shoot tip will likely turn upwards again, but the crotch angle will be set at a much stronger 90° angle to the leader. If the extending shoot becomes too upward (which is usually related to being closer to the leader apex as well as being dependent on

Fig. 7.2. Tree trunks are painted with a white latex paint to reduce the potential for winter injury (courtesy of G.A. Lang).

the variety's growth habit), twine can be used to tie it down to a trellis wire, a support post or soil clip, or to the lower trunk.

If trunk guards were not installed, the trunk should be painted white with a latex paint in the fall to protect it from potential winter injury. Such injury can occur on sunny cold days when the bark on the southwest (in the northern hemisphere; or the northwest in the southern hemisphere) side of the tree can heat up during the day and then be subjected to severe low temperature freezing and splitting overnight (Fig. 7.2).

For the second and third years, the promotion of another tier or whorl of lateral shoots should be repeated on the portion of the leader that grew the previous summer. Lateral branches that developed in the previous season should be headed about 80 cm (30 in) from the trunk, just above two side buds (versus top or bottom buds). This will promote further lateral branching, rather than vertical or pendent shoot formation. Any vertical or pendent shoots, and any branch directly above a lower branch such that shading may occur, should be removed with thinning cuts. The goal over the first 3 years is to develop tiers of well distributed branches around the central leader, such as either two to three distinct tiers or a continuous spiral whorl, so that a maximum amount of sunlight is intercepted and distributed among the branches with minimal dense shading.

By the fourth year, significant fruiting should have begun, especially for trees on precocious rootstocks, and promotion of an additional tier of lateral shoots in the uppermost portion of the canopy that grew in the third year is generally not needed. Pruning of the fruiting branches should focus on good light interception and distribution, and removal of any areas of densely spaced nodes, as these will later form dense spur fruit clusters. This management of future crop loads can be accomplished by heading the apical portion of previous season shoot growth where nodes are more densely spaced, or by manual removal of some of the spurs, either selectively to reduce positional density, or non-selectively as by rubbing the pruning loppers upward against the bottom of branches to knock off the very bottom spurs that would have the poorest exposure to light.

7.2.2   Vogel central leader, Zahn spindle and modified Brunner spindle

The Vogel central leader (VCL) is a free-standing, single leader tree that fruits on a structure of renewable scaffold branches developed in tiers or whorls around the leader, creating a conical spindle or 'Christmas tree' shape of longer lower scaffolds progressively diminishing in length up to the top of the leader (Fig. 7.3). In most ways, VCL trees are similar to other spindle training systems such as the Zahn spindle and modified Brunner spindle. F. Zahn and T. Vogel, and T. Brunner and K. Hrotko, were tree fruit scientists and/or advisers in Germany, and in Hungary, respectively. Hrotko developed the modified Brunner spindle from Brunner's initial work on spindle canopies. This is indicative of how canopy training concepts, such as the spindle, have been developed in ways suitable for individual growing regions and available rootstocks.

Good quality knip nursery trees are useful for rapidly developing the first tier or whorl of fruiting scaffolds for precocious yields. Training and growth of the fruiting scaffolds is relatively horizontal, which also promotes precocity. If good quality knip trees are not available, lateral branch induction techniques (see Chapters 5 and 6, and Section 7.1: 'General principles of orchard training systems') are important for establishing enough lateral branches to moderate their individual growth and develop fruit buds the following year. Heading is avoided for inducing branching on the leader and is discouraged for the scaffolds as well. Pruning of initial lateral branches is limited to primarily thinning cuts to remove excessive primary laterals or overlapping, vertical or pendent secondary shoots. Fertilization of young trees should be minimal, especially on semi-vigorous to vigorous rootstocks, to promote a moderate growth rate. Training of future scaffolds with clothes pins to set wide crotch angles and/or tying down to maintain

Fig. 7.3. Vogel central leader (VCL) tree (Oregon) (courtesy of L.E. Long).

flat growth can be labor intensive, but it favors precocity and lowers the maintenance needs of the mature structure.

For VCL and Zahn spindle trees, the central leader is the only permanent structural wood, with secondary scaffolds giving rise to fruit-bearing tertiary and quaternary growth. Modified Brunner spindles may retain a low permanent whorl of four to five scaffolds. At maturity, one or two of the largest non-permanent scaffolds are annually cut back to stubs to be regrown. The conical tree shape promotes good light interception and distribution by the canopy. Due to the single leader nature of this canopy training system, a dwarfing or semi-dwarfing rootstock is necessary to help maintain

Fig. 7.4. The lateral branches in the upper canopy (red arrows) tend to be more vigorous and dominant than branches in the lower canopy (green arrow and below) (courtesy of G.A. Lang).

tree height at 3.0–3.7 m (10–12 ft) at maturity, creating a semi-pedestrian orchard. The vertical orientation of the permanent structural leader can eventually result in upper canopy vigor becoming a major challenge, at the expense of fruitwood renewal (Fig. 7.4). That is, the best fruiting zones tend to move upward in the canopy over time. Delayed spring and/or summer pruning of the upper portion of the canopy can help devigorate it, and dormant pruning of the lower canopy can help invigorate that portion. Another technique to encourage uniform vigor throughout the tree is to bend upper branches in the tree top to near horizontal in the first year of growth, either by tying or through the use of toothpicks (Fig. 7.5) and spreaders (see Fig. 6.4 in Section 6.4.2.3: 'Stubbing and Thinning').

Fig. 7.5. Toothpicks were used on newly formed branches to keep them flat and to help manage branch vigor (Oregon) (courtesy of L.E. Long).

Depending on rootstock and site vigor, trees are typically spaced 1.8–2.75 m (6–9 ft) in the row and 4.0–4.9 m (13–16 ft) between rows.

7.2.3 Tall spindle axe and fusette

Central leader cherry tree canopies that are taller and narrower than VCL or Zahn spindles include the tall spindle axe (TSA, from the USA) and the fusette (from Italy). The tree canopy is characterized by a continuous whorl of lateral branches of modest vigor that are developed by activation of selected buds (via bud scoring, bud removal or PGR application), usually on dwarfing to semi-dwarfing rootstocks unless the orchard soils are weak. The goal is to achieve ten or more lateral shoots per year of initial canopy development; the more laterals that form, the more moderate will be their vigor. Fruiting is primarily on spurs on those laterals, with about one-third borne on flowers at the base of previous season shoot growth. The increased height increases yield potential and the narrower width is facilitated by the lack of permanent scaffolds, which are replaced by weak secondary lateral branches where fruitwood is developed and regularly renewed, reducing canopy density and improving light distribution. TSA training is further differentiated from VCL and Zahn or modified Brunner spindle canopies by annual heading of lateral shoots to manage future crop loads and maintain a favorable balance in LA:F ratios.

The TSA training system favorable for precocious fruiting due to minimal structural pruning of the leader, rapid formation of many laterals for fruiting, and the use of precocious rootstocks that promote basal fruiting of the laterals in the year after formation. At maturity, the crop load is a favorable mix of spur-borne fruit to increase quantity and basal fruit to increase quality, with annual partial renewal assuring that fruiting spurs remain relatively young. As in other central leader canopies, the vertical orientation of the permanent structural leader can eventually result in greater upper canopy vigor compared to the lower canopy, at the expense of fruitwood renewal. Delayed spring and/or summer pruning of the upper portion of the canopy can help devigorate it, and dormant pruning of the lower canopy can help invigorate that portion. Depending on rootstock and site vigor, trees are typically spaced 1.25–2.0 m (4–6.5 ft) in the row and 3.5–4.0 m (11–13 ft) between rows.

7.2.4 Solaxe

The solaxe (SLX) is another modified spindle canopy architecture in which the permanent primary tree structure is limited to the central leader, ultimately achieving a tall conical architecture that differs from TSA and fusette in several key ways. Similarities include a continuous whorl of lateral branches of modest vigor that are developed by activation of selected buds (via bud scoring, bud removal or PGR application), usually on dwarfing to semi-dwarfing rootstocks unless the orchard soils are weak. However, key to SLX is the tying down of laterals below horizontal to slow growth and promote fruit bud formation, and the removal of the most basal 40 cm of buds on all laterals arising from the leader to create an open 'chimney' in the center of the canopy for light distribution (Fig. 7.6). Thinning cuts are used to remove any laterals on the main branches that are greater than one-third of the diameter of each main branch, to minimize the forking of fruiting branches and the removal of any thin lateral shoots back to a fruiting spur.

The main branches are never headed, to maintain narrow, somewhat pendent columns of fruit and avoid stimulating growth. These techniques promote a higher proportion of spur fruiting sites and a lower proportion of basal fruiting sites (due to the reduction in lateral shoots) compared to TSA training. Consequently, spur 'extinction' (selective removal of spurs) is imposed to manage crop loads and improve LA:F ratios. Also, the leader terminal is rarely headed at maturity to prevent problematic invigoration of the upper canopy; rather, the leader is tied down to a modest angle to slow growth without triggering watersprout formation. Annual pruning costs are reduced compared to other spindle systems, but this is somewhat of a trade-off since labor is needed for tying down of branches and the

Fig. 7.6. Lateral branches tied below horizontal to promote fruit bud formation (Chile) (courtesy of L.E. Long).

leader terminal, as well as spur extinction. Due to the lack of shoot re-newal, many SLX cherry growers have eventually experienced declines in fruit quality, especially on precocious, vigor-controlling rootstocks, leading to imposition of some heading cuts to renew fruiting branches and modify the typical SLX canopy.

7.2.5 Tabletop central leader

The tabletop (TTP), also known as 'sistema de renovación permanente' (SRP) is a three-dimensional canopy architecture in which the permanent primary tree structure is limited to the central leader with two tiers of horizontal, triennially renewed fruiting branches. The TTP central leader training system was developed in The Netherlands as a unique technique for optimizing annual renewal of fruiting wood, balancing crop loads with adequate leaf area, and reducing the eventuality of excessive vigor in the top of the central leader canopy and shade/low vigor in the lower canopy. It was designed for high density orchards using precocious, vigor-limiting rootstocks like Gisela 5. However, even Gisela 5 can be too vigorous for this system in some situations. To overcome this problem, only shoots of about 1 m (~3 ft) in length and 8–10 mm (3–4 in) in diameter are selected; these are broken and tied to the table wires to heal. By selecting the right shoots, more uniformity is obtained. For very productive cultivars, flower buds are removed to better balance yields.

A double 'T'-trellis system with 1.0 m (~3 ft) wide, low and high cross-arms is required to provide structural support for the two tiers of side wires that help guide the canopy formation of the two 'tabletops' of continuously renewed fruiting wood. These two tiers of fruiting wood are generally at heights where lateral branching naturally occurs, the first being a bit below the terminus of the nursery tree (a height of about 1.1 m or 3.5 ft) and the second being just below the terminus of the first year's orchard growth of the nursery tree leader (about 2.8 m or 9 ft) (Fig. 7.7a and b). Knip nursery trees at about 1.8 m (6 ft) tall, having at least six branches at the desired first tier height, are preferred for early fruiting potential. If whip nursery trees must be used (or appropriate lateral branches are insufficient on the knip tree), the tree terminal should be scored with a light saw cut and broken about 20 cm (8 in) above the lower tabletop wires, or buds should be activated by scoring or with PGRs, to induce lateral branch formation without pruning of the leader. The breaking of the leader should be done in the early spring before bloom of the second year. Once broken, it should be tied upright to the pole. This reduces growth on the leader and stimu-lates more growth in the tree base, guaranteeing new shoots for renewal on the table wire.

In August or September (in the northern hemisphere; or February–March in the southern hemisphere), after the lower tier of branches has formed, each should be scored with a small saw cut directly in front of

Fig. 7.7. The two tiers of fruiting wood can be clearly seen on this mature fruiting tree and on these recently pruned trees under the tabletop (TTP) training system (courtesy of R. Vermeulen).

a vegetative bud on the top side of the branch, about 10–15 cm (4–6 in) from the central leader and then broken (Fig. 7.8). The broken branch is then clipped or tied to the lower side-arm wire for support while healing. Late summer breaks help to keep the vigor low, heal more quickly than in spring, and the risk of bacterial canker is less. The broken branch should never be headed for the duration of its existence. The purpose of cracking and breaking the branches is to initiate: (i) fruit bud formation on the longer portion of the broken branch; and (ii) to grow a replacement shoot from the basal portion of the broken branch near to the leader.

Ideally, this will create a new future fruiting branch with a balanced LA:F ratio of first-year fruiting spurs and solitary flower buds at the base of its terminal extension shoot growth, plus the emergence of a replacement shoot at its base. The year following breakage, the initially broken branch will fruit and then be removed with a pruning cut where it was broken (in front of its replacement shoot); the replacement branch will be broken and tied down; and it will generate another replacement shoot at its base near the leader. If any lateral shoots form more terminally on the broken branches, they should be removed with thinning cuts, maintaining relatively narrow horizontal columns of fruit. The goal is to develop a continuous

Fig. 7.8. Under the tabletop (TTP) training system, lateral branches are broken and left in place, in front of a top bud to induce fruit bud formation and encourage a replacement branch to grow from the top bud (courtesy of R. Vermeulen).

cycle of four to six fruiting branches per tier, with an equal number of replacement branches that are forming future fruiting spurs plus an equal number of new shoots that will become replacement branches. Thus, other than the permanent central leader, the canopy should be comprised of only three populations of branches: new shoots, year-old branches, and first-year fruiting spur branches.

The second tabletop tier of fruiting should be developed the same way, and any laterals arising from the central leader in-between the tiers should be pruned back to the most basal vegetative bud, promoting the formation of a 'window' that maintains good light exposure throughout the lower tier. On precocious rootstocks, the intermediate shoots in the window will have basal flower buds that fruit and are supported by the shoot that regrows from the retained vegetative bud (similar to the standard pruning of SSA trees, described later in this chapter). These non-tabletop shoots help reduce vigor of the central leader terminal. If too many lateral shoots form for the second tier (thereby increasing the risk of shade for the lower tier), some of them should be pruned back to their most basal vegetative bud, to open a temporary light channel while promoting regrowth as a future fruiting branch to replace the nearest current fruiting branch. If any of the lateral shoots are too vigorous, they should be removed entirely rather than pruned back to a vegetative bud. If lateral shoot emergence is inadequate to form the second tier target of four to six fruiting branches, the leader can be partially cut and broken from budswell to bloom at about 2 m (6.5 ft) to promote new lateral formation. The broken leader should be tied back upright to heal and resume growth (Fig. 7.9).

At maturity, if the number of renewal laterals begins to fail at either height, a double chainsaw cut into the trunk (one on each side, separated vertically by about 25 cm or 10 in) above the tiers can shock the leader into activating latent buds for the formation of new (replacement) shoots. This also weakens vigor of the shoots above the chainsaw cuts, countering the natural acrotonic growth and apical dominance of sweet cherry. The typical maximum mature tree height is 3.25 m (10.5 ft), tree spacing is 1.0–1.5 m (3 to 5 ft) and typical row spacing is 4.0 m (13 ft). On more dwarfing rootstocks or poorer soils, spacings may be slightly closer, and about 0.5–1.0 m (~2–3 ft) farther apart on more vigorous rootstocks. If growth in the upper canopy is too strong, the leader can be broken to slow growth without stimulating many new shoots as would occur with a heading cut. Chainsaw cuts into the trunk have also been used to reduce vigor, rather than breaking or root pruning. This, however, is an emergency tool to control the growth, but is usually not needed if the leader is properly broken at the right time. It is always advisable to avoid chainsaw cuts where risk of bacterial canker infection is significant. Another potential advantage of the TTP system in hot climates is the shading of the root zone that is provided by the lower tier of branches in mid- and late summer, maintaining soil temperatures for good root function.

As soon as the leader reaches the designated height, it is cut or broken. The tree will respond with three to five new upright shoots which will be broken again in August. This procedure is repeated every year. By selecting eight shoots each year and cutting out eight branches each year in August, this becomes a very simple system with a guarantee of production.

Fig. 7.9. The leader is broken and then tied back to vertical to reduce vigor in the tree top and promote lateral formation in this zone, under the tabletop (TTP) training system (courtesy of R. Vermeulen).

7.3   Free-Standing (3-Dimensional) Multiple Leader Trees

Multiple leader training systems generally develop free-standing trees, since: (i) more vigorous rootstocks are utilized with multiple leader cano-pies; and (ii) the steps taken in pruning to develop the multiple leaders and secondary fruiting structures allow the root system to become better established before the upper canopy is fully developed and bearing a crop load. Due to the increased dimensions of multiple leader canopies, orchards trained to multiple leader systems are typically planted at low to moderate tree densities.

7.3.1 Open vase/goblet

The most traditional orchard training system for sweet cherry trees is probably the open vase or goblet (OVG) canopy architecture, which is still common in many parts of the world, including the otherwise highly advanced Pacific Northwest (PNW) US sweet cherry industry. Its popularity is likely due to its low cost of establishment and its distribution of the natural vigor of cherry among the multiple leaders, facilitating good yields with canopies having relatively moderate heights at maturity. The disadvantages of this system include a relatively long time to reach full production; high labor costs (not to mention increased insurance and worker safety concerns) due to the inefficiencies of pruning and harvest that require ladders; and a complex, non-uniform canopy that may yield fruit of variable quality depending on fruit location within the large canopy.

The OVG canopy architecture includes a permanent primary tree structure that is limited to a very low trunk with three to five vertically spreading permanent secondary structural leaders, from which fruit-bearing tertiary and quaternary growth is developed (Fig. 7.10). These structural leaders are initiated by heading of the nursery tree at planting, about 60–75 cm (24–30 in) above the ground, choosing the three to five new leaders for equal distribution around the trunk, and using clothes pins to flatten the crotch angle when the new shoots are about 10–15 cm (4–6 in) long. To promote earlier cropping (as on non-precocious rootstocks), the leaders should not be headed; lateral branch formation on the leaders will occur naturally just below the terminal every spring. When precocious rootstocks are used, the initial multiple leaders can be headed to an outer bud to promote both a wider spread of the leaders and lateral branching just below the point of the heading cut. Very little additional pruning (other than thinning cuts to remove overlapping, weak or dead shoot growth) is required until the tree has mostly filled its allotted orchard space and cropping has begun. Once cropping becomes significant, lateral shoots can be headed by about 20–30% to balance cropping potential with leaf area, and thinning cuts can be used to remove overlapping or poorly-oriented lateral shoots and/or to reduce the shading of lower fruiting wood.

The typical maximum mature tree height is 4.0–4.5 m (13–15 ft), tree spacing is 2.5–4.5 m (8–15 ft) and typical row spacing is 4.0–4.5 m (13–15 ft), depending on rootstock vigor. The vertical orientation of the permanent secondary structural leaders means that despite initial diffusion of the typical tree vigor into multiple leaders, eventually managing excessive

Fig. 7.10. Traditional multiple leader, open vase/goblet (OVG) cherry trees on vigorous rootstock (Washington) (courtesy of G.A. Lang).

upper canopy vigor can become a challenge, at the expense of retaining and renewing lower fruiting wood, resulting in the best fruiting zones moving to the top of the canopy over time. Often, pruning of the upper canopy, especially where cuts into large diameter wood are required, is accomplished in late summer after harvest to minimize regrowth and to remove leaf area when carbon and nitrogen reserves are building. Pruning of the lower canopy and detail pruning for refining potential crop loads is accomplished during dormancy to invigorate those portions of the canopy.

7.3.2   Steep leader

The steep leader (SL) training system is a modification of OVG canopies that was developed in the PNW region of the USA and can best be described as a free-standing tri- or quad-axe multiple leader tree with horizontal scaffold branches projecting from the base of each leader. The permanent primary tree structure is limited to a very low trunk, its vertically-oriented permanent structural leaders that are relatively close together, and its low tier of horizontal scaffolds. The center of the canopy (between the leaders) is maintained free of shoot growth, creating a 'chimney' of light similar to the idea of the open vertical cylindrical area around the central leader of the SLX system. Each leader mimics a one-sided spindle tree. From this permanent structure, temporary lateral fruiting shoots and spurs are formed and periodically renewed. This canopy architecture creates a

Fig. 7.11. The steep leader tree consisting of four upright leaders and a permanent bottom whorl at its base (courtesy of G.A. Lang).

broadly pyramidal shape that facilitates relatively good light distribution throughout the canopy (Fig. 7.11).

Due to the limited number of leaders, SL trees on full-size rootstocks can grow to heights of 5.5–6.1 m (18–20 ft) if not contained by relatively extensive annual pruning. Pruning at maturity consists of regular dormant renewal of a portion of all fruiting wood and limiting tree height, generally after harvest.

Fig. 7.12. The Spanish bush (SB) system produces a small tree that allows fruit to be harvested from the ground or small ladders (Spain) (courtesy of L.E. Long).

7.3.3 Spanish bush

The Spanish bush (SB) (also sometimes called 'Italian bush') training system is a variation of OVG consisting of usually a greater number of permanent leaders of relatively shorter stature, on which are developed numerous small lateral fruiting shoots that are renewed periodically. The SB and the similar Italian bush were developed in generally hot, arid Mediterranean locations having relatively poor soils. The multiple leaders and more bush-like canopy, compared to OVG trees, were a way to produce a semi-pedestrian orchard that could be harvested efficiently from the ground and small ladders (Fig. 7.12). Tree formation is easy, consisting of repeated heading cuts to multiply the potential leaders until an adequate number can be selected that are evenly spaced and relatively uniform. No tying, bending, clipping, scoring, breaking, etc., is required. Mature pruning also is relatively quick and simple, consisting of periodically renewing a portion of the horizontally-oriented fruiting wood on the permanent vertical scaffolds. The SB utilizes vigorous or semi-vigorous rootstocks, and the number of vertical leaders established should be proportional to tree vigor.

7.3.4 Kym Green bush

Just as the SB is a variation of the OVG, the Kym Green bush (KGB) is a variation of the SB. Initial orchard establishment is similar to the SB, with repeated dormant heading cuts over the first 2–3 years to form even more potential leaders than the SB. The most upright and uniform are selected after each round of heading cuts and those that are weak and non-upright are eliminated, while the most vigorous may be subjected to a second heading during the growing season to diffuse their vigor further. This is repeated until a target number (proportional to rootstock vigor and tree spacing) are achieved, and these are then allowed to extend, mature and fruit without any further heading. However, the leaders are temporary, unlike SB leaders, and lateral shoots are removed as they form, rather than allowed to fruit as for SB. Thus, KGB trees maintain an ever-renewing cycle of vertical columns of spur-bearing wood, which facilitates good light distribution between the many leaders (Fig. 7.13).

KGB was developed in a relatively vigorous growing environment with Lapins, a very upright-growing and spur-bearing cultivar, on a vigorous rootstock, Mazzard. Due to the diffusion of all this vigor by the many leaders, and the limber character of the leaders due to their lack of being headed, KGB tree training creates a fully pedestrian orchard, as the leaders can be pulled down for fruit harvest and for removal of lateral shoots without use of ladders or orchard platforms. The fruiting leaders are renewed as they become too stiff to pull down (usually after five to six years) which maintains relatively young populations of productive, high quality fruiting spurs.

Fig. 7.13. These Kym Green bush (KGB) trees on Gisela 6 rootstock have nearly 20 moderately vigorous leaders per tree (Moldova) (courtesy of L.E. Long).

Fig. 7.14. In this example of whole tree renewal, after several years of cropping, all branches of this Kym Green bush (KGB) tree were stubbed back to renew the fruiting wood of the entire canopy (courtesy of L.E. Long).

The KGB training system creates the only three-dimensional canopy architecture in which the permanent primary tree structure is limited to a very low trunk, and the many fruit-bearing secondary leaders grow entirely vertically, with the removal or severe reduction of any tertiary structural growth. The vertical orientation of the fruit-bearing secondary structures and the overall vigor of the system components facilitate successful renewal, and the leaders are replaced before each leader's terminal vigor becomes excessive and difficult to manage. Another challenge that is more common on vigor-limiting rootstocks is achieving adequate sprouting and regrowth of renewal leaders when they are in competition with heavy crop loads on mature trees. This can lead to a diminishing number of fruiting leaders over time and has led to some growers eventually transitioning their mature KGB trees to SB training. However, a recent solution to this situation is the imposition of whole tree renewal (WTR; see Chapter 6, 'Sweet Cherry Pruning Fundamentals') on a proportion of the orchard each year so that after a cycle of five to seven years, all trees have been renewed and those with the oldest leaders are subject to WTR again to begin the next cycle of renewal (Fig. 7.14; see also Figs. 6.8, 6.9, and 6.10). A secondary favorable outcome of WTR is the opportunity to reset the target number of leaders in the tree and their uniform development, as compared to KGB

Fig. 7.15. Sylleptic branches (red arrows) have arisen from one-year-old regrowth and will need to be removed on this renewed Kym Green bush (KGB) tree (Oregon) (courtesy of L.E. Long).

trees that undergo annual partial renewal, and therefore each canopy is a mixture of fruiting leaders at every stage of development and fruitfulness. A challenge of WTR is that the vigorous rate of regrowth that often occurs as a response to this method can lead to sylleptic (secondary) shoots even as the renewal (primary) leader is regrowing (Fig. 7.15). Monitoring for excessive regrowth vigor and imposing a second heading or a PGR to the most vigorous shoots may be necessary to optimize shoot development and spur formation.

Since KGB is a specialized training system tailored to higher vigor rootstocks and spur-bearing cultivars with upright growth, imposing KGB training on some cultivar/rootstock combinations can be problematic. It has been observed that Gisela 5 and 6 rootstocks tend to impart a more horizontal growth habit to many cultivars, and thus a significant number of the KGB leaders that develop on these rootstocks may lack sufficient verticality to remain upright, especially when laden with fruit. Some cultivars tend to readily produce lateral shoots, which reduces the desirable spur-fruiting sites on KGB leaders and increases labor costs for lateral shoot removal. Some, like Regina and Kordia, produce a significant proportion of fruit on basal non-spur flower buds on previous season lateral shoots, which are eliminated in KGB training. Consequently, KGB is not recommended for non-spur type cultivars or those that form a lot of lateral shoots.

7.4   Trellised Planar (2-Dimensional) Single Fruiting Walls

Trellised planar single fruiting wall canopy architectures are designed to be semi-pedestrian for training and harvest labor, as well as suitable for partial mechanization of some tasks with mobile platforms for labor and tractor-based hedging or thinning technologies. Trellises for planar canopy architectures certainly provide tree support, but primarily are used for tree and fruiting wood orientation and precision in development. If the orchard management plan is to do some of the pruning by hedging, the trellis facilitates precise placement of the fundamental canopy structure for most effective hedging. Planar canopy architectures with vertically-oriented fruiting wood may utilize as few as two wires (sometimes with intermediate twine or bamboo supports for alignment) or as many as five wires, depending on height. For example, SSA trees may only require a bottom wire for precise tree orientation at planting and a top wire for leader orientation at maturity, whereas UFO trees require a bottom wire for cordon orientation plus intervening wires (in lieu of twine) at distances above the cordon appropriate for catching and orienting the annual growth of the upright fruiting offshoots until they reach the top wire. Horizontally-oriented planar canopy architectures may utilize as many as seven wires, depending on trellis height, for orienting the lateral fruiting branches at precise distances from each other.

To reduce the potential for bacterial canker from wounds caused by the rubbing of tree tissue against the trellis wires, several options have been used. High tensile polypropylene plastic wires or plastic-coated steel wires have been shown to reduce the potential for bacterial canker infection from trellis wires. Some growers have used old drip tape tubing to slide over galvanized steel trellis wires where tree tissue may contact the wire. Similarly, rubber tubing used to tie plant tissue to steel (or plastic) wires can first be tied to the wire to create a cushion, then tied to the branch or shoot (Fig. 7.16). With angled dual plane trellises, upward growing shoots or branches can be tied down (or clipped when using plastic training clips) near to the wires such that they pull up and away from the wire. When bamboo is used to bridge wires, the plant tissue can be tied to the bamboo rather than a wire. Vertically-oriented shoots that need to be attached to wires should be tied on the leeward

Fig. 7.16. Plastic tubing is used to tie branches to trellis wires so that there is a barrier between plant tissue and wire (courtesy of G.A. Lang).

side, so that the prevailing winds blow them away from, rather than into, the trellis wires.

7.4.1 Palmette, tri-axe/trident and candelabra

The palmette (PLM), tri-axe/trident and candelabra canopy architectures are all variations of two-dimensional planar training systems that are created with at least three vertical leaders to diffuse vigor. Consequently, these are planted at moderate to high densities on semi-dwarfing to semi-vigorous rootstocks. The permanent primary tree structure is limited to a very low trunk which is split into three (tri-axe/trident, sometimes PLM) to five (candelabra, sometimes PLM) vertically-oriented permanent planar leaders, spaced apart at equal distance. The leaders of PLM and candelabra trees form a palm- or fan-shaped primary structure (Fig. 7.17). Establishment of the primary leaders is accomplished by heading a whip nursery tree so that each new leader begins growth from relatively similar points on the trunk.

For each of these training systems, fruit-bearing renewable tertiary shoot growth is developed on the permanent leaders (Fig. 7.18). The vertical orientation of the permanent structural leaders tends to diffuse terminal vigor in proportion to the number of leaders, and, together with good light distribution through the narrow canopy, this somewhat reduces the difficulty of managing upper canopy vigor and maintaining fruit-bearing capacity lower in the canopy. The arrangement of fruiting branches off the main structural leaders can be irregular (filling space as lateral shoots arise) or highly structured (with lateral shoots induced at specific points and/or tied to trellis wires for precise spacing).

The challenge with the three-leader canopy structures is to prevent the middle leader from becoming significantly more vigorous than the two leaders on either side of it. Similarly, the test with more than three leaders is to achieve a relatively uniform vigor among each. Horticultural strategies to achieve such uniformity include the selective use of plant growth inhibitors such as prohexadione-calcium (P-Ca) on the more vigorous leaders as they form, or repeated selective pinching of the growing points of the

Fig. 7.17. Palmette (PLM) tree development (Italy) (courtesy of G.A. Lang).

Fig. 7.18. These tri-axe trees consist of three upright permanent leaders and weaker, fruit-bearing laterals (Kazakhstan) (courtesy of L.E. Long).

more vigorous leaders. Alternatively, a single selective heading back of the more vigorous leader(s) to several basal buds, when they reach about 45 cm (18 in), will usually initiate two replacement shoots and thereby diffuse individual shoot vigor that will be more in balance with the other non-headed leaders. At the end of the growing season, the extra leader of each replacement pair can be removed entirely, retaining the one that is most in balance with the other non-headed leaders. Consequently, significant attention must be devoted to creating a balanced permanent structure during the first 1–3 years, to most effectively establish efficient use of space and uniformity that will simplify future canopy maintenance and renewal decisions.

7.4.2 Drapeau marchand

For drapeau marchand canopy training, the nursery tree is planted at a 45–60° angle in the tree row, and the structural framework is developed

Fig. 7.19. Vertical, single plane espalier (ESP) trellised cherries (New Zealand) (courtesy of G.A. Lang).

by three to five secondary leaders, arising from the top side of the primary leader at 90°. This angled primary and secondary tree structure helps reduce vigor, since no vertical structural development is allowed. Thus, the primary leader is inclined to the north or south, while the secondary leaders are inclined in the opposite direction (south or north). The angled growth must be attached to trellis wires to maintain the angles, otherwise new extension growth will tend to revert to vertical. For this reason, trellis posts cannot be spaced too far apart, otherwise the trees furthest from each post will pull the trellis wire upwards, changing the angle relative to trees nearer the post. Weaker tertiary, renewable fruiting shoots are developed on the main angled structures. This fruiting wood can be trained in the trellis row to maintain a very narrow planar canopy, or trained out into the tractor alleys to make a somewhat wider canopy.

7.4.3 Espalier

The espalier (ESP) is a narrow planar canopy architecture in which the permanent primary tree structure is limited to a central leader with ten or more horizontally-oriented permanent scaffold branches tied to trellis wires, from which fruit-bearing renewable tertiary growth arises (Fig. 7.19). The vertical orientation of the permanent single leader can result eventually in too much vigor at the top of the leader, though the many horizontal branches help partition the vigor to the fruit-bearing tertiary fruitwood renewal and replacement.

Successful establishment of ESP trees requires good promotion of lateral shoots each year on the central leader until the mature tree height is reached. Lateral shoot formation is promoted by girdling, application of PGRs such as Promalin®, or bud selection/removal to initiate laterals precisely near their trellis wire and eliminate extraneous potential laterals between wires or oriented perpendicular to the trellis (out into the tractor alley). Heading of the leader is not used to promote laterals, since although this can precisely initiate laterals at desired heights, it usually results in an insufficient number of laterals that grow too vigorously.

The main challenges with ESP training include: (i) the formation of lateral horizontal scaffold branches to fill each wire; (ii) the eventual renewal of any horizontal scaffold branches if productive fruiting sites become too sparse; and (iii) the prevention of new vertical shoots from arising on the horizontal scaffold branches. The latter is best accomplished by the use of plant growth inhibitors (where allowed) such as paclobutrazol or P-Ca.

7.4.4 Super slender axe

The super slender axe (SSA) is a very high density orchard system of semi-free-standing single leader trees at up to 5000 trees per hectare (2000 trees per acre), usually requiring dwarfing rootstocks and a two-wire trellis. The permanent primary tree structure is limited to the central leader with short spreading renewable secondary scaffolds that bear fruit (Fig. 7.20). The production habit of SSA trees is significantly different from other canopy training systems, with the majority of the fruit grown on solitary flower buds produced at the base of one-year-old branches that are pruned severely at the beginning of the fruiting season (Fig. 7.21). This limits crop load and results in a generally favorable LA:F ratio, resulting in low yields per tree of high quality fruit. Consequently, high tree densities are needed to achieve adequate yields. However, cultivars differ in their inherent traits for basal bud formation, so productivity can vary significantly by cultivar. Research in the Emilia-Romagna region of Italy found the best SSA yields with Grace Star, Ferrovia, Sylvia and Giorgia; yields were fairly poor with Regina, Summit, Kordia (Attika), Sweet Early, Early Star, Early Bigi and Black Star (Musacchi *et al.*, 2015). The most important characteristics for varieties suitable for SSA training are the capacity to produce fruit on basal buds of one-year-old shoots and the ease with which lateral shoots can be initiated.

Mature pruning consists of 100% annual renewal, such that every lateral branch is pruned severely ('short pruning'), thereby requiring simplified but extensive annual labor. The majority of each lateral shoot is removed, retaining only the basal solitary (non-spur) flower buds plus one

Fig. 7.20. Young super slender axe (SSA) trees with fruiting sites at the base of each lateral branch (Chile) (courtesy of L.E. Long).

to three vegetative buds for regrowth. This short pruning is most easily done during budswell to differentiate between basal floral and vegetative buds. Since the solitary flower buds will fruit once and then become blind nodes, as each fruiting lateral shoot is subsequently short pruned to regrow and fruit, it creates small segments of blind wood with the basal fruiting sites extending further and further away from the leader. The first year that each new lateral shoot arising directly from the central leader is short pruned, four to five vegetative buds should be retained, as the lower buds will become spurs while one to three of the distal buds will become the new

Fig. 7.21. Super slender axe (SSA) limb after pruning (courtesy of G.A. Lang).

shoots for fruiting the next year. The presence of several spurs retained close to the leader provides potential sites for renewing the entire lateral after it is has fruited for 4–5 years and the additive blind wood segments extend too far from the central leader.

Successful establishment of SSA trees requires good promotion of lateral shoots each year until the mature tree height is reached. Lateral shoot formation is promoted by girdling and/or application of PGRs such as Promalin®. The more laterals that are induced on previous season leader growth, the more moderate will be their vigor, which tends to result in good subsequent fruitfulness. Laterals are never induced with bud selection/removal, which results in the loss of too many future growing points, or with heading, which usually results in an insufficient number of laterals that grow too vigorously.

As the number of fruiting laterals increases (i.e. when two vegetative buds are left annually, each previous lateral shoot subsequently divides into two new lateral shoots), individual shoots that are poorly positioned (i.e. pendent, vertical or may cause shading) may be removed entirely with thinning cuts. More of these secondary and tertiary laterals may be left in the lower third of the central leader canopy, while more thinning cuts and fewer vegetative buds should be left in the upper third of the canopy to reduce its vigor and potential shading.

At such high densities, root competition contributes to significant vigor control. However, even with root competition, dwarfing rootstocks are critical for the success of SSA production. The vertical orientation of the single permanent structural leader means that management of vigor in the upper part of the canopy can eventually become a major challenge. Excessive upper canopy vigor occurs at the expense of the fruit-bearing secondary fruitwood renewal and replacement, with the best fruiting zones moving upward in the canopy over time. In a fertile soil in Michigan, Benton trees on dwarfing Gisela 3 yielded well but still added 1–1.5 m of annual terminal leader growth even at 10 years old; trees on semi-vigorous Gisela 6 performed very poorly with extremely low yields due to the excessive vigor of most laterals and as much as 2 m or more of annual terminal leader growth at 10 years old (G. Lang, personal communication). Root pruning or plant growth inhibitors (where allowed) may be required to correct excessive vigor. Hedging in late summer can help reduce vigor somewhat and improve light distribution to the developing basal flower buds that otherwise may become too shaded. Planting of an SSA orchard on more vigorous rootstocks should only be attempted if the orchard soils are very low in fertility and water holding capacity.

The high initial tree cost for SSA orchards can be reduced by planting dual leader (bi-axe) nursery trees or by low heading of a single whip nursery tree at planting to develop dual leaders in parallel down the tree row. This allows the orchard to be planted at half the tree density as with single leader trees, but the wider tree spacing thereby creates less root competition, losing some of the vigor control that occurs through closer spacing.

7.4.5   Upright fruiting offshoots

The most quantifiable and precisely structured sweet cherry canopy architectures are the planar fruiting walls created by ESP and upright fruiting offshoot (UFO) training. Whereas ESP canopies are comprised of horizontally-oriented, primarily spur-based fruiting wood borne on a single permanent vertical leader, UFO canopies are comprised of vertically-oriented, primarily spur-based fruiting wood borne on multiple leaders ('upright fruiting offshoots') that arise from a single or a dual permanent horizontal cordon. While the general profiles of these two narrow planar systems are similar, there are a number of implications of the multiple leader vertical orientation of the UFO canopy structure compared to the single leader horizontal canopy structure of the ESP. First, the multiple leaders can be a significant advantage for diffusing vigor and reducing the tendency for upper canopy vigor to become more difficult to manage over time. Consequently, UFO training can utilize a wide range of rootstocks simply by increasing or decreasing the number of leaders in proportion to the vigor of the rootstock (and site and growing climate). Second, vertical growth is the most natural habit for most sweet cherry cultivars, so that the development of canopy structure to efficiently fill (and renew) vertical fruiting space tends to be more easily accomplished than filling, and especially renewing, horizontal fruiting space (particularly maintaining and renewing fruiting space lower in the tree canopy/fruiting wall).

UFO training creates a fully trellised canopy architecture that optimizes labor efficiency and fruit quality by creating a narrow fruiting wall that is

Fig. 7.22. Vertical, single plane upright fruiting offshoot (UFO) orchard formed by planting the whip at a 45° angle (Washington) (courtesy of G.A. Lang).

precocious and easy to harvest and prune. Like KGB training, it produces fruit on renewable vertical fruiting units which arise from a low horizontal trunk, similar to a grape cordon. The advantages of this semi-pedestrian system include: (i) high early and mature yields; (ii) easy and labor efficient training and harvest; (iii) potential mechanization of some tasks like flower thinning and summer hedging; (iv) good uniformity of light distribution (and minimization of shade) to promote uniform, high quality fruit throughout the canopy; and (v) sufficient air movement to minimize disease incidence and achieve good canopy coverage with spray applications of pesticides, foliar nutrients or protectants, and/or PGRs. Although the canopy is relatively easy to prune and maintain at maturity, initial establishment of the trellised upright fruiting leaders is more intensive, time-consuming, and costly than some of the other canopy architectures like KGB.

UFO trees generally consist of six or more leaders, or upright fruiting offshoots, that arise from one or two horizontal cordons. Single cordon UFO trees are best developed by planting the nursery tree at a 45° angle and tying it to the bottom trellis wire to maintain the initial angle (Fig. 7.22). After vertical shoots begin developing following budbreak, the portion of the leader above the bottom wire is bent horizontally and tied down to the wire, maintaining a slight angle upwards towards the apex (avoiding a hump at the bend that would be higher than the terminal of the leader). The advantages of a single cordon UFO include earlier production, since the nursery tree serves as the cordon and development of the upright fruiting offshoots can begin immediately, and where the tree rows are established going up and down slopes, the single cordon can always be

Fig. 7.23. Vertical, single plane upright fruiting offshoots (UFO) formed by heading the whip after planting (courtesy of G.A. Lang).

oriented uphill. In windy sites, the cordon can be oriented so that the fruiting offshoots develop into the wind. In hot sunny locations, the cordon can be oriented towards the sun (towards the south in the northern hemisphere or towards the north in the southern hemisphere), to better shade the cordon from potential sunburn. The disadvantages of a single cordon UFO include potentially greater difficulty in initiating a full complement of offshoots in the first year due to nursery tree transplant shock, and potentially weak vegetative growth as the root system becomes established. Also, it can be more difficult to develop uniformity of the offshoots due to the variability in offshoot formation in the establishment year and the gradient of stronger to weaker vigor from the most basal offshoots (closest to the rootstock) to the most terminal offshoots (at the leader apex).

Dual cordon UFO trees are developed by planting dual leader (bi-axe) nursery trees if available, or planting a normal whip single leader nursery tree and heading the tree above three good lateral buds located below the first trellis wire. An advantage of either type of dual cordon UFO tree is the ability to plant the nursery trees vertically as normal (rather than at a 45° angle) (Fig. 7.23). Planting a dual leader nursery tree achieves the same earlier production advantage as a single cordon UFO, with the added advantage of symmetry in development due to a more balanced gradient in vigor, since the distance from rootstock to terminal is half that of a single cordon UFO. However, planting dual leader nursery trees has the same potential disadvantages regarding transplant shock and less uniform early development of the offshoots. Planting and heading a single leader nursery tree to develop the leaders that will become cordons at the end of the first growing season has two advantages. The first is initiating three leaders from which the two most uniform can be selected (with removal of the third once the two most uniform are identified). The second advantage is the delaying of upright offshoot development until the second year, by which time the root system has become well established and the initiation of upright offshoots tends to be more extensive and vigorous, often in excess of the number required, so that the most uniform complement can be selected, with excessively weaker or stronger ones removed. The disadvantage of developing a dual cordon UFO from a single leader nursery tree is a one-year delay in fruiting on upright offshoots, albeit with potentially greater uniformity for future years. A disadvantage of dual cordon UFO trees developed from either type of nursery tree is in orchards with significant slopes where the rows are oriented up and down hill, rather than terraced with the hill, since one of the tree's dual cordons will be oriented uphill and the other oriented downhill.

Uniformity in offshoot development should be prioritized during the first 2 years of establishment by monitoring vigor of the developing upright offshoots and intervening early where vigor appears to be excessive. For example, when the most vigorous new offshoots are 40–45 cm (16–18 in) long (preferably before the summer equinox), if the variation is great between the largest and the smallest, the largest can be head pruned back to about 10–12 cm (4–5 in) and several low buds, while the less vigorous are left alone. This will usually promote two new shoots to elongate from the heading cut, thereby partitioning the inherent high vigor of the original offshoot into two less vigorous replacements that will end the season more in balance with the remaining unpruned offshoots. At the end of the summer, if both regrown offshoots are needed to fill the allotted space, both are retained; if only one is needed, the other should be removed entirely with a thinning cut. In the case of single cordon UFO trees, if the most basal offshoot is significantly more vigorous than the others, it can be pruned back to a few low buds with a heading cut as described above, to diffuse its vigor into multiple replacement upright offshoots. Conversely, if the cordon from the adjacent tree does not completely fill the linear space of the trellis, the vigorous basal offshoot from one tree can be bent back at the end of the first growing season to form a second cordon, and upright offshoots can be developed on it during the second growing season.

To optimize planar canopy light interception and fruiting capacity, precision positioning of the uprights can be achieved by placing a clip every 20 cm apart for each upright on the second wire, allowing it to be moved into the defined position regardless of where it originated on the cordon. Alternatively, twine can be used for precise training to continuously bridge the gap between the top and bottom wires (intersecting with at least one intermediate trellis wire for stability), which makes upright shoot orientation quicker and cheaper than clipping, by periodically intertwining each

Fig. 7.24. Precision spaced twine for quick winding and orienting of upright fruiting offshoots (UFO) (twine should be tied where it crosses the middle wire, to prevent shifting) (courtesy of G.A. Lang).

shoot with the twine as its growth extends (Fig. 7.24). The twine should be tied to the top wire at 20 cm intervals, wrapped twice (once on each side of the twine to prevent it from sliding) around the intermediate wire, and then to the bottom wire (Fig. 7.25). This also allows fewer trellis wires to be used. Although this is a greater investment in labor in the first year, it reduces future training labor and cost of clips during the filling of vertical canopy space, and makes upright removal during planned renewal quicker since twine can be cut and replaced rather than unclipping from each wire.

Finally, if less than 50% of the upright offshoots have formed during the year allotted for their formation (the first year for single cordon UFO trees and dual cordon UFO trees formed by dual leader nursery trees, or the second year for dual cordon UFO trees formed by heading the single leader nursery tree in the first year), it is often more advantageous in the long term to remove all of the offshoots that have formed with a heading cut back to two basal vegetative buds. These offshoots are allowed to regrow

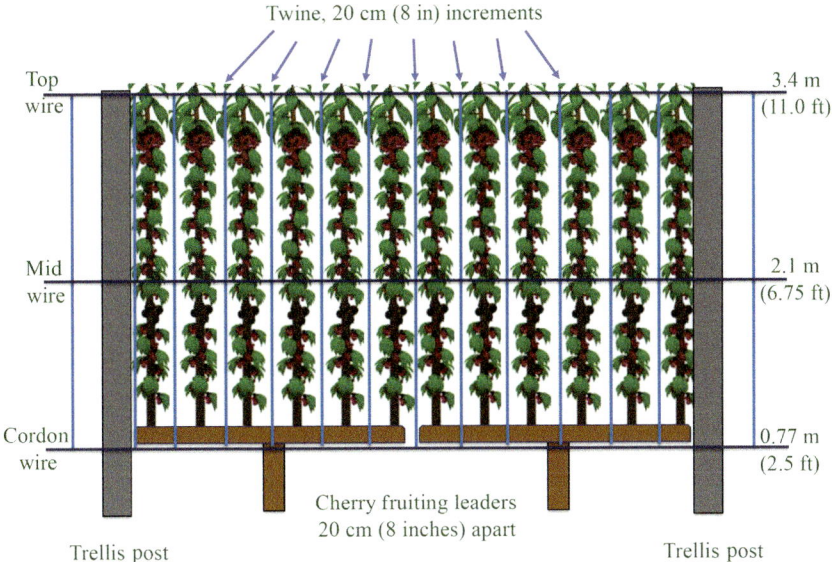

Fig. 7.25. Single plane upright fruiting offshoots (UFO) spacing diagram of trellis wire and twine dimensions (courtesy of G.A. Lang).

the following season in parallel with the other new offshoots to fill the missing spaces, resulting in a fuller complement and greater uniformity as all offshoots would be the same age at the end of that year. Although this may delay initial fruiting for 1 year, the greater uniformity will improve tree management and cropping consistency for years to come. Promotion of buds to elongate into new offshoots can be done as with any of the methods used in central leader systems to promote lateral shoot formation: scoring, bud selection and removal, or use of high levels of PGRs (e.g. Promalin®) at green tip or lower levels (e.g. 600 ppm) at initial leaf emergence.

Since sweet cherry cultivars differ in their inherent propensity to produce fruit on spurs (desirable for UFO training) and form lateral shoots (undesirable for UFO training), suitability to maintain the upright fruiting offshoots with minimal lateral branches can vary significantly by cultivar. When annual lateral shoot formation is greater than desired, the laterals may be removed entirely if the cultivar is highly productive (that is, adequate spurs remain for good yields), or they may be removed with SSA-type short pruning to leave only basal flower buds plus one vegetative bud, to increase the yield potential. The greater the lateral branching, the greater the need for summer hedging to maintain a favorable uniform light environment at the spurs on the upright offshoots, which may or may not be followed up with dormant hand pruning to either remove them entirely or short prune them.

Traditional renewal of the upright fruiting offshoots is similar to that for KGB vertical leader renewal – annual renewal of the largest one to

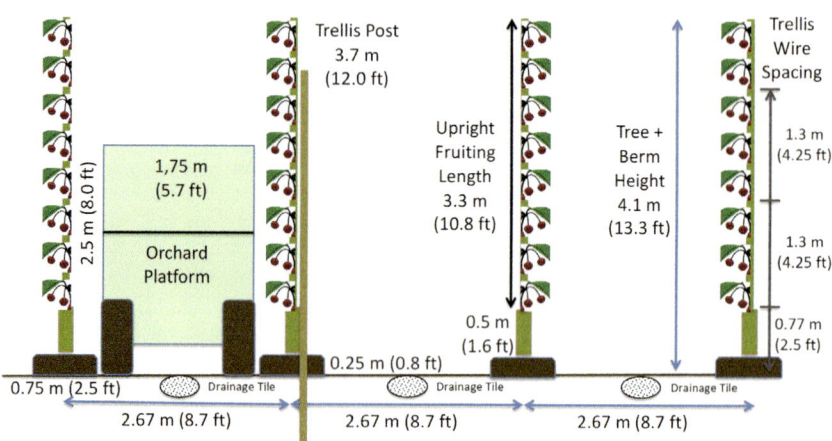

Fig. 7.26. Single plane upright fruiting offshoots (UFO) spacing diagram of row and trellis dimensions (courtesy of G.A. Lang).

three offshoots per tree (or ~12–15% of the total number) by cutting them back to a basal vegetative bud to be regrown, or a~10–15 cm (~4–6 in) stub to promote activation of one or more latent buds to generate a new shoot. This annual percentage of renewal assures that no upright fruiting offshoot is older than 6–8 years, maintaining a relatively young canopy regardless of cordon age. The natural vertical growth orientation of the fruit-bearing structures facilitate their easier renewal compared to horizontal fruiting structures in spindle and ESP canopies, and uprights are replaced before the upper vigor of each leader becomes a major challenge (upon reaching their mature height, terminals are cut back to a weak lateral replacement until the entire upright is renewed). Since the UFO canopy is so narrow, with some light diffusing through between uprights, fruiting upright height above the cordon can be extended as much as ~20–25% higher than the row width (Fig. 7.26).

UFO-trained tree canopies also are conducive to WTR, such that every upright fruiting offshoot on every tree in one entire row can be cut back to stubs just above the cordon, promoting uniform renewal of the entire row all at once (Fig. 7.27). This can be done to 12–15% of the rows in the orchard, resulting in no row with structural fruiting wood older than 6–8 years, and the ability to manage renewal by entire rows so that non-bearing rows can be skipped for some fruit-related spray applications such as pre-harvest gibberellic acid or brown rot protection. The same attention to promoting uniform vigor of WTR uprights applies as when the trees are first developing their initial uprights. The commitment to row-by-row WTR should begin as early as the second or third year, once the full complement of uprights has been achieved across the orchard, so that from that point onward, there is always 12–15% in the first stage of renewal, and therefore

Fig. 7.27. Whole tree renewal (WTR) of upright fruiting offshoot (UFO) trees (Washington) (courtesy of G.A. Lang).

the final rows to be renewed for the first time are no older than 8 or 9 years. It is also possible to partially mechanize WTR, using a tractor-mounted circular saw to stub back each upright in the row.

In conclusion, the UFO canopy architecture simplifies training, pruning and crop load management, with the potential for implementing partial mechanization of some of these tasks. The extremely narrow canopy structure optimizes light uniformity throughout the fruiting wood while minimizing shade. The upright orientation of the fruiting offshoots utilizes sweet cherry's natural growth habit, and vigor can be managed across a wide range of rootstocks by varying the number of offshoots in proportion to the vigor of the rootstock, soil conditions and growing season climate.

7.5 Trellised Planar (2-Dimensional) Dual Fruiting Walls

Dual angled plane, two-dimensional trellised fruiting walls are referred to as 'Y'-canopies when each tree has alternating portions of its canopy trained into each angled plane (Y = one trunk, two angled planes of fruiting structure) (Fig. 7.28). They are referred to as 'V'-canopies when separate trees are alternated to form each angled plane (V = two trunks, two planes of fruiting structure) (Fig. 7.29). V-canopy dual angled fruiting walls can be developed by planting trees in a single row, alternating their incline east and west, or in a narrow double row of offset trees, with one row inclined east and the other inclined west. Although all orientations of Y- and

V-planar canopies facilitate harvest from the tractor alley, the narrow double row allows greater access by labor to the middle of the V-canopy for
pruning.

Although Y- and V-canopies require more complex trellis structures
for tree structural training and support, they also facilitate greater light
interception per orchard area (and therefore greater yield potential) compared to vertical single plane fruiting walls. Thus, the achievement of higher yields is somewhat at the expense of slightly reduced labor efficiency for
picking and training (angled fruiting walls take more time to harvest from
compared to vertical fruiting walls) and slightly less ease/efficiency for partial mechanization by hedging and thinning.

Almost any trellised planar single fruiting wall training system can be
adapted to a dual angled planar fruiting wall canopy architecture. Perhaps
the first significant adoption of such canopy architectures for tree fruit
production was by Bas van den Ende at the Tatura Research Centre in
Australia, resulting in the Tatura trellis. As adopted for sweet cherry production, Tatura trellis cherries usually create the Y- or V-canopy at a 60–72°
angle (relative to the ground) with a permanent single (V) or dual (Y)
leader tree having horizontal lateral fruiting branches like an ESP. Annual
pruning and management is similar to that for single vertical fruiting wall
ESP trees. Similarly, Y- and V-dual angled plane canopies can be developed
under PLM-type training utilizing three to five permanent leaders; with
UFO-type training utilizing cordons and a variable number of vertical

Fig. 7.28. Y-trellis espalier (ESP) or Tatura trellis (Washington) (courtesy of G.A.
Lang).

Fig. 7.29. V-trellis, dual planar upright fruiting offshoot (UFO) orchard (courtesy of G.A. Lang).

renewable upright fruiting offshoots; or with SSA-type training utilizing a single (V) or a bi-axe (Y or V, depending on orientation of the dual leaders) tree.

With the extremely narrow planar UFO canopy training that results in some light diffusion through the narrow wall (canopy light porosity), a further potential exploration of dual fruiting wall training systems is the use of double UFO (U2FO) rows oriented vertically rather than at an angle (Fig. 7.30). That is, dual tree rows spaced ~1 m (3 ft) apart alternating with tractor alleys of 3.1–3.5 m (10–11.3 ft) provide increased light interception and fruit-bearing wood, yet retain the vertical orientation that best accommodates efficient

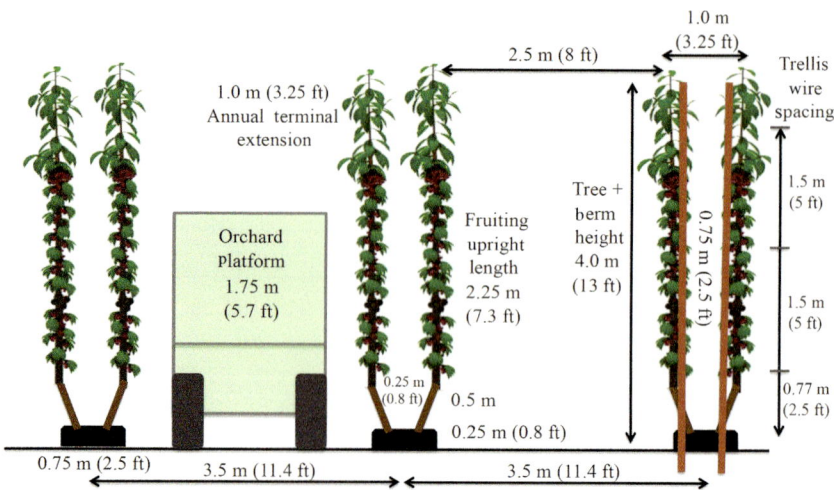

Fig. 7.30. Dual parallel plane upright fruiting offshoot (U2FO) spacing diagram of double row and trellis dimensions (courtesy of G.A. Lang).

harvest and partial mechanization. The U2FO double tree row spacing alternating with the wider tractor alley spacing accommodates access by labor and over-the-top hedging implements between the tree rows, as well as standard narrow tractors and spray equipment in the tractor alleys, and the light porosity of the precisely spaced U2FO uprights increases both light and sprayed materials into the interior of the double row. It remains to be determined whether the traditional 20 cm (8 in) upright spacing should be increased to perhaps 25 cm (10 in) to increase light porosity in double rows. Certainly, such double rows also accommodate partially mechanized WTR strategies more easily than a V- or Y-angled dual fruiting wall.

7.6   Matching Training Systems with Rootstocks and Cultivars

When selecting a training system, it is always important to consider the rootstock and cultivar. Free-standing multiple leader systems, such as the KGB, must be grown with a rootstock that is strong enough to provide adequate vigor to all of the leaders. This means that vigorous and semi-vigorous rootstocks are best for this canopy architecture. However, it must also be remembered that vigorous rootstocks tend to be less precocious than semi-vigorous or semi-dwarfing rootstocks. Most higher density training systems rely on precocious cropping to help mitigate tree vigor. When cropping is delayed until the fifth leaf, excessive vigor can delay cropping even more. Training systems with fewer multiple leaders, such as the free-standing SL or the trellised PLM, are suitable for semi-vigorous rootstocks.

The trellis support of the multiple leader UFO allows it to be adapted to a wide range of rootstock vigors, with tree spacing and upright fruiting off-shoot number being varied in proportion to rootstock vigor.

Conversely, rootstock selection is just as important with single leader trees such as the VCL, SL, TSA, TTP and SSA. Vigorous rootstocks will promote very tall single leader trees or trees with excessive upper canopy growth at the expense of lower fruitwood development and increasing shade. Usually, semi-dwarfing to dwarfing rootstocks are recommended for VCL, TSA and TTP trees, depending on the site vigor factors (soil fertility plus climate). Due to the very close in-row spacing of SSA trees, dwarfing rootstocks are critical for successful production unless soil conditions are very poor. Also, some rootstocks promote more upright growth (such as Mazzard, Colt and Gisela 12) and some rootstocks promote more open branching (such as Gisela 3, 5 and 6). The latter rootstock traits are beneficial for single leader trees that are developed with many lateral branches for fruiting, but are problematic for free-standing KGB trees in which the multiple leaders should have minimal laterals and grow upright with vertical rather than spreading angles.

It is also important to take the fruiting pattern of cultivars into consideration when choosing a training system. Some cultivars, such as Regina and Kordia (aka Attika) are considered non-spur type bearers since they produce fewer spurs and crop a significant amount of fruit at the base of one-year-old lateral shoots. Lateral shoots are eliminated in both the KGB and UFO trees, so these training systems are more problematic for non-spur type cultivars. The majority of the SSA crop is produced at the base of one-year-old shoots, so non-spur type cultivars are fine unless they produce too many nodes with basal fruit. This creates longer sections of blind wood (no leaves or fruit) that extend the fruiting area of the canopy away from the leader too much, losing some of the advantages of being a planar fruiting canopy. Most cultivars bred in North America are spur type, but several European cultivars are non-spur types.

References

Long, L., Lang, G., Musacchi, S. and Whiting, M. (2015) Cherry training systems. Pacific Northwest extension (PNW) publication 667. Available at: https://catalog.extension.oregonstate.edu/pnw667 (accessed 19 June 2020).

Musacchi, S., Gagliardi, F. and Serra, S. (2015) New training systems for high-density planting of sweet cherry. *HortScience* 50(1), 59–67.

Managing the Orchard Environment

8

Managing the routine aspects of the orchard environment, such as irrigation, fertility and weed competition, are key factors for successful sweet cherry production. Additionally, climate-related environmental factors can significantly influence cherry cropping (see Chapter 4, 'Planning a New Cherry Orchard'). Sweet cherries are not grown as widely as apples or peaches largely because of climatic limitations, such as low temperature injury to buds and other plant tissues in winter, low temperature injury to blossoms and young fruits in spring, rain-induced fruit cracking, the effects of excessive heat, rain-disseminated diseases, insufficient winter chilling to break endodormancy, etc. Growers can manage some orchard environmental factors more readily than others, but all must be optimized as much as possible for sustainable sweet cherry production.

8.1 Soil Moisture Monitoring

As soon as possible after planting, some type of soil moisture monitoring system should be installed. For greatest accuracy, in newly planted trees, the monitoring instrument should be placed in the rooting zone of the tree, e.g. the top 45 cm (18 in) of the soil. If drip irrigation is to be used, the drip emitter for a newly planted tree needs to be located in this area as well. In a mature orchard, the monitoring probes can be moved away from the tree but still should be in the root zone wetted by the drip emitter, or between trees in the row in a sprinkler-irrigated orchard. Tree roots, however, can extend to 60–95 cm (24–36 in) deep, so some monitoring should be done at these levels as well. By monitoring at 30, 60 and 90 cm (12, 24 and 36 in), the soil moisture profile will reveal how quickly the surface zone will become dry and need replenishment, but will also monitor moisture in the lower zones to ensure that it does not become saturated or depleted. For example, such monitoring may help to schedule frequent short duration irrigations to replenish the zone of most active water uptake, but

Table 8.1. Determining soil texture by hand (modified from Thien, 1979).

	Forms a weak ribbon <2.5 cm (1 in)	Forms a ribbon 2.5–5 cm (1–2 in)	Forms a ribbon >5 cm (2 in)
Ribbon feels gritty	Sandy loam	Sandy clay loam	Sandy clay
Ribbon feels equally gritty and smooth	Loam	Clay loam	Clay
Ribbon feels smooth	Silt loam	Silty clay loam	Silty clay

will occasionally identify when a longer duration irrigation is needed to replenish moisture in the lower depths.

To properly monitor soil moisture, it is first necessary to determine the soil texture in the orchard, which may vary within the orchard and at different soil depths (see Chapter 4 for guidance). A simple way to determine soil texture is to moisten the soil and compress it between your thumb and forefinger, forming a ribbon (see Table 8.1).

The definition of several terms for interpreting soil moisture is useful. 'Field capacity' is the amount of soil moisture or water held in the soil after excess water has drained away, and the rate of downward water movement has become minimal. This usually takes place 2–3 days after rain or irrigation in soils of uniform structure and texture. 'Available water' is the range of water that can be stored in soil and is available for growing crops. 'Allowable depletion' is the proportion of available water that can be used before irrigation is needed. For cherries, the allowable depletion is about 50% of the total amount of water available in the soil. Clay soils retain the most water and sandy soils the least (Table 8.2).

If the effective rooting zone of a cherry tree on dwarfing rootstock is 90 cm (36 in), the soil texture in each rooting subzone, 0–30 cm (0–12 in),

Table 8.2. Water holding capacity of soil based on texture (National Center for Appropriate Technology, 2009).

Soil texture	Available water capacity (mm/cm)	Available water capacity (in/ft)
Coarse sand	0.2–0.7	0.2–0.8
Fine sand	0.6–0.8	0.7–1.0
Loamy sand	0.7–1.1	0.8–1.3
Sandy loam	0.9–1.3	1.1–1.6
Fine sandy loam	1.0–1.7	1.2–2.0
Silt loam	1.5–2.1	1.8–2.5
Silty clay loam	1.3–1.6	1.6–1.9
Silty clay	1.25–1.7	1.5–2.0
Clay	1.1–1.5	1.3–1.8

30–60 cm (12–24 in) and 60–90 cm (24–36 in), must be determined. Use Table 8.2 to find the water holding capacity of each soil layer in the rooting zone. For example, if the top 30 cm (12 in) of soil is determined to be a silt loam, then the available water holding capacity of that soil layer is 1.5–2.1 mm/cm (1.8–2.5 in/ft). If the next soil layer is a silty clay loam, then the water holding capacity at that depth is 1.3–1.6 mm/cm (1.6–1.9 in/ft). By adding the water holding capacity of each layer, it is possible to determine the water holding capacity of the total rooting zone. Knowing the water holding capacity of the soil will help to prevent overwatering and leaching of nutrients.

There are several ways to make relatively accurate measurements of soil moisture, though none are perfect. The system that is right for a grower will depend on orchard size, soil type, resources and management strategies.

8.1.1 Dielectric sensors

Dielectric soil moisture sensors measure the dielectric constant of the soil to determine its moisture content. Small changes in soil moisture cause a significant change in the dielectric constant. The most common dielectric devices are capacitance sensors or frequency domain reflectometry (FDR) and time domain reflectometry (TDR) sensors. FDR devices consist of two electrodes separated by a dielectric material that does not conduct electricity. Once installed in the soil, a frequency is applied to the electrodes. Since the moisture in wet soil resonates at a lower frequency than drier soil, it is possible to determine soil moisture content by reading the resonant frequency. A calibration equation is then used to determine soil moisture content.

Some FDR instruments, such as the EnviroSCAN, can measure continuously and provide trends over time that can be quite useful. However, there are some limitations to this method, as the accuracy of these devices can be affected by soil type, salinity and temperature. In addition, they can be very expensive.

TDR devices consist of two or three parallel steel rods, known as 'waveguards', that are inserted into the soil. An electrical pulse is sent through the rods. Since moist soil reflects the signal at a different speed than dry soil, the time that it takes to reflect the signal back indicates the dielectric constant of the soil. A calibration equation is then used to determine soil moisture content. TDR devices can be quite accurate and are not sensitive to temperature or soil type. However, they can be complicated to use and the signal analysis equipment is expensive.

8.1.2 Granular matrix sensors

One of the simplest methods to measure soil moisture is by electrical resistance blocks, such as gypsum blocks or their improved version, the granular matrix sensor (GMS) (Fig. 8.1). GMS sensors are blocks filled with a

Fig. 8.1. Granular matrix sensors (GMS) are easy to install and give a fairly accurate reading of the soil moisture levels in the immediate vicinity of the probe (courtesy of L.E. Long).

granular material above and below a gypsum wafer. Gypsum is a reasonable conductor of electricity, as is water.

Installing GMS sensors is a simple process. Prior to installation, each sensor is soaked for several minutes in water. A soil probe is used to make a hole in the soil that should be the exact diameter of the sensor. The first sensor should be installed at 20–30 cm (8–12 in), the second around 60 cm (24 in) and the third around 90 cm (36 in). The active region of the sensor is 2 cm (0.8 in) above the tip, so this additional span should be added to the probe hole to obtain the correct depth. Accurate readings depend upon good contact of the sensor with the soil, so expect a tight fit. A dowel will help to insert the sensor into the hole. For coarse soils that will hold their shape, pour a small amount of water (e.g. 60–90 ml or 2–3 oz) into the hole before inserting the sensor. For silt soils that can easily lose their shape, insert the sensor into the hole first and then add the water. Soil should be backfilled in the hole, making sure that no air pockets exist.

When the soil is moist, the sensor fills with water and resistance to an applied electrical current is low. A meter is attached to two leads connected to the block to measure the electrical resistance of the soil. A value near zero indicates a wet soil. As the soil dries, resistance increases and the readings rise. This is a very inexpensive way to measure soil moisture levels,

Fig. 8.2. Neutron probes are extremely accurate, but require a special license to operate (courtesy of L.E. Long).

providing useful soil water potential trends that are directly related to plant water needs. However, accuracy is less precise than other instruments, the measurements reflect a very small localized area, and results are affected by temperature and salinity. Resistance sensors are less effective in very coarse and sandy soils, and they will overestimate soil moisture in saline soils.

8.1.3   Neutron probes

The neutron probe is one of the most accurate devices available to measure soil moisture (Fig. 8.2). Access tubes are inserted into the ground and remain present throughout the season. The probe, which is portable, is then lowered into the ground where the instrument is activated and emits neutrons that interact with soil moisture. The density of the neutron flux varies depending on the amount of water in the soil. The density of the neutron flux is monitored by a gauge on the instrument. Neutron probes are not influenced by temperature and are affected only slightly by salinity.

Fig. 8.3. These tensiometers are reading soil moisture at two different depths (courtesy of C. Kaiser).

In addition, these instruments sample a greater volume of soil at several depths compared to the very limited soil volume monitored by other methods.

Due to the radioactive nature of the source material in the probe, a license needs to be obtained to use this instrument. In addition, a film badge, similar to those worn by X-ray technicians, must be worn by the operator to detect any radioactive exposure. This and the fact that the instrument is expensive tends to limit its use to professional irrigation consultants.

8.1.4 Tensiometers

The tensiometer consists of a vacuum gauge connected by a glass tube to a porous ceramic tip. The tube is filled with water which saturates the ceramic tip. The tube is inserted into the soil in the root zone. In a non-saturated soil, water will flow out of the tip creating a vacuum which is read by the gauge (Fig. 8.3). Unlike GMS sensors, tensiometers detect soil moisture tension instead of the electrical resistance of the soil. As the soil dries, the suction increases and more water is pulled from the instrument causing a higher reading on the gauge. In a saturated soil, the pull of water from the tensiometer will be weak, mirroring the ease by which roots are able to extract moisture from the soil. Thus, this soil water tension measurement is directly related to the force necessary for plant roots to extract soil water. Although it is more expensive than GMS, it is very commonly used and it is not affected by salinity. However, it reflects only a very small sample area of the root zone, can lose accuracy if not well maintained, needs to be protected from freezing and works best in coarse textured soils.

Typically, two tensiometers should be installed at 20–30 cm (8–12 in) and 45–60 cm (18–24 in) for each soil type within the orchard. Tensiometers are available for the exact soil depth desired. If soils are homogeneous, then one set of instruments should be installed for each irrigation block. They should be located in the active rooting zone, within 15 cm (6 in) of a drip emitter with the porous tip located at the edge of the wetted area. In a sprinkler-irrigated block, tensiometers should be placed within the irrigation pattern under a tree. Before installation, the tensiometer should be filled with clean pure (distilled) water and allowed to stand in a vertical position for at least 30 min so that the ceramic tip becomes saturated. Once saturated, refill the tube. Use the vacuum pump that comes with the unit to remove air bubbles in the gauge, fill to capacity and cap the tube. To install in the soil, make a hole with a soil probe that is similar in diameter to the tube. A soil probe with a 2.2 cm (0.875 in) diameter will be perfect. The hole should be no deeper than the desired measurement depth. There should be no space between the tube and the soil, as this will affect the readings. After installation, the top of the tensiometer should be 5–8 cm (2–3 in) above the soil surface.

Tensiometers should be read early in the morning at the same time each day. Tensiometers need to be monitored at each reading to make sure that the water level is maintained within the tube. If there is more than 0.6 cm (0.25 in) of air below the cap, it should be refilled with distilled water. Tensiometers should be removed from the orchard each fall to prevent freezing.

8.1.5   Automated monitoring systems

It is possible to obtain useful soil moisture monitoring information from any of the instruments described above. However, it takes time and dedication to monitor soil moisture throughout the growing season, which can be challenging as harvest approaches. Consequently, a number of manufacturers have developed automated systems to monitor soil moisture. In fact, some systems provide both automated soil moisture monitoring and automated systems that schedule and apply irrigation based on tree water demand. All of these systems can be monitored remotely on desktop computers, but the most advanced can be monitored and accessed on smartphones and computer tablets from anywhere in the world.

These systems typically support a network of sensors strategically located throughout an orchard block to continuously monitor soil moisture levels. This provides the information necessary for growers to make informed irrigation management decisions without the need to monitor each sensor individually and then record and graph the results. This can enable significant savings in time and money and such automation ensures that monitoring will continue throughout the growing season.

Where automated soil moisture monitoring is coupled with a completely automated irrigation system, irrigation can be scheduled as needed, controlling pumps and valves remotely at preset soil moisture values and for specific durations. It is also possible to schedule filter flush cycles automatically, by operation time or pressure differential. Fertilizer injectors can be added to provide fertigation, all controlled through an app on a smartphone or tablet. Similarly, such irrigation automation can be used to turn on sprinklers for frost protection in spring or for evaporative cooling (such as to improve chilling in winter, delay budbreak in spring or protect from excessive heat in summer). Although automated systems are expensive investments, they provide continuous feedback even during the busiest times of the season; provide ready access to the detailed information necessary to make well informed irrigation scheduling decisions; and can significantly reduce labor costs.

8.2   Irrigation Strategies and Scheduling

8.2.1   Irrigation to meet plant water demand

For consistent commercial yields of high quality sweet cherries, all orchards should have the capability to be irrigated. Even in locations where rainfall is abundant during the growing season, dry periods of a week or more during critical times for fruit growth can limit cherry fruit size and quality. Where rainfall is too abundant during ripening, the risk of fruit cracking may require the installation of orchard covers (see Chapter 5, 'Orchard Establishment and Production') to protect fruit from contact with excessive rain, and hence the root zones in protected orchards may dry out. Furthermore, it is recommended that any orchard that utilizes dwarfing rootstocks must be irrigated, since the size of the canopy generally reflects the size of the root zone, meaning that root zones of small trees tend to be shallow and limited, and therefore subject to water stress more rapidly than vigorous or semi-vigorous rootstocks. Even some semi-vigorous rootstocks, like Gisela 6, can exhibit water stress without supplemental irrigation to bridge dry periods. Ideally, the orchard environment should be managed to provide water to the root zone whenever it is needed during the growing season, while preventing excessive water in the root zone from rain and preventing contact of the fruit with rain when it is susceptible to cracking (during Stage III of fruit growth, beginning with the green to straw color change through fruit expansion and ripening; see Chapter 9, 'Fruit Ripening and Harvest'), and preventing exposure of the tree canopy to rain that can disseminate diseases like cherry leaf spot or bacterial canker (see Chapter 11, 'Managing Orchard Pathogens and Disorders').

For the best establishment of newly planted orchards, cherry trees need to be irrigated within hours of planting. If the irrigation system is not installed prior to planting, it is important to water trees in by hand with 11–19 l (3–5 gal) of water per tree. This will provide moisture around the roots, settle the soil in good contact with the roots, and eliminate air pockets that may be present after planting the tree. The trees should continue to be watered by hand, as needed, based on temperature and soil texture until the irrigation system is installed. Due to the very limited root zone of newly planted cherry trees, frequent irrigation of short durations will be needed. A good estimate of the irrigation need at this time is 12.5 mm (0.5 in) of water per irrigation, but this should be verified by making sure that the water is penetrating to the lowest point of the root system. Initially, a soil probe can be used to confirm adequate irrigation. Check several trees 24 h after the irrigation is turned off. During this time, gravity will move the water through the soil profile. Be careful that the watering is thorough, but avoid over-irrigation.

Like most tree crops, mature sweet cherries require about 100 cm (~40 in) of annual rainfall during the growing season. This demand varies significantly as leaf area increases in spring and temperatures rise in the summer. Water demand in commercial orchards in The Dalles, Oregon were found to be 10 cm (4 in) in May, 15 cm (6 in) in June, 20 cm (8 in) in July, 18 cm (7 in) in August, and 10 cm (4 in) in September, and this demand varied higher and lower across years by as much as 10% (J. Le Roux, personal communication). Thus, it is easy to see how water in the root zone is likely to become limiting during critical periods of growth when rainfall quantity does not keep up with tree demand.

Other factors to consider when comparing rainfall with tree water use is surface runoff versus infiltration rate of the soil, evapotranspiration (ET) of any weeds or grasses in the root zone and internal drainage through the soil. Designing irrigation application rates and timing should also take these factors into account. As a rough rule of thumb, 2.5 cm (1 in) of precipitation or applied irrigation water will move 25 cm (10 in) deep in the soil. Since the majority of feeder roots are located within the top 45 cm (18 in) of soil, an adequate level of irrigation should be applied to wet that soil zone. Consequently, long irrigation applications are to be avoided except to flush salts out of the root zone if there has been a salt build-up.

Whether the trees are newly planted or well established, determining when to irrigate and for how long are important questions that need to be answered in order to grow healthy, productive trees. Too often, growers settle for calendar-based irrigation scheduling that can over-irrigate trees in the spring when temperatures are cool and rains may be more prevalent. This keeps the soils cold late into the spring, slowing nutrient uptake and growth, or can directly kill the tree through root asphyxiation or indirectly

Table 8.3. Guidelines for irrigating using tensiometers and resistance blocks to determine soil moisture tension and 50% depletion in different soil types (after Werner, 2002).

Soil type	Soil moisture tension at field capacity (kPa)	Soil moisture tension value to begin irrigation (kPa)	Soil moisture tension at 50% depletion (kPa)
Loamy sand/sandy loam	10–15	20–30	45–60
Loam/silty loam	10–20	30–40	50–70
Clay loam/clay	15–25	35–45	60–100

lead to tree death by weakening it and increasing its susceptibility to diseases such as bacterial canker (*Pseudomonas syringae*) or collar rot (*Phytophthora* spp.). Over-irrigation will also leach critical nutrients from the soil profile, potentially causing groundwater pollution or runoff into surface water sources as well as squandering costly resources. Conversely, as harvest approaches, calendar-based scheduling may under-irrigate trees during the critical fruit expansion stage. Rapid cell expansion begins immediately after pit hardening and continues through final ripening. Any stress to trees during this stage of development can reduce fruit size. In most years, this stage is associated with rising temperatures and an increased need for water. Careful soil moisture monitoring is needed to assure optimal growth conditions.

To determine the amount of water to apply at each irrigation, estimate the amount of water needed to fill the soil profile as determined by the available water holding capacity of the soil (Table 8.2). Apply only enough water to return the soil water content to field capacity. Soil moisture tension values of 0–10 kPa (1 kPa = 1 centibar, cb) indicate a saturated soil. In a loam or silt loam soil, field capacity generally is reached at 10–20 kPa (Table 8.3). Irrigation is not needed at 0–20 kPa. The efficiency of the irrigation system will affect the amount and duration of water that is actually applied. Unless the lower root zone depths are already saturated, the soil moisture sensors at the lower depths should detect the added water within 24 h of irrigation.

The goal of many cherry growers is to simply maintain the level of moisture in the soil between field capacity and the point of allowable depletion. However, allowing the soil moisture to fluctuate between these extremes can cause fruit swelling and shrinkage that can increase the incidence of microcracks in the fruit epidermis and predispose the fruit to rain-induced cracking. To reduce this potential and minimize tree stress, one management strategy is to maintain soil moisture close to field capacity throughout the growing season. This ensures that water is always readily available so that the trees and the developing fruit are never under stress. Research with cherries on dwarfing rootstocks in a coarse soil showed that four brief irrigation sets per day produced more growth and higher yields than one longer irrigation every other day, even though the same amount of water was applied in both treatments (Neilsen *et al.*, 2014).

Maintaining soils at or near field capacity can be done with microsprinklers or drip emitters; however, soil moisture should be monitored frequently and recorded. In a loam or silt loam soil, to maintain soil moisture at or near field capacity, irrigation should be applied when the tensiometer soil moisture value at 30 cm (12 in) is 30 kPa (Table 8.3). Irrigation initiation values will vary depending on soil type. Since the majority of the feeder roots are in the top 30 cm (12 in) of soil, this is the zone that needs to be kept evenly moist. The tensiometer at 60 cm (24 in) can be used to determine how long to irrigate. When field capacity has been reached in this zone, irrigation should be stopped. Irrigating beyond this point only wastes water and potentially leaches nutrients and chemicals below the main root zone. Typically, when the crop is within 10 days to 2 weeks of harvest, growers will change their irrigation regime from providing 80-95% of the soil water holding capacity to 80-90%. In neither case is the tree stressed, but keeping the water holding capacity to a narrower range as harvest approaches helps to concentrate the sugars in the fruit. Also, when irrigating with microsprinklers that may contact low hanging fruit during ripening (which could be susceptible to cracking), maximum irrigation duration is reduced to 2-4 h. In general, postharvest irrigation strategy should be to maintain soil water above 50% available water, and to be especially careful about not inducing too much stress if the orchard is planted on lightly textured soils or on shallow rooted rootstocks such as Gisela 5 or 6.

Another irrigation strategy is to replace the water lost the previous day by ET, the combined process of water transpired from the tree canopy and that evaporated from the soil. In the spring, most soils are filled with moisture from winter precipitation, which allows irrigation to be delayed for some time based on ample available water in the root zone profile. As the canopy leaf area develops, depletion of soil moisture will eventually accelerate and increased soil moisture tension values will indicate that it is time to begin replacing water lost to ET. In many commercial fruit growing regions, local weather stations may provide estimates of daily ET, which can guide how much irrigation will be needed to replace that water lost during the previous day. Maintenance of full ET replacement from pit hardening to ripening will result in good fruit growth and high quality fruit at harvest. However, sweet cherry trees are relatively drought tolerant, so that after harvest, irrigation rates can be reduced below replacement ET. Thus, quality cherries can be produced when only 80% of the seasonal ET is replaced, and some growers have successfully produced good yields of quality fruit with only 50–60% of the seasonal ET, especially in cases of mature trees grown on deep rooted rootstocks (i.e. Mazzard), provided the trees are not water stressed in the month prior to harvest. One caveat, though, is that in locations with hot summer climates, this postharvest reduction in available soil moisture may increase the risk of fruit doubling the following year. Reduced ET can lead to higher canopy temperatures, and developing floral meristems may become abnormal if exposed to temperatures above 37°C (99°F) in July and August (in the northern hemisphere).

Some automated irrigation systems in commercial fruit production regions can utilize climatic data from nearby weather stations and estimate irrigation needs, automatically initiating irrigation while receiving feedback to close valves once the soil profile becomes saturated. These systems allow growers the ability to monitor and track crop moisture needs when desired, but also allow for automation to reduce workload during labor intensive times like harvest. The greatest limitation to this option is cost, as subscriptions can be expensive, but this may equate to labor designated for soil moisture monitoring and irrigation scheduling.

Finally, irrigation services from consultants are available to growers in some areas, that can monitor and interpret soil moisture data without the large initial investment of purchasing a system or maintaining labor to monitor each orchard block. This option keeps labor needs low as the service provides the personnel necessary to monitor each block and interpret the data. In addition, many of these services are licensed to monitor soil moisture with a neutron probe to provide extreme accuracy.

In summary, universal goals of cherry growers should be to avoid stress from over-irrigation in early spring during cell division, and to avoid stress later in early summer from under-irrigation during the rapid cell expansion stage that occurs after pit hardening. Applying the appropriate amount of water to trees when needed can be done most effectively by consistently monitoring soil moisture levels. Basing irrigation decisions on monitoring will keep the trees healthy and strong, and produce the premium quality fruit for which domestic and foreign markets pay well.

8.3 Irrigation System Use for Non-Irrigation Purposes

8.3.1 Irrigation system use for evaporative cooling

Over-tree microsprinklers are sometimes used not for irrigation (due to potentially increasing the incidence of canopy diseases or fruit cracking), but for evaporative cooling. Evaporative cooling of the tree canopy can have several valuable effects on production, depending on when and how it is scheduled. In locations with mild sunny winter climates, evaporative cooling of endodormant buds on sunny winter days can improve the chilling needed to transition physiologically from endodormancy to ecodormancy, promoting stronger bloom and foliation in spring that otherwise might be poor due to lack of adequate winter chilling. Evaporative cooling of buds after the transition to ecodormancy can slow acquisition of the heat units required for budbreak, thus delaying bloom to avoid or reduce the risk of spring frost. Evaporative cooling of the tree canopy during summer

Fig. 8.4. Over-tree microsprinklers in this orchard are located 8.5 m (28 ft) apart and are used to reduce fruit doubling (courtesy of L.E. Long).

can reduce heat stress, fruit or tree sunburn, and/or abnormal flower bud development that would result in fruit doubling the subsequent spring (Fig. 8.4).

8.3.2   Enhancing chilling unit accumulation or delaying budbreak

Evaporative cooling of buds during endodormancy (to enhance chilling) or ecodormancy (to delay budbreak) utilizes computer control of the microsprinkler system to operate on a pulse basis, wetting the surface of the bud, then turning off while the moisture evaporates. The computer records inputs of wind, radiation and air temperature through a leaf wetness sensor, to estimate how long it will take until complete evaporation has occurred, then it reactivates another brief irrigation to rewet the buds and start the cooling cycle again. This provides a very efficient use of minimal water to achieve cooling, and prevent saturation of the soil or cause issues with pathogen dissemination. This method is not recommended, however, where water quality is poor, as the repeated evaporation cycles can leave salt deposits on the buds if the irrigation water is not relatively pure. The same fundamental operational parameters for cooling are used for either endodormant or ecodormant buds, with the difference being at what point during winter the cooling cycles are begun.

8.3.3 Mitigating sunburn

Sunburn in cherries is generally the result of a combination of excess heat and light. Sunburn symptoms can be induced in the absence of light when temperatures reach 52°C (125°F). In the laboratory, fruit exposed to ultra-violet (UV) light at an intensity of 12,500 kW/m² (equivalent to 48 h of full midday sun) exhibited sunburn symptoms when temperatures exceeded 40°C (104°F) for 8 h (Kaiser *et al.*, 2012). In contrast, at temperatures below 35°C (95°F), fruit exposed to 8 h of UV at the same intensity did not exhibit sunburn damage. This suggests that both UV and temperature are the determining factors in causing sunburn and that a certain minimum total radiant energy is required to produce sunburn symptoms. In the orchard, visible light also contributes to the total radiant energy equation, and sunburn on cherry tree limbs is usually expressed on south-facing exposed surfaces (in the northern hemisphere). This is typically the case when major limbs are pruned during the summer months, exposing shaded bark that is unconditioned to intense heat and UV light. It is always a good idea to paint exposed bark on the southern side of the tree trunk with undiluted white latex (water soluble) paint to reflect visible and UV light.

Sunburn may be managed by using over-tree evaporative cooling during the hottest summer months, typically early July through early August in the northern hemisphere. As with cooling during winter, computer control of the microsprinkler system facilitates operation on a pulse basis, wetting the canopy, then turning off while the moisture evaporates. The differences are in setting an air temperature threshold for initializing the microsprinklers and in the duration, since a fully foliated tree requires the irrigation system to run longer to fully wet the canopy compared to a bare tree in winter.

8.3.4 Mitigating fruit doubling

Cherry doubling (usually two fruit fused together with one stem) is a fruit abnormality that can cause significant financial losses when it occurs, since doubled cherries are treated as cull fruit on the packing line. The conjoined fruit may develop to produce two smaller than normal fruit, or one of the conjoined fruit may shrivel, resulting in an undersized fruit with a vestigial growth (sometimes referred to colloquially as a 'bird beak') attached along the suture. Doubling is a result of excessive heat during the summer months when meristematic flower differentiation takes place. This critical period of susceptibility to doubling is related to each cultivar's harvest timing, beginning about one month after harvest for a duration of approximately two weeks.

Doubling can be managed with over-tree microsprinkler evaporative cooling during the susceptible period of bud differentiation. Doubling is worse if air temperatures are consistently above 37°C (99°F) throughout the day. Cycling over-tree sprinklers through: (i) 30 min on–off intervals; or (ii) cycles of 20 min on, followed by 10 min off, have reduced doubling by up to 50%. Computer control of the microsprinkler system to operate on a pulse basis (as with the other evaporative cooling strategies described above) to wet the canopy, then turn off while the moisture evaporates, can increase efficiency of water use while possibly reducing doubling further than the simple clock cycle thresholds.

Some growers use postharvest applications of kaolin (e.g. Surround®), with or without evaporative cooling, to reduce bud temperatures and the incidence of doubling, especially for cultivars that are highly susceptible. This can also reduce the possibility of sunburn in the canopy, though kaolin is not used while fruit is present since it is difficult to remove from the fruit surface before sale. Canopy training systems with a lot of horizontal fruiting wood, like Vogel central leader (VCL), espalier (ESP), tall spindle axe (TSA), and tabletop (TTP) are more likely to have an increased incidence of doubling than systems with vertical fruiting wood, like Kym Green bush (KGB) and upright fruiting offshoots (UFO).

8.3.5   Irrigation system use for frost and fruit cracking protection

Impact sprinkler or microsprinkler irrigation systems can also be utilized in sweet cherry production for spring frost protection and/or to reduce the risk of rain-induced fruit cracking. For frost protection, water must be available and the system must be designed to be operable in late winter, since the freezing of water in above-ground irrigation lines could damage the system. The most useful irrigation system design for protecting trees from frost damage utilizes under-tree sprinklers or microsprinklers. The system can be operated to add heat to the orchard when frost conditions are predicted. The amount of heat released will depend on the temperature of the irrigation water, which will release heat into the orchard as the water freezes (Fig. 8.5). Under-tree sprinklers can be operated in conjunction with wind machines (see later in this chapter) to help pull the heat released by the water back down into the orchard as it rises. Typically, wind machines are operated as a first line of defense against low temperatures in a radiative frost, with the addition of the irrigation system operation if temperatures will be below the effective range of the wind machine. The sprinklers do not have to be operated during the entire low temperature event, only when additional heat is required. This strategy in conjunction with wind machines can help reduce the amount of water applied and reduce the potential for soil saturation.

In past decades, over-tree sprinklers were also utilized as a frost protection strategy. This technique requires turning on the sprinklers before the

Fig. 8.5. Ice builds up on the ground as water from under-tree sprinklers freezes, releasing heat to the orchard (courtesy of L.E. Long).

air temperature falls below 0°C (32°F) so that as water begins freezing on the tree, the latent heat of freezing releases heat directly next to the plant tissues in danger, which keeps the flower bud temperatures from falling any lower than 0°C (32°F). The sprinklers must be operated continuously until the air temperature rises back above freezing and the accumulated ice on the tree melts. It does not provide significant protection in windy conditions, which can lead to evaporative cooling and a potential increase in freeze damage. This technique is rarely utilized today because it requires large quantities of water, can saturate the orchard soils making it difficult to operate spray equipment in the orchard, and can promote increased infections of diseases such as bacterial canker.

However, over-tree microsprinklers have been used to reduce rain-induced fruit cracking by operating them similarly to their use for evaporative cooling. This rain-cracking prevention strategy utilizes an injection of calcium chloride into the irrigation line, and computer control of the microsprinkler system to operate on a pulse basis to cover the tree with a layer of calcium (Ca)-infused water during rain events. The Ca lowers the osmotic potential of the otherwise pure rainwater, slowing its uptake into the fruit by reducing the osmotic differential between the water on the outside of the fruit and the sugar-infused juice in the fruit flesh. The computer records inputs of rainfall to initiate the first pulse, wetting the canopy with Ca solution, then turning off while the continuing rain dilutes the Ca layer on the fruit. After a threshold of additional rain, the computer reactivates the next pulse and the cycle continues until the rain bucket indicates the event is over. This provides a very efficient use of water and calcium chloride to achieve osmotic protection, adding minimal amounts of water to the root zone. As will be discussed in Chapter 9, cherry fruit cracking is often a function of both water contact with the fruit (osmotic uptake) and water uptake by the root system, so this protection technique only addresses the first cause of cracking.

8.4   Plant Nutrient Demand, Monitoring and Scheduling

8.4.1   Orchard nutrient analyses

The conventional methods for determining nutrient deficiencies are regular soil and foliar tissue analyses. For new plantings, it is recommended that soil samples be analyzed prior to planting (Chapter 5) and every 2–3 years afterwards until trees reach maturity. For established orchards, soil sampling intervals can be reduced to every 3–4 years. It is important to keep in mind that there is no reliable test for soil nitrogen (N) and that soil phosphorus (P), potassium (K), Ca, and magnesium (Mg) cannot reliably predict tissue levels of these elements. Consequently, leaf samples should be taken at least every 2 years to provide recommendations on maintenance fertilization, unless a problem appears that needs diagnosing. These foliar analyses are usually taken just after harvest, when most growth is finished and leaf nutrient levels are somewhat stable, and are sent to commercial laboratories for analysis. The disadvantage of this timing is that any necessary adjustments to the nutrient program are too late to affect the current crop.

In recent years, a tissue analysis process called 'sap analysis' has been developed to determine the level of nutrients in the tree by evaluating the nutrient content of the sap. This intensive nutrient management program relies on bi-weekly sap analysis for macro- and micronutrients. Results of these analyses provide information for targeted foliar remediation sprays that are used to adjust nutrient levels in near real time throughout the growing season. This serves to minimize nutritional deficits and maximize fruit yield and quality. This approach is used in commercial greenhouse operations in Europe, but is relatively new to tree fruit production, especially in the USA. Only a few labs in the world are currently providing this service, but it is becoming more available.

There is clear evidence that altering the mineral nutrition of plants (either via soil or leaves) can influence insect pest and disease severity. For example, plant mineral content and nutrition explained 40–57% of the variation in Colorado potato beetle abundance (Alyokhin et al., 2005), and plant nutrition impacted insect pests present in maize (Phelan, 1997). The book *Mineral Nutrition and Plant Disease* (Datnoff *et al.*, 2007) describes in detail the relation of specific nutrients to plant physiological function, and includes examples of nutrients that reduce susceptibility to plant disease. Nutrients influence the plant, pathogens, and overall microbial growth (in the soil and potentially on the leaf surface) in complex ways that have not been fully understood or exploited in orchard management. It is believed by some that increasing 'plant health' leads to lower pests and diseases, and consequently can lead to the use of fewer pesticides.

8.4.2 Nutrient disorders

Most growers do not think about nutrient deficiencies in cherries until they start seeing symptoms. The problem with this approach is that by the time symptoms are visible, the problem already exists and it may be too late to avoid reductions in growth, fruit quality and/or yield. In addition, many other factors can produce symptoms that are similar to nutrient deficiencies or excesses. Diagnosis also can be complicated when one or more elements are deficient at the same time.

The most common reason for nutrient deficiencies is soil pH that is outside the ideal range for cherries. Cherries grow best at a pH of around 6.5–7.0. If pH levels are significantly outside this range, trees may not be able to utilize some nutrients even when present in the soil. Thus, soil pH should always be checked and, if necessary, modifications made prior to planting a new orchard (see Chapter 5). Check soil pH every 3–4 years after planting. The use of some fertilizers can quickly change the soil pH, especially in the root zone where the fertilizer is applied (see Table 8.4).

Soil nutrient availability varies from region to region. N is very mobile in the soil and can become limiting anywhere in the world. K deficiencies are relatively common in the Willamette Valley of Oregon, but not in the Mid-Columbia production region of Oregon. P is usually not deficient in the Pacific Northwest (PNW); however, in the volcanic soils near Mt. Hood, Oregon, there have been significant growth responses to added P. Other deficiencies occasionally found in the PNW cherry production region include boron (B), Mg, sulfur (S) and zinc (Zn). In Michigan cherry production areas, N and K deficiencies are common. P, Mg, B, manganese (Mn) and Zn shortages occur occasionally, while shortages of Ca, S, chlorine (Cl), copper (Cu), iron (Fe) and molybdenum (Mo) are rare (Hanson, 1996).

8.4.2.1 Nitrogen

Nitrogen (N) deficiency in conventionally managed sweet cherry orchards is rare. However, in organic orchards, or other orchards where manures are used as the main source of fertilizers, N can become deficient. N deficiency results in weak or little to no new growth. Young trees with less than

Table 8.4. Characteristics of nitrogen (N) fertilizers (Hanson, 1996).

Fertilizer	% N	Other nutrients present	Soil pH reaction[a]	Limestone equivalent[b] (kg CaCO$_3$/kg N)
Ammonium sulfate (AMS)	21	S (24%)	acidic	5.3
Ammonium nitrate	32	none	acidic	1.8
Calcium nitrate	16	Ca (19%)	basic	1.3
Diammonium phosphate (DAP)	17	P$_2$O$_5$ (50%)	acidic	4.1
Monoammonium phosphate (MAP)	11	P$_2$O$_5$ (48%)	acidic	3.5
Potassium nitrate	13	K$_2$O (44%)	basic	2.0
Urea	46	none	acidic	1.8

[a]acidic = reduces soil pH; basic = increases soil pH
[b]limestone equivalent = amount of lime that is equivalent to the reaction of 1 kg N applied to the soil (ratios are the same for lb CaCO$_3$/lb N)

60 cm (24 in) of growth may need supplemental N applied to boost vigor. Terminal growth may cease in mature trees with inadequate N. In this case, additional N and severe pruning may be needed to promote adequate growth. Well balanced young cherry trees should have 60–90 cm (24–36 in) of shoot extension growth throughout the tree. Mature, bearing trees should have 60 cm (24 in) of shoot extension growth throughout the tree, not just in the tree top.

Since N is mobile in the tree, symptoms first appear on the older leaves, which will remobilize N to new growth. These generally become pale green to yellow. However, pale green leaves in cold, wet spring periods may simply be an indication of cold soils that are limiting N uptake, but leaves will become darker green once the soils warm. Trees low in N may produce more spurs and flower buds than normal, typical of weak growth in cherry. Leaves may drop earlier than normal in the fall. Trees will lack vigor and shoots and twigs will be of small caliper. Fruit set may be light and/or there may be a heavier than normal June drop (aborted fruit). Fruit size will be small and fruit may mature earlier than normal.

Interpreting the results of N foliar analysis requires caution, as it is possible to have low leaf levels of N, yet have adequate shoot extension growth >60 cm (24 in). In this case, the low leaf N levels should be ignored and the current fertilizer program maintained. It is also possible to have high leaf N but low tree vigor. Increasing N fertilization and more aggressive pruning may be appropriate in this situation. In general, leaf N levels should be maintained between 2.3–3.3%.

Excessive N levels can delay production, reduce yields, lower fruit quality, delay the onset of cold acclimatization, and make trees more susceptible to diseases, such as powdery mildew and bacterial canker, and insects such as San Jose scale.

An important aspect of N nutrition in sweet cherry production is the role of N stored in the tree over winter for remobilization during budswell and early spring growth. Cherry trees on dwarfing rootstocks have smaller root systems and trunks for storing N, yet often more flower buds per woody tissue than trees on vigorous rootstocks. Consequently, it has been shown that, in addition to soil-applied N in the spring, high density cherry orchards may benefit from a fall foliar spray of low biuret (i.e. containing low levels of impurities) urea of around 34–50 kg/ha (30–45 lb N/acre), applied over two applications about 7–10 days apart, in very late summer or early fall (September in the northern hemisphere). The N taken up by the foliage is remobilized to storage tissues in the fall, mainly to spurs. This is then available for the initial flush of spring growth, including the development of the very important spur leaves. Larger spur leaves in the spring produce more carbohydrates for fruit growth, improving the leaf area-to-fruit (LA:F) ratio. Some marginal leaf burn might occur at higher N spray rates.

8.4.2.2   Phosphorus

Except in a few locations, phosphorus (P) deficiencies are rare in tree fruits. If foliar P values are low, it is often best to establish a small trial to determine if trees will respond to supplemental P applications. Since P is not mobile in the soil, any additional P applied to the trees will need to be added in the root zone for the tree roots to quickly access it. Low P values often are associated with diseased or stressed trees and very low (<6.3) or very high (8.3–8.7) soil pH. Correcting other problems is often more useful than applying additional P fertilizers. P levels that are above normal or excessive may result in marginal Zn levels. Leaf P levels should be maintained between 0.23–0.38%.

Symptoms of P deficiency include weak, slender shoot growth and dark green leaves that turn bronze or reddish purple. Leaves develop a characteristic leathery feel. As symptoms progress, petioles and young shoots turn reddish. Finally, leaf size is affected and early defoliation may occur, beginning with the oldest leaves.

8.4.2.3   Potassium

Potassium (K) deficiency can be a problem in most stone fruits. Leaf margins that turn upward and yellow are the initial symptoms, with more severe conditions exhibiting a bronze color and leaf margins eventually become necrotic (Fig. 8.6). Since K is mobile within the tree, symptoms will first appear on older leaves, but in severe cases, young leaves may exhibit symptoms as well. As fruit ripens, K will move out of leaves and into the fruit, so

Fig. 8.6. Potassium (K) deficiency on sour cherry (from Hanson, 1996).

symptoms may be most severe during the later stages of ripening in a heavy cropping year. Deficiencies can be found in any soil types, but they are most often associated with poorly drained, fine textured soils. Leaf K levels should be maintained between 1.0–1.9%.

When deficiencies exist, trees respond well to fertilizer additions. Unfortunately, adding K to the soil is expensive, so amendments should be considered carefully. Low K readings can be the result of a high crop load, or in the case of non-irrigated orchards, a particularly dry year in which surface roots are less active. High K levels in the soil are also of concern as this may inhibit Mg or Ca uptake and induce deficiency symptoms of these elements. If K soil levels are high, the Mg level should be checked; if K is high but Mg is normal, this is not a problem.

8.4.2.4   Calcium

Cherry trees deficient in calcium (Ca) have shortened internodes and re-duced growth. Defoliation and dieback often follow in more severe cases. On young trees, light brown to yellow marks may appear on the leaves. Leaves may become ragged with numerous holes. As symptoms progress, leaf blades may roll inward and upward and develop large chlorotic patch-es before abscising. Normal leaf Ca levels range from 1.6–2.6%.

Since Ca does not readily move in the plant, most growers apply mul-tiple applications of foliar Ca throughout the growing season. Higher Ca concentrations in fruit result in potentially less cracking, longer storage capacity, less pitting, better luster and improved flavor after prolonged cold storage. Care must be taken, however, as high levels of Ca applications can reduce fruit size. Calcium chloride and calcium citrate forms of Ca are most prone to affecting fruit size, whereas calcium chelate has not nega-tively impacted fruit size and is the safest form to use.

8.4.2.5   Boron

Boron (B) is often the most commonly applied micronutrient on sweet cherries. B plays an important role in good fruit set, with benefits even when foliar B levels are in the adequate range. Typically, growers will apply B as a foliar spray in the fall or delayed dormant period of development so that the element will be present at bloom. Deficient B levels lead to severely reduced new shoot growth and dieback of shoot tips (Fig. 8.7). Vegetative buds may fail to open in the spring, or they may open and then die. Shoots on deficient trees are short, thin and may die by late summer. Leaves are glossy, small, narrow, and pinched at the base with enlarged, distorted mid-ribs. In addition, flower buds may be absent or fail to open, and fruit yields can be reduced severely.

Excessive B is toxic. Toxicity symptoms are very similar to deficiency symptoms, including twig dieback often accompanied by gumming. In

Fig. 8.7. Boron (B) deficiency on sour cherry showing delayed bud opening and small, narrow leaves (from Hanson, 1996).

severe cases, this gumming may extend to main limbs and trunks. Leaf size and shape is normal, but necrotic areas may develop along main veins. Flower buds may fail to open and fruit set is impaired. There is a narrow range between deficiency and excess, but at maintenance application amounts, it is unlikely that toxicity levels will be reached. Nevertheless, it is important that B levels in the tree are monitored on a regular basis. Sufficient leaf B levels range between 30–50 ppm.

8.4.2.6   Iron

Symptoms of iron (Fe) deficiency include interveinal chlorosis, in which the leaf veins remain green, but the tissue between the veins is yellow (Fig. 8.8). The newest leaves are affected first. Deficiencies caused by low levels of Fe in the soil are rare, but symptoms of iron chlorosis are not. High soil pH is a leading cause of iron chlorosis symptoms. Cold and wet soils in the spring can exacerbate this condition. Since Fe deficiency in the soil is rare, supplemental Fe applications are seldom suggested except in severe cases and as a stop-gap measure until the underlying cause can be mitigated. Iron chlorosis symptoms are effectively dealt with by moderating the soil pH or eliminating waterlogged soils. Symptoms can be temporarily alleviated with foliar applications of Fe chelate. Also, MaxMa 14 and some Mahaleb rootstocks perform better than others in high pH soils, eliminating symptoms of iron chlorosis. Leaf Fe levels should range between 75–150 ppm.

Fig. 8.8. These iron (Fe) chlorosis symptoms include interveinal chlorosis on the newest leaves and are due to high soil pH (Uzbekistan) (courtesy of L.E. Long).

8.4.2.7 Magnesium

In cherries, magnesium (Mg) deficiency often occurs in sandy soils where nutrient levels are low. Heavy applications of K in the spring can also cause Mg deficiency. The most common symptoms include yellowing along the leaf margins and between the veins (Fig. 8.9). As symptoms develop, the yellow regions turn brown and die, leaving a Christmas tree-shaped green area in the leaf center. The most severely affected leaves often will drop. Since Mg is mobile within the tree, symptoms typically develop on older leaves first, generally around mid-season or later.

Most often, Mg deficiencies are associated with low pH soils. In these cases, application of dolomitic limestone can help. Mg deficiencies usually develop slowly, with a gradual decline over several years. Since Mg competes with Ca and K, when leaf Mg levels are above normal, K levels should be checked for sufficiency. Normal foliar Mg levels range between 0.49–0.65%.

8.4.2.8 Manganese

Manganese (Mn) deficiency is expressed as a yellowing of the older leaves, which starts at the leaf margins and progresses inward between the veins

Fig. 8.9. Magnesium (Mg) deficiency on sweet cherry (from Hanson, 1996).

(Fig. 8.10). The veins and the tissues immediately next to the vein remain green. Spur leaves are the first to be affected, but shoot leaves develop symptoms as the deficiency progresses. In the most severe cases, young leaves also may be affected. Fruit tend to be small and lack juice. As with Fe deficiency, the most common cause of Mn deficiency is high soil pH and

Fig. 8.10. Manganese (Mn) deficiency on sweet cherry (from Hanson, 1996).

wet, poorly drained soils. Foliar Mn applications can relieve symptoms, but more permanent control is only possible by lowering soil pH and eliminating wet soils. Foliar Mn levels should range between 15–40 ppm.

8.4.2.9 Sulfur

Where sulfur (S) is applied as a control agent for fungal diseases, S deficiencies are rare. However, in other regions, S deficiencies can be common. S deficiency causes poor growth and newer leaves that are uniformly yellow. Above normal S concentrations do not seem to be a problem. Foliar S levels should range between 0.16–0.29%.

8.4.2.10 Zinc

Without supplemental zinc (Zn) applications, Zn deficiencies would be common in many cherry producing areas of the USA, like the PNW and Michigan. Typical symptoms include a rosette-type growth at the shoot tip, in which the internodal region of the shoot is greatly reduced causing a proliferation of leaves. In some cases, the spacing of the leaves may be normal, but the leaves are smaller and more narrow than normal. Leaf chlorosis also is a symptom of Zn deficiency, especially in the area between the veins. Zn deficiency is most common in high pH soils. Zn is not very mobile within the tree, so multiple applications of foliar Zn during the growing season are needed to provide adequate Zn for new leaves as they develop. Normal leaf Zn levels range between 15–40 ppm.

8.4.3 Fertilization techniques

Like sweet cherry training and production systems in general, fertilization techniques and strategies have changed dramatically since the turn of the century. These changes have been in response to both higher prices for fertilizers, especially N, leading to economic incentives for more efficient use, as well as in response to improved environmental stewardship, leading to reductions in total use and improved timing of applications to minimize the amount of nutrients that can leach beyond the root zone or run off from the soil surface. As emphasized in the earlier part of this chapter, current fertilization practices should be driven by soil and tissue sampling to determine what are the actual plant needs for each nutrient. Estimates can be made of the nutrients removed each year through cropping, pruning and leaf drop, but crop loads vary from year to year, and some nutrients are recycled through leaf decomposition and perhaps mowing/chopping of orchard pruning brush. If weeds or grasses are growing in the tree root zone, they will compete for fertilizer. Organic matter (OM) and soil microbial systems can significantly influence the availability of nutrients to trees.

Annual leaf nutrient analyses provide the most precise strategy for building an understanding of orchard nutritional needs, block by block.

Standard fertilization practice in low density orchards of large, vigorous cherry trees used to be to broadcast the nutrients needed for an entire season, usually just before bloom. Large trees have extensive, deep root systems that access nutrients throughout a large volume of the soil profile. As orchard densities increased to distinct tree rows, sometimes the fertilizer was banded into the row, to better place the nutrients in the tree root zone, reduce the promotion of tractor alley groundcover growth and minimize potential surface runoff of nutrients. With the advent of vigor-controlling rootstocks and high density orchard training systems, the effective root zone also has decreased comparably, narrowing the orchard floor area where nutrients need to be applied for use by the tree. Similarly, the now relatively shallow depth of root zones of trees on dwarfing or semi-dwarfing rootstocks means the volume of soil to store nutrients until they can be taken up has greatly decreased. Depletion of nutrients in the effective root zone can occur much more quickly in higher density orchards.

Consequently, fertilization strategies are influenced by the identification of plant needs through tissue analyses, and by rootstock vigor, tree density, and orchard training systems. Other major factors include soil type and climate. Sandy or rocky soils have poor nutrient holding capacities compared to loamy or clay soils, so fertilizer must be applied in small increments, more frequently on the former soil types to prevent leaching of nutrients beyond the root zone. Fertilizer applied in the fall on loamy soils may be available to roots for uptake in spring, but fall applications on sandy soils may be leached beyond the root zone by spring. Rainfall frequency also affects nutrient movement through the root zone: dry fertilizer applied to the soil surface may be washed off to some extent if rainfall is excessive; moved into the root zone efficiently if rainfall is regular and moderate; or may fail to move into the root zone if rainfall is rare, as in the arid regions that otherwise can be quite advantageous for cherry production.

Finally, an area of precision orchard nutrient management that is still being refined through physiological research is to determine how the needs of the tree for each critical nutrient element may vary through the growing season. As modern fertilization strategies tend towards more frequent applications of small amounts of specific nutrients, this seasonal knowledge can inform whether and how the make-up of nutrient delivery should change from application to application. For example, research has shown that N is advantageous for early season leaf and shoot growth, providing the photosynthetic capacity to support fruit growth, but too much N promotes excessive shoot growth that can lead to excessive shading, with detrimental effects on both fruit quality and flower bud initiation for the next year's crop. Furthermore, K and Ca are necessary nutrients, but uptake of too much K can reduce the availability of Ca for developing fruit, possibly leading to increased fruit cracking, pitting or softer fruit. While nutrient needs have yet to be elucidated for each stage of cherry tree and fruit development, scientists are making progress in this area and so growers should plan their orchard nutrient delivery techniques to be able to adapt to such advances.

Thus, modern fertilization methods tend to utilize strategies to provide multiple nutrient applications throughout the growing season, possibly utilizing banding of dry fertilizer, application of soluble fertilizer through irrigation water ('fertigation'), and/or foliar applications of specific nutrients (and sometimes growth promoting materials not strictly considered nutrients, often with plant hormonal activities). An increasingly attractive strategy is the use of weekly fertigation applications, which provides small, easily altered amounts of nutrients to the roots that have developed in the soil profile wetted by the irrigation emitters. In arid climates, this fertigation may be accomplished via drip irrigation emitters; in relatively rainy climates, the preferred irrigation emitters may be microsprinklers since rainfall can promote root growth (and thus nutrient uptake activity) beyond the typical wetting zone of a drip emitter. If the cherry orchard is protected by a rain covering system (e.g. row covers, high tunnels, greenhouse-like structures), the potential impact of the covering system on rain or irrigation water movement through the soil, and hence on the zone of root activity, must also be taken into account when planning irrigation and fertigation techniques.

Fertigation provides a method to quickly alter nutrient delivery to the soil profile where root uptake can respond relatively quickly. So, if nutrient analyses during plant growth identify an insufficient amount of a specific nutrient, often that can be corrected quickly through fertigation. Foliar application of specific nutrients can act even more quickly to remedy nutrient deficiencies, provided the nutrient formulation can be readily absorbed by leaves. Another potential advantage of foliar nutrient applications includes cool spring periods, when budbreak and initial leaf and shoot development have begun, then the weather turns cool and cold soil temperatures inhibit root uptake of nutrients. Trees may begin to look N deficient, which can be rectified by a foliar N application until the weather turns warm and soil nutrient uptake begins. Similarly, in the event of a loss of cherry crop due to winter or spring low temperature damage, soil applications of N can be reduced or avoided with only occasional foliar applications, to prevent excessive compensatory growth in the absence of crop nutrient demand.

Bypassing the root uptake and transport pathway also may have some physiological advantages for certain nutrient demands. Research has shown that tree N reserves in the fall are important for cold acclimatization and cold hardiness during winter, as well as driving flowering and initial spur leaf growth in spring (since roots tend to not take up N until the soil warms and leaves emerge to begin driving uptake through transpiration). While low N reserves can occur in any tree, this situation can be more problematic in high density trees on vigor-limiting rootstocks, since the mass of trunk and root tissues for accumulating reserves is considerably less than in large vigorous trees. Yet, to increase N reserves in the fall, N applied to the soil may be converted to amino acids for transport to growing points that may promote renewed shoot growth that could fail to cold-acclimatize and be subject to low temperature damage during winter. Research has shown that foliar-applied N (in the form of low biuret urea) absorbed by the leaves is converted to amino acids for transport to storage tissues, thereby providing a precision technique for remedying low reserves without adversely affecting cold acclimatization.

8.5   Orchard Floor Management

Ideally, orchard floor management strategies should contribute to producing the highest quality fruit possible while maintaining and sustaining the ecosystem in which the tree is grown. Most growers expect their orchard floor management decisions to provide a healthy environment for tree growth. Unfortunately, over time the orchard floor ecosystem can break down due to soil compaction, restricted water infiltration, soil erosion or other factors. The ideal ecosystem would resist this degradation while providing an environment that limits competition from weeds, rodents, pests and disease organisms. Orchard floor management strategies vary greatly around the world and each has its own strengths and weaknesses.

8.5.1   Cover crops

Orchard grass is the most common cover crop used in orchards in the PNW. Other sod-type cover crops include hard fescue, sheeps fescue, chewings fescue, creeping red fescue and perennial rye grass. Other cover crop mixtures have included crimson clover, annual ryegrass/hard fescue and berseem clover. There is considerable ongoing research in the use of legumes in orchards. These naturally provide N to the trees, and also can serve as a home for predatory mites throughout the year. Unfortunately, due to the N fixed and released by leguminous cover crops, it is difficult to control nutrient inputs as fruit ripen and as trees approach dormancy in the fall. Also, there is increased potential for rodent damage with legumes.

Planning of cover crops for orchard floor management should begin with soil preparation for planting the orchard. Lime or other needed amendments should be incorporated into the soil before seeding the cover crop. Late summer is often the best time for seeding as the soil is warm and the seedlings grow quickly. To prevent soil erosion during winter, mix grass seed with 66 to 101 kg/ha of wheat (60 to 90 lb/acre). If late summer establishment is not possible, plan on a mid-March to mid-May planting window (in the northern hemisphere). At establishment, 22 to 44 kg/ha (20 to 40 lb/acre) of actual N should also be applied to promote a dense, healthy cover. Seeding the entire orchard floor, rather than just the alleys, also will help protect the soil surface from erosion.

Once the seedlings are well established, define the tree rows by creating a planting strip with an herbicide spray or use an organic means of killing the seedlings. This should be done in the fall to avoid driving equipment on wet spring soil. The dead cover crop strip will help to protect the soil during the winter from erosion. In many production regions, cover crops will need supplemental irrigation to survive summer droughts. In sprinkler-irrigated orchards, this is not a problem, but where drip irrigation is utilized, many species of cover crops will die out in the summer. For this reason, some orchards install a dual irrigation system, with drip emitters providing irrigation for the trees, and microsprinklers supplying just enough water to the cover crop to keep it alive.

Grass cover crops need to be mowed several times per year, generally when the grass blades are 7.5–10 cm (3–4 in) long. To reduce competition for bee pollination activity during bloom, it is a good idea to control flowering weeds in the cover crop. In addition, some broadleaf weeds may be a host for Western X phytoplasm. Although cherry blossoms compete well with dandelions for bee pollination activity, it is still a good idea to control competing flowers during bloom. These can be controlled either with herbicides in the fall or mowed as trees come into bloom. However, eliminating weed bloom at the same time that the orchard is coming into bloom could potentially move thrips from the weeds to the cherry flowers, causing damage to the fruit. Be sure to mow again prior to harvest to facilitate labor and equipment access. Finally, it is important to mow just before winter so the cover crop does not serve as a refuge for rodents. Some growers are averse to fertilizing cover crops due to cost and the potential increase in mowing requirements. However, 22 to 44 kg/ha (20–40 lb/acre) of actual N applied in the spring will help promote a dense cover that can withstand orchard equipment traffic throughout the year.

8.5.2 Grass alleys

In most tree fruit orchards in the US, a well managed orchard grass is established and maintained in the tractor alley, with trees planted in strips that are usually kept weed free with herbicides (Fig. 8.11). This weed free zone is commonly 1.5–1.8 m (5–6 ft) wide in high density orchards. In fruiting wall or ultra-high density training systems, such as UFO or super slender axe (SSA) orchards, weed free zones may be as small as 0.75–0.9 m (2.5–3 ft) across. In Europe, herbicide strips are small when present, constituting no more than 20% of the total orchard floor. The majority of the tree roots can be found in this area, and these tend to grow close to the surface. Tractor-based mowing implements are now available that blow the grass clippings into the tree row when the alleyway is mowed. This can provide significant inputs of nutrients and OM that increase tree vigor and yields.

Fig. 8.11. Grass alleys are being established in this newly planted orchard (Oregon) (courtesy of L.E. Long).

The grass alley helps provide some resistance to soil compaction from orchard equipment and erosion, and aids in soil aeration and water infiltration. It also will provide a cooler environment that buffers air temperature changes, which can delay harvest by several days, a situation that may or may not be advantageous. Where powdery mildew is a problem, the higher humidity maintained by the grass alley can significantly increase risk of infection. Grass or cover crop tractor alleys can increase the risk of rodent or frost damage.

8.5.3   Solid vegetation cover

In some parts of the world, orchards may have cover crops that encompass the entire surface of the orchard, with vegetation growing up to the tree trunks. To implement this strategy, the areas directly surrounding the tree must be kept vegetation free until the tree becomes mature. The cover crop is mowed regularly, with cut vegetation returned to the orchard floor. This method reduces soil erosion and is especially advantageous for reducing erosion where slopes are severe. Solid cover crops also keep the orchard cooler, reducing air temperature by up to 0.5–1°C (1–2 °F), they help keep the soil aerated, and help moderate tree growth due to root competition. A potential disadvantage is increased frost potential, and if the vegetation out

Fig. 8.12. Clean cultivation is a common practice in this Spanish bush (SB) orchard in the Ebro Valley of Spain (courtesy of L.E. Long).

competes the tree's root system, growth and fruit quality can be affected. In addition, rodent damage can be a potential problem and *Pseudomonas syringae*, the causal agent of bacterial canker, can live epiphytically on cover crops which may provide inoculum for infections under certain conditions (see Chapter 11).

8.5.4   Clean cultivation

The most common alternative to a well manicured vegetated orchard floor is open, bare ground (Fig. 8.12). In drier climates, where drip irrigation is used, there is not enough precipitation to support a grass alley or cover crop, thus the ground is kept bare mostly from lack of adequate moisture, although in some situations, herbicides are also used. In drip irrigated orchards, the soil becomes compacted and serves as a surface for all orchard practices. Where orchards are non-irrigated, cover crops would compete with the trees for scarce water. In these situations, the ground is commonly tilled to a depth of 30 cm (12 in) so that a 'dust mulch' is formed that helps to reduce soil moisture loss. Orchards with bare ground, whether tilled or hard packed, are as much as 2°C (3.5 °F) hotter than those with cover crops, which will advance ripening, a potential advantage in some situations. This can also be an advantage during spring frosts, although it could

Fig. 8.13. In this replant site, old trees were chipped up and returned to the orchard as mulch under these newly planted trees (Oregon) (courtesy of L.E. Long).

also lead to earlier bloom (and therefore could increase frost risk). The downsides to a vegetation free orchard floor include the potential for lower fruit quality, including smaller/softer fruit, an increase in sunburn potential, and increased depletion of soil OM. In addition, using equipment on bare ground during or shortly after rains can be slippery and cause severe soil compaction. The negative effects of soil erosion also must be considered, as wind and rain can quickly remove topsoil. Nitrate leaching is also greatest with this management strategy.

8.5.5 Mulches and composts

Orchard mulching has been used by some growers to reduce soil moisture loss, increase soil OM and improve soil biology (Fig. 8.13). Scientists have documented many of the benefits of mulching in orchards, including modifying the soil environment, buffering roots from moisture and temperature

Fig. 8.14. A freshly applied layer of straw mulch will conserve moisture and add to the biodiversity of the soil under these mature trees (Oregon) (courtesy of L.E. Long).

extremes, adding carbon for soil biology and improving physical properties related to water intake and storage. However, even with these known benefits, mulching is still a practice that is not widely adopted. Limiting factors related to the use of mulch are the availability of appropriate and consistent mulch materials, and equipment to efficiently apply the mulch.

However, as the importance of soil biology in orchard systems has become better known, interest in composts and mulches has increased. Some of the more popular mulches include straw, wood chips, chipped prunings and geo-textile fabrics. All of these will conserve moisture, and the organic mulches can help stimulate biodiversity in the soil (Fig. 8.14). However, rodent control needs to be aggressively pursued when mulches are used, and all of the organic mulches will tie up soil N. Additional N needs to be added to avoid a deficiency.

Composts can be added to the organic mulches, either applied just prior to the mulch or mixed with the mulch to help remediate all or some of the N deficit (see Table 5.2 and the discussion in Section 5.2.3.5: 'Remediation of organic matter'). The compost/mulch combination also can enhance soil biology and improve soil health. Scientists have shown that the soil rhizosphere is teeming with microbial life. This biodiversity, which has been called the 'soil food web', consists of bacteria, fungi, protozoa, arthropods, nematodes, earthworms, insects and more. How these micro-organisms affect the health and growth of trees, and ultimately the fruit quality, is not well understood. However, the benefit of these amendments is supported by scientific research. In a commercial organic orchard in Washington, a wood chip mulch led to a yield increase and positive return on investment of nearly $10,000/ha ($4,000/acre) over three seasons (Granatstein *et al.*, 2014). Another study on cherries in Oregon with straw mulch found reduced irrigation water usage and increased fruit size at an economically meaningful level (Yin *et al.*, 2012).

Compost from municipal waste or other sources is often mixed with a source of high carbon mulch (such as wood chips, straw, corn cobs or other organic sources) in a 1:2 ratio along with lime or gypsum. Some growers attempt to further enhance biological diversity with applications of fish emulsion and other 'activators' applied bi-weekly via tree row ground sprays or fertigation to improve biological activity in the rhizosphere while improving mineralization, increasing OM, aeration and water infiltration in the soil.

In order to maintain the benefits derived from the mulch, reapplication is necessary on an intermittent basis since the product decomposes and is incorporated into the rhizosphere. This is especially true during the first year after application, as the OM is rapidly broken down. Nevertheless, many growers believe that the benefits are worth the cost.

Some growers also use geo-textile fabrics to serve as a permanent mulch in the tree row (Fig. 8.15). This is a woven synthetic fabric, usually black, that is positioned in the weed free zone under the trees at planting. The best materials can last 10 years or more, providing weed control while reducing herbicide usage and irrigation needs. In a study by Yin *et al.* (2007), woven fabric had annual gross returns greater than annual costs in the fourth year after planting, by $8,181/ha ($3,131/acre) compared to herbicide-treated strips. By the sixth year after planting, the geo-textile fabric had $17,796/ha ($7,201/acre) net returns greater than the total costs of all previous years.

Another class of manufactured 'mulch' materials is reflective fabrics or films that are applied temporarily for a specific period of time, usually to cover the tractor alley during fruit development. These can be white extruded plastic or woven geo-textile type materials (such as Extenday), or silver metallicized plastics (such as Mylar) that are rolled out on the orchard floor to reflect sunlight back up into the tree canopy (Fig. 8.16). These materials can be advantageous in high density orchards wherever there is a significant amount of sunlight not being intercepted by the tree canopy (therefore, wasted radiant energy for fruit production) as with vertical fruiting wall canopy architectures, and/or particularly where lower portions of the canopy are relatively shaded, as with Y- or V-trellises (provided there is a gap for sunlight between the tops of the tree canopies). If there is mostly shade on the orchard floor throughout the day, there will

Fig. 8.15. A geo-textile fabric has been used as a mulch in this newly planted upright fruiting offshoot (UFO) block. Benefits include reduced weed control, conservation of moisture and precocity (Oregon) (courtesy of L.E. Long).

be little benefit to using reflective mulches. Similarly, growing regions with lower light intensity or increased cloudiness may have proportionally less of a plant response than in higher light, sunnier regions. In orchards with continuous protective covers, the amount of direct sunlight reaching the orchard floor is reduced and thus reflected light will also be limited. In orchards with row covers that have gaps over the tractor alley, direct sunlight penetrating between the cover gaps can potentially be reflected into the tree canopies.

Decisions on the use of reflective mulches should also consider the potential production benefits and the durability of the materials. The photosynthetically active wavelengths of sunlight that are reflected back up into the canopy can increase leaf photosynthetic activity, improving fruit ripening uniformity and sweetness. The UV wavelengths of sunlight can increase red pigment formation (such as anthocyanin biosynthesis) on the fruit surface, which can be especially valuable for developing an attractive blush on yellow-fleshed cherries like Rainier. Beyond fruit benefits, reflected light into parts of the canopy that are less exposed to sunlight, can improve flower bud formation (or, seen in another way, reduce flower bud abortion), thereby increasing potential productivity. Similarly, such reflected light can help retain vegetative buds and improve the potential for renewing future

Fig. 8.16. Reflective mulches can increase fruit quality and provide higher yields (courtesy of G.A. Lang).

fruiting shoots. In terms of material durability, woven fabrics tend to arrest potential tears in the material better than extruded films. Thicker products are usually more durable than thin, and can potentially be reused for more production seasons. Some materials may have reinforced edges or even grommets for use of hook-ended bungee cords to help keep them in place during the season. Films are sometimes kept in place by shoveling small amounts of soil on to the material edge every 1–2 m (3–6 ft). Of course, when using reflective materials for more than one season, as they become dirty, the quantity of light reflected is reduced proportionally.

8.6   Weed Management

Cherry trees, especially when young, do not compete well with weeds (or cover crops) for water and nutrients. Such competition forces cherry roots to grow at lower soil depths, relegating the best and most biologically active soil to the weeds or cover crop. Weeds can interfere with sprinkler mechanisms and water distribution, and also shelter rodents that can damage tree trunks. When these weeds grow up to and around the trunk, they allow mice or voles to feed on the tree bark without fear of predators. A careful weed management program can significantly reduce potential rodent damage.

Weeds also can shelter some insect pests and diseases, allowing populations to build up before spreading to the trees. *Pseudomonas syringae* can grow epiphytically on grasses and weeds before populations reach infectious levels to cause bacterial canker in the tree. Pests such as *Lygus* bugs can multiply on legumes and plants in the mustard family, then spread as adults to developing fruit. This can also happen with thrips and spider mites when weeds are mowed in preparation for harvest.

In newly planted orchards, weeds must not be allowed to compete with young trees. Weeds can quickly stunt trees and inhibit yield potential. Careful hoeing, hand pulling or herbicide treatments are all options to control weeds. To prevent phytotoxicity, glyphosate should be avoided during the first 2 years after planting. Careful application with a contact herbicide, such as paraquat, can help reduce weed competition. A tree guard placed around each trunk is recommended to protect the tree from potential hoeing, mowing and herbicide damage.

8.6.1 Types of weeds and control strategies

Annual weeds complete their life cycle from seed germination to flowering, seed production and dying all in one growing season. Examples include pigweed, prostrate knotweed and wild buckwheat. Since these weeds produce seeds shortly after germination, it is important to control them early in their life cycle. Ideally, the best way to control annual weeds is to use pre-emergent herbicides to prevent them from germinating in the first place. Post-germination, annual weeds can be controlled with a contact herbicide or hoeing/cultivation.

Biennial weeds complete their life cycle in two growing seasons. These grow vegetatively the first year, then flower, produce seeds and die in the second year. Examples include common mullein, burdock, bull thistle and wild carrot. In the first year of growth, a contact herbicide or hoeing/cultivation may be enough to kill a biennial weed. Herbicides are generally most effective against biennials if applied before the weed has bolted and produced a seed stalk.

Perennial weeds establish a large underground root structure that allows them to survive and produce seeds or spread clonally (as by runners or rhizomes) for many years. Examples include Johnson grass, field bindweed and Canada thistle. Due to their well established root systems, mechanical control measures, such as flaming and tilling, are ineffective since roots tend to regenerate new growth. Systemic herbicides that translocate to kill the crown and root are usually the most effective methods of control.

8.6.2 Types of herbicides and use

The most common method to control orchard weeds is with herbicides. This can be challenging in newly planted orchards since young trees have thin bark that is easily penetrated by herbicides, with the potential for significant damage. For example, glyphosate damage can occur in young orchards due to bark contact or spray drift that contacts the foliage on low branches. Since glyphosate moves systemically within the tree, damaged trees can be stunted to the point that removal is necessary. It is possible to protect the trunk with non-porous wraps, grow tubes or waxed cardboard containers, or by carefully painting the trunk with a white latex paint. Some growers avoid glyphosate use for the first 2 years, instead using a contact herbicide, such as paraquat. Sometimes, 2,4-D amine is mixed with paraquat to improve efficacy. However, even paraquat is absorbed by tree bark, causing a chemical burn on the trunk that can predispose it to bacterial canker or other infections. The use of residual herbicides also should be avoided in the first 2 years after planting. Young trees have shallow root systems and cracks in the soil around the newly planted trees can be an avenue for herbicides to travel into the root zone.

The soil residual herbicides simazine, diuron, dichlobenil, and terbacil are not recommended for newly planted orchards since they are potentially mobile in the plant and can be absorbed by roots. For this reason, special care should be taken in orchards with sandy or rocky soils with low OM. The soil residual herbicides indaziflam, rimsulfuron, oryzalin, pendimethalin, oxyfluorfen, and pronamide are not particularly mobile within the tree. Due to their greater level of safety, they are generally the most widely used residual herbicides today. For increased efficacy, they often are used in combination to control a broader spectrum of weeds.

Systemic contact herbicides include glyphosate and 2,4-D. These are powerful broad spectrum herbicides that require extra care for use due to their systemic mode of action. Glyphosate is active against many difficult weeds, both grasses and broadleaf. Mixing glyphosate with water containing high levels of minerals can significantly reduce its efficacy. In hard water situations, concentrating the glyphosate by using up to two-thirds less water will increase its activity. The efficacy of glyphosate also can be improved by adding ammonium sulfate (AMS) to increase uptake into the plant and prevent cations, such as calcium sulfate, from combining with glyphosate. The most common symptoms of glyphosate injury in cherry are strap-like leaves, but limb dieback and stunting, and even death of young trees, can occur (Fig. 8.17). Damaged trees may exhibit symptoms and remain stunted for several years. Recent evidence suggests that prolonged use of glyphosate can cause trunk cankers and cracked bark. In addition, it appears that glyphosate may accumulate in both the plant and soil, so damage can become significant with even minor drift incidents. 2,4-D is active only against broadleaf weeds and is most effective when plants are

Fig. 8.17. Glyphosate damage includes elongated, strap-like and crinkled leaves (courtesy of L.E. Long).

young. It is absorbed by foliage and bark, and can move through soil to be absorbed by roots. 2,4-D damage in cherry includes twisting and abnormal development of newly formed leaves.

Non-systemic contact herbicides include paraquat, gluphosinate-ammonium and pyraflufen-ethyl. These often are used to control weeds in young orchards, although applications should still avoid contact with green bark, especially with gluphosinate-ammonium. As noted above, even localized bark injuries can lead not only to temporarily reduced growth, but also infections by serious pathogens. A wetting agent should be used to improve contact herbicide coverage. Repeat applications will be necessary for season-long control.

All herbicide use in cherry orchards, at any stage, should follow label directions carefully. Improper herbicide applications can cause significant damage, especially to young cherry trees. Depending on the type of herbicide, damage can be local or systemic. The response of the tree to the damage depends on the rate, dosage, tree age, vigor and stress factors. To reduce the potential for herbicide injury, applications should be made on wind free days at pressures no greater than 200 kPa (30 psi). Herbicides may be applied with hydraulic sprayers at pressures ranging from 130–200 kPa (20–30 psi). Most cherry orchards are sprayed with a boom-type sprayer, with three or four flat fan nozzles placed about 30 cm (12 in) apart. For complete coverage, the spray pattern from each nozzle should overlap by about one-third on both sides. Avoid nozzles that produce a fine mist that can easily drift.

8.6.3 Precision weed management

New technological advances in weed management utilize advanced camera/sensor optics and computer circuitry to identify the presence of weeds and target pulse herbicide applications to only the weeds, from directed spray nozzles. This saves significant money on herbicides, and reduces the environmental impact. Currently, scientists are developing technologies to determine the minimum amount of herbicide needed to kill a weed. These microdoses would significantly reduce or eliminate drift and leaching of chemicals into root zones and ground water. Similarly, as machine vision and image processing technologies improve, increasing the precision of automated weed identification, individual weeds may be targeted with the most effective control method available. For example, if an identified weed is resistant to glyphosate, another chemical or method will be used to eliminate it. Autonomous, battery-powered tractor sprayers guided by global positioning satellite (GPS) coordinates may soon be available to patrol orchards on a periodic basis to spot-spray weeds to minimize their establishment and competition with the trees.

8.7 Low Temperature Management

Plant responses to low temperatures are discussed at length in Chapter 4. Management strategies to reduce the potentially negative impacts of low temperature events in established orchards focus on the type of event being managed. A distinction must be drawn between potential damage from fall/winter (advective) freezes and spring (radiative) frosts. Freeze damage occurs in the late fall or winter when temperatures drop below the critical temperature at which plant tissues (buds, cambium, etc.) are cold-acclimatized, resulting in cellular damage and tissue death. Plant tissues become acclimatized to increasingly lower temperatures in the fall and winter, but this acclimatization is dynamic and can fluctuate depending on the temperatures that precede critical temperature exposures. For example, several days of warm weather in midwinter can cause tissues that previously were cold-acclimatized to $-20°C$ ($-4°F$), to de-acclimatize and become susceptible to damage at $-16°C$ ($3°F$).

This can occur even on a diurnal basis, such as south-west or north-west winter injury (in the northern or southern hemispheres, respectively) to cherry tree trunks. On cold sunny winter days in the northern hemisphere (usually a result of a cold air mass originating in the Arctic), the south-west side of the tree trunk exposed to afternoon solar radiation becomes

Fig. 8.18. Southwest injury has occurred to this young tree during the preceding winter (courtesy of L.E. Long).

much warmer than the air temperature, and loses some of its cold hardiness (Fig. 8.18). If the overnight low temperature plunges to values below its modified level of cold acclimatization, the cells will freeze, causing death in the vascular cambium and ultimately cracks in the bark in the spring as the dead tissue fails to expand with growth of the surrounding live tissue.

This contrasts with radiative freezes in fall or spring (also known as frosts). In these situations, the potentially damaging low temperatures are not the result of a cold air mass across the region, but rather the loss of heat through radiation to the atmosphere on a calm clear night. In frost situations, day temperatures are typically well above 0°C (32°F). At night, heat from the earth and other objects radiates into the atmosphere, resulting in warm air from the ground rising above upper colder air layers, forming a temperature inversion so that lower air temperatures drop below 0°C (32°F) (Fig. 8.19). The inversion layer of air warmer than at the orchard surface is typically about 11 m (~36 ft) off the ground. When inversions exist, the use of a wind machine (a motor-driven large propeller on a tower) can pull some of the upper layer of warm air back down into the orchard and mix it with the heavier colder air (Fig. 8.20). Thus, wind machines are most effective when the temperature differential between the inversion layer at 11 m (36 ft) and the orchard at 1.5 m (5 ft) above ground is >2–3°C

Fig. 8.19. These flowers and fruitlets exhibit the typical damage caused by a radiative spring frost (courtesy of G.A. Lang).

(~4°F). Persistent winds in the range of 8–16 kmph (5–10 mph) will prevent the formation of an inversion.

Besides installing wind machines, other strategies to avoid or reduce frost damage caused by a radiative freeze include:

1. Maintain soil moisture near field capacity and wet the top 30 cm (12 in) of the entire soil surface 2–3 days in advance of a predicted frost event. It is not necessary to wet the entire profile since the top layer will insulate and protect the soil lower in the profile from losing heat. Orchard floor management has an impact when bare firm moist ground is warmer than an orchard floor covered with a tall grass or cover crops that can restrict air drainage. This can make a difference of as much as 3–4°C (6–8 °F).

2. Mow the orchard grass or cover crop as close to the ground as possible in late winter before budbreak and the first risk of spring frost.

3. As noted earlier in this chapter, under-tree sprinkler irrigation systems can be operated to add heat to the orchard.

4. Other sources of supplemental heat can be used, with or without wind machine operation, such as burning propane gas (Fig. 8.21). In past decades, other fuel sources included diesel, kerosene, wood, straw or

Fig. 8.20. Wind machines help to protect this valley from spring frosts (Oregon) (courtesy of L.E. Long).

Fig. 8.21. Propane heaters are used to protect this orchard from extreme winter cold and spring frost (Washington) (courtesy of G.A. Lang).

tires, but these have negative environmental consequences and have lost favor, if not been banned outright, in many areas. Propane (or other) heaters should be distributed more heavily on the perimeter of the orchard, especially the windward side so that if there is any wind, the heat is carried into the orchard; the remaining heaters should be distributed evenly within the orchard.

5. Pathogen management can also contribute to protection from frost damage. Ice-nucleating bacteria, including *Pseudomonas syringae, P. fluorescens, P. viridiflava, Erwinia herbicola* and *Xanthomonas campestris* var. *vesicatoria* are ubiquitous in nature. These bacteria promote the formation of ice crystals at higher temperatures than normal, for example, commercial ski slope snow-making operations use *P. syringae* to produce snow at −2.7°C (27°F) versus pure water at −9.4°C (15°F). Thus, reducing the populations of these bacteria, such as with copper hydroxide applications prior to a predicted frost event, can minimize ice nucleation and frost damage as long as temperatures do not drop below −3.2°C (25°F).

For More Information

Black, B., Hill, R. and Cardon, G. (2008) *Orchard Irrigation: Cherry.* Utah State University, Logan, UT.

Shock, C.C., Wang, F.X., Flock, R., Feibert, E., Shock, C.A. and Pereira, A. (2013) Irrigation Monitoring Using Soil Water Tension. Oregon State University EESC. *EM 8900.* www.catalog.extension.oregonstate.edu/sites/catalog/files/project/pdf/em8900.pdf (accessed 19 June 2020).

References

Alyokhin, A., Porter, G., Groden, E. and Drummond, F. (2005) Colorado potato beetle response to soil amendments: a case in support of the mineral balance hypothesis? *Agriculture, Ecosystems & Environment* 109(3-4), 234–244. DOI: 10.1016/j.agee.2005.03.005.

Datnoff, L., Elmer, W.H. and Huber, D.M. (eds) (2007) *Mineral Nutrition and Plant Disease.* APS Press, St. Paul, MN, pp. 278.

Granatstein, D., Andrews, P. and Groff, A. (2014) Productivity, economics, and fruit and soil quality of weed management systems in commercial organic orchards in Washington State, USA. *Organic Agriculture* 46, 197–207. DOI: 10.1007/s13165-014-0068-0.

Hanson, E. (1996) *Nutrition of Fruit Crops.* Michigan State University, MSUE Bulletin E-852.

Kaiser, C., Christensen, J.M., Long, L.E., Over, S.M. and Erm, K. (2012) Reduction of sunburn in 'Golden Delicious' apple (*Malus x domestica* Borkh.) fruit using a hydrophilic biofilm. *American Society of Horticultural Science Annual Conference*, Miami, Florida, Jul 31-Aug 3.

National Center for Appropriate Technology (2009) *The Pacific Northwest Irrigator's Pocket Guide.* Butte, Montana.

Neilsen, G.H., Neilsen, D., Kappel, F. and Forge, T. (2014) Interaction of irrigation and soil management on sweet cherry productivity and fruit quality at different crop loads that simulate those occurring by environmental extremes. *HortScience* 49(2), 215–220. DOI: 10.21273/HORTSCI.49.2.215.

Phelan, P.L. (1997) Soil-management history and the role of plant mineral balance as a determinant of maize susceptibility to the European corn borer. *Biological Agriculture & Horticulture* 15(1-4), 25–34. DOI: 10.1080/01448765.1997.9755179.

Thien, S.J. (1979) A flow diagram for teaching texture by feel analysis. *Journal of Agronomic Education* 8(1), 54–55.

Werner, H. (2002) Measuring soil moisture for irrigation water management. South Dakota State University Extension. *FS* 876.

Yin, X., Seavert, C.F., Núñez-Elisea, R., Núñez-Elisea, R. and Cahn, H. (2007) Effects of polypropylene groundcovers on nutrient availability, sweet cherry nutrition, and cash costs and returns. *HortScience* 42(1), 147–151.

Yin, X., Long, L.E., Huang, X.-L., Jaja, N., Bai, J. *et al.* (2012) Transitional effects of double-lateral drip irrigation and straw mulch on irrigation water consumption, mineral nutrition, yield, and storability of sweet cherry. *HortTechnology* 22(4), 484–492. DOI: 10.21273/HORTTECH.22.4.484.

Fruit Ripening and Harvest 9

Cherry fruit exhibit a three-stage, double sigmoid growth pattern. Stage I is characterized by rapid growth and final cell division following successful pollination and fertilization of the flower. Fruit also photosynthesize during this stage, contributing to their own growth resources. During Stage II, growth of the fruit slows as cell division ceases, the embryo develops and the pit hardens. Fruit photosynthesis also occurs during Stage II. Stage III is marked by the beginning of the change in color, from green to straw-yellow to various depths of red (in red-fruited cultivars) at maturity. Fruit photosynthesis declines and growth becomes exponential as the fruit flesh swells due to cell elongation. Fruit sugar levels increase during Stage III, which lowers the osmotic potential in the fruit. This rapid increase in weight and volume can lead to significant mechanical stress (tension and strain) in the cuticle that thins out as expansion occurs, often resulting in microcracks. Some studies have shown that irregular water supply to cherry trees during Stage III can increase the number of cuticular fractures.

9.1 Fruit Ripening and Maturation

Achieving good postharvest fruit quality begins long before harvest. Preharvest factors that affect postharvest fruit quality include crop load, light intensity, fruit calcium (Ca) levels and stress factors. Fruit grown in a balanced tree canopy, with the proper leaf area-to-fruit (LA:F) ratio, will be larger, have reduced tendency to pit, and have more sugar and higher acidity than fruit on an overset tree. These fruit will also be firmer after storage. This all translates to higher quality, better flavor and a superior eating experience for the consumer. Heavy crop loads not only affect fruit size negatively, but also increase the potential for pitting during the postharvest period as well as decrease fruit firmness, stem pull force, and fruit sugar levels. Heavy crop loads delay ripening and harvest date. Some self-compatible varieties, such as Lapins and Sweetheart, are more likely to set heavy crop loads, even on less productive rootstocks such as Mazzard or Colt.

Although it is expensive, many growers in Chile, Spain and even the Pacific Northwest (PNW) of the US will hand-thin trees with excessive fruit sets to improve fruit quality. Hand-held string thinners also are used commercially to thin flowers in the PNW. Workers typically target the terminal sections of two-year and older shoot growth, where the flower clusters are most dense. Unfortunately, thinning of flowers or fruitlets with spray applications of abscission-inducing chemicals is still in the experimental stage and not yet reliable commercially. By far the easiest and least expensive way to achieve balanced tree crop loads is with pruning. To prevent the formation of heavy clusters in highly productive varieties or varieties that overset on highly productive rootstocks, annual head pruning of the terminal section of previous season shoot growth can remove the future dense cluster sites before they form fruiting spurs. Approximately one-third of the terminal shoot growth should be removed each year to better balance fruit bud formation with supporting leaf area (see Chapter 6, 'Sweet Cherry Pruning Fundamentals'). Without crop load management, varieties that set dense clusters of fruit can be difficult to harvest gently. Dense fruit clusters make finger access to pick the fruit by the stem difficult, increasing the likelihood of damage during harvest. Furthermore, fruit in dense clusters usually have lower dry matter levels and are smaller compared to more widely spaced, less dense fruit clusters.

Fruit firmness is related to variety, growing practices, temperature and light interception. Choosing varieties with firm fruit and growing cherries on a tree with a balanced crop load and good light distribution throughout the canopy will greatly improve the potential to export fruit to distant markets. Firm fruit is better able to withstand picking, processing and shipping with minimal damage. Fruit firmness can be easily measured by a hand-held durometer in the field (Fig. 9.1) or a Firmtech II instrument in the packing house (Fig. 9.2). A firmness value of 70 to 75 on the durometer or 275 g/mm on the Firmtech are minimum values for export quality fruit.

As fruit maturation approaches harvest, good light distribution to the fruit-bearing sections of the tree canopy becomes more important for achieving premium quality cherries. A general guideline is that 30–40% sun exposure is optimal for fruit quality. Sugar levels are lower and fruit is softer at harvest and after cold storage, when light levels are not properly managed. In addition, fruit Ca levels, important for storage, stem color, fruit luster and many other quality factors, are also lower for shaded fruit. Some training systems, such as the upright fruiting offshoots (UFO), inherently provide good light penetration due to their planar canopy architecture. Other training systems, however, can benefit from detailed pruning in the winter for establishing a good canopy structure and balanced cropping potential, and summer pruning (prior to harvest) to remove suckers and

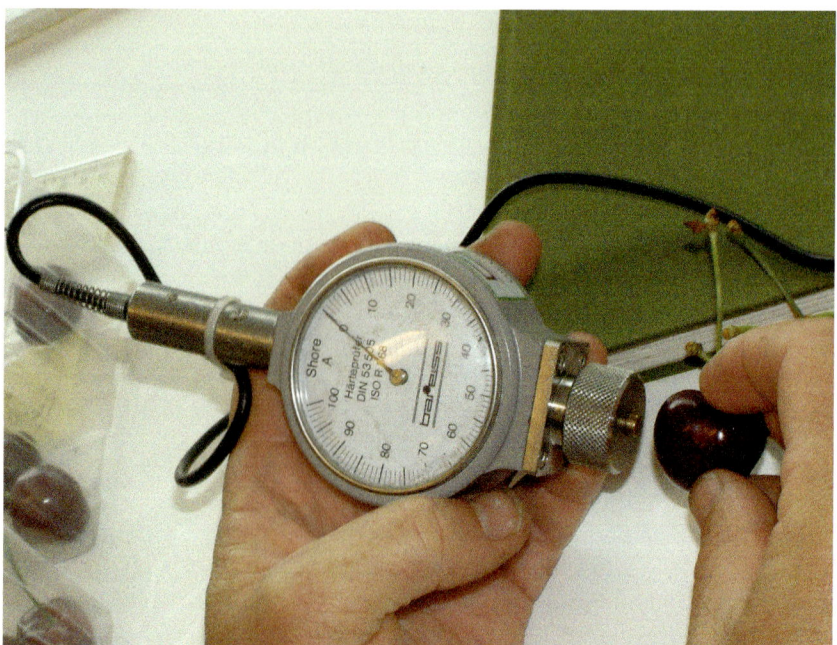

Fig. 9.1. The durometer is used widely throughout the world to determine cherry fruit firmness. These instruments can be used either in a laboratory or taken to the field (courtesy of C. Kaiser).

other extension shoot growth that interferes with canopy light infiltration during ripening.

Plant stress, due to inadequate irrigation or other factors, can cause fruit chilling injury in storage that is manifested as internal browning, wooliness or bleeding around the pit. Some years are worse than others for the expression of these defects. Therefore, avoiding any form of transient stress during the growing season is important.

Ca plays an extremely important role in both fruit cell wall and cell membrane development. The level of Ca in cell walls affects fruit firmness, physiological disorders and decay, while its presence in cell membranes determines respiration activity and even flavor. Fruit Ca levels are directly proportional to fruit firmness; the higher the Ca, the higher the firmness. In addition, fruit with higher Ca levels have reduced respiration levels, less stem browning and decay, greater resistance to pitting, better luster and higher acidity. High fruit Ca levels will also improve resistance to rain-induced fruit cracking and heat stress.

However, fruit is often deficient in Ca due to its unidirectional movement in the plant to areas of higher transpiration, such as leaves. Compounding this issue are environmental and cultural factors that further hinder Ca movement to the fruit, such as acid soils, high temperatures,

Fig. 9.2. The laboratory-based Firmtech instrument is used mainly in North America and Chile to determine fruit firmness (courtesy of L.E. Long).

water stress, excessive pruning, low crop load and uptake/translocation competition with nitrogen (N) and potassium (K).

Since Ca moves so poorly within the tree, efforts to increase soil Ca levels will have only limited advantages. Ca sprays, applied directly to the fruit, can be effective. Unfortunately, most growers do not apply enough Ca to significantly impact fruit quality.

Effective application rates are 0.1–0.15% using chelated calcium (either amino acid or organic acid chelated products), with the amount of Ca product used varying proportionally with the Ca concentration in the product. A liquid acid chelated product containing 8% Ca would be mixed at 12 l per 940 l of water per hectare (5 quarts per 100 gal of water per acre). Due to the high Ca concentration, applications should be made only after pit hardening, in order to prevent phytotoxicity to young, developing leaves. In addition, Ca should not be applied when temperatures are greater than 27°C (80°F). If applications are not tank mixed with a pesticide, a non-ionic wetting agent should be added to the spray tank to enhance Ca uptake and reduce the potential for leaf margin burn. Chelated calcium (0.05% Ca), sprayed six times at weekly intervals from pit hardening to harvest, was found to be more effective than other Ca sources for improving fruit quality (Wang *et al.*, 2015). Treatments of calcium chloride and calcium citrate negatively affected fruit size, but other Ca sources did not.

Although the research was conducted with six applications, some preliminary data indicates that four applications at the indicated rate will provide similar benefits.

Fruit Ca levels can also be enhanced postharvest (by adding Ca^{2+} (in the form of calcium chloride, $CaCl_2$) in the hydrocooler water (Wang et al., 2014). This will impart many of the same benefits obtained through preharvest foliar sprays, including an increase in fruit firmness, less stem browning, reduced pitting susceptibility, reduced postharvest splitting, reduced respiration that helps to maintain acid and flavor levels, and an overall reduction in fruit decay. Add Ca^{2+} to the hydrocooler water at a rate of 0.2–0.5%. Cherries should be treated for 5 min. As the cherries will adsorb the Ca^{2+}, levels of Ca^{2+} should be constantly monitored. Higher rates of Ca^{2+} will damage the pedicels, so care should be taken to maintain Ca^{2+} levels at the indicated rate (Wang *et al.*, 2014).

9.2   Cherry Fruit Cracking

Rain-induced cracking of sweet cherry fruit can have severe economic impacts, and is a major determinant of where economically viable fresh market sweet cherry production has been sustained around the world. Cracking of 25% or more generally renders the crop uneconomical to pick and pack, unless damaged fruit can be removed autonomously in the packing house using optical sorting equipment and software. Since cracking can occur anytime from several weeks to only a day or two before harvest, the potential always exists for a high value crop to become progressively or instantly worthless, depending on the number and duration of rain events. This can lead to a loss of long-term market opportunities if marketing contracts cannot be fulfilled on a consistent basis. Understanding the causes of cracking is critical to devising effective strategies and/or implementing effective technologies to reduce their incidence.

Although cherry fruit cracking is affected by morphological, physiological, environmental and genetic factors, there are two major practical driving forces: (i) water on the surface of the fruit that is absorbed through the cherry cuticle; and (ii) water that moves into the fruit internally through vascular connections following uptake by the root system (Fig. 9.3).

During rain events, there are three areas of the fruit surface that tend to remain in prolonged contact with rainwater: (i) the 'bowl' of the fruit where the pedicel is attached tends to collect water; (ii) where two fruits touch side-to-side, surface tension tends to hold a small quantity of water between the fruits, whereas water runs off the remaining sides of the fruits; and (iii) at the tip of the fruit, where water drips off but the droplet formed there provides prolonged contact with the cuticle. The tip of the fruit is also the site of the scar from the style of the flower, which is less elastic than the normal cuticle and therefore more prone to cracking, especially in

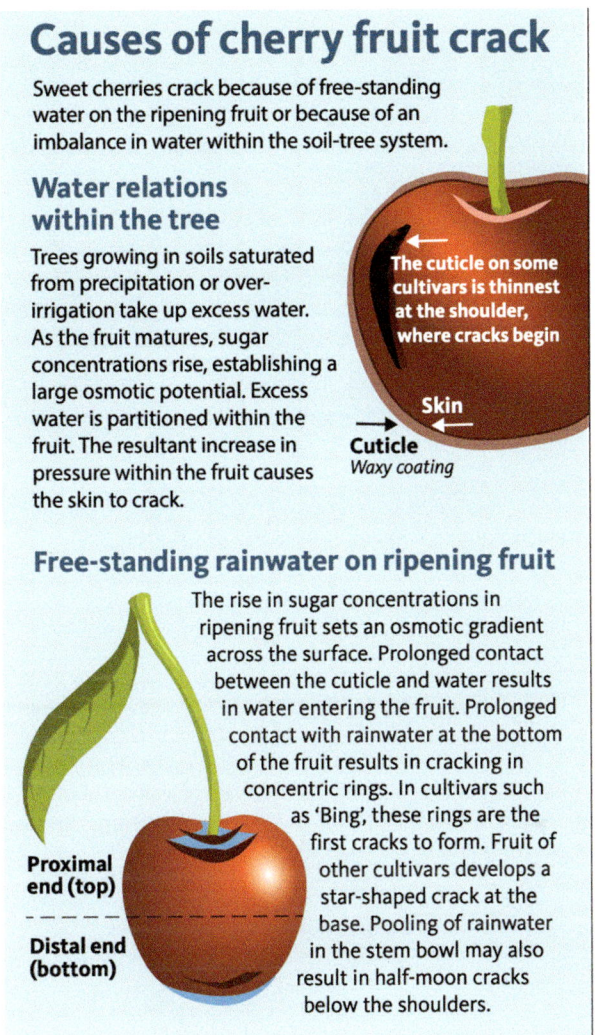

Fig. 9.3. There are two potential causes of rain-induced fruit cracking (courtesy of C. Kaiser).

those cultivars with large stylar scars. Each of these sites of prolonged water contact promote localized water uptake through microcracks in the cuticle, causing localized swelling until the cuticular cracks enlarge and fruit flesh begins splitting open, resulting in bowl cracks, tip cracks, or less frequently, shoulder cracks. The greatest prevalence of microcracks occurs at the stylar end of the fruit. A study in the northern hemisphere found that fruit on the southern side of the tree developed more microcracks than on the northern side, presumably due to greater fruit shrink–swell activity on the sunnier side of the canopy (Hovland and Sekse, 2003).

Fig. 9.4. This photo illustrates the three main types of rain-induced cracking in cherry fruits (courtesy of R. Bastias, University of Concepcion).

Site for bowl cracking

Site for root uptake side-cracking

Site for fruit-to-fruit side cracking

Site for tip cracking

Fig. 9.5. Typical sites for the formation of rain-induced cracking in sweet cherry fruits (courtesy of G.A. Lang).

When water moves into the fruit internally through the vascular connections, following uptake by the root system, the localized swelling tends to be along the sides of the fruit, resulting in side cracks (Figs 9.4 and 9.5). Consequently, the incidence of fruit cracking can vary between rain events, and each event may result in fruit cracking from water absorption through the cuticle or uptake through the vascular system. Furthermore, no two

Table 9.1. Sweet cherry cultivar susceptibility to fruit cracking, based on natural rainfall events in The Dalles, Oregon (2008–2019), and international observations.

High	High to moderate	Moderate	Low
Bing	Benton	Chelan	Big Star
Brooks	Sandra Rose	Coral Champagne	Black Pearl
Early Robin	Sweetheart	Cristalina	Black Star
Rainier	Tieton	Ebony Pearl	Blaze Star
Royal Edie		Garnet	Burgundy Pearl
Royal Helen		Kordia (Attika)	Tulare
Santina		Lapins	
Selah		Radiance Pearl	
Skeena		Regina	
Utah Giant		Royal Hazel	
Van		Samba	

rain-cracking events are likely to be similar, since cracking susceptibility can vary based on stage of fruit development, temperatures during and after each event, wind and cloudy versus sunny conditions following the rain, soil moisture profile preceding and following the rain event, and root zone wetting and drying patterns (affecting continuous versus episodic fruit growth) preceding the rain event.

Consequently, excessive rainfall events (>40 mm or 1.5 in of rain per event) and soil moisture profiles near field capacity before a rainfall event, can increase cracking. Plant internal water relations are directly influenced by rootstock and soil water relations, as well as atmospheric conditions (temperature, wind, relative humidity, vapor pressure deficit) that affect transpiration. Rootstock trials in The Dalles, Oregon, found Mahaleb to generally be the most cracking-susceptible rootstock, followed by Mazzard, Gisela 5, Gisela 6, MaxMa 14, Krymsk 5 and Krymsk 6. In Australia, trees on Mazzard were more susceptible to fruit cracking than trees on Mahaleb, but both were more susceptible than Stockton Morello. Some cherry cultivars are more crack-resistant than others due to genetics, which controls cuticle morphology, composition and thickness. For example, Bing, Skeena and Brooks are highly susceptible to rain-induced fruit cracking, while Regina and Black Pearl is somewhat tolerant (Table 9.1).

Irrigation and plant nutrition also are linked to fruit cracking and must be carefully managed. High leaf K concentration has been correlated positively with fruit cracking; since K competes with Ca uptake, high K levels may indicate that Ca levels in the fruit are correspondingly low. Crop load has an effect on fruit cracking; light crops of larger fruit are more prone to cracking than heavy crops of smaller fruit. Gibberellic acid (GA_3) applications at straw fruit color may increase fruit cracking. GA_3 reduces transpiration at the time of application, which may lead to a transient increase in fruit cracking susceptibility. However, transpiration rates recover within 48 h of application. Training systems that provide better leaf cover over the fruit can reduce cracking. Examples include the spur-bearing vertical fruiting wood canopies in UFO or Kym Green bush (KGB) training systems, compared to canopies with more horizontal fruiting wood and exposed non-spur fruiting sites.

Orchard management strategies and technologies to reduce rain-induced fruit cracking should always be developed with the predominant type of potential cracking in mind, that is, cracking due to prolonged fruit contact with rainfall or cracking due to excessive water in the root zone.

9.2.1   Orchard covers

Row or orchard covering systems to protect the fruit during cracking-sensitive stages of the growing season are increasingly being used in many areas of the world, particularly where rain during ripening is almost guaranteed. Typically, orchards are covered from 3 weeks prior to harvest or when the fruits start to change color from green to straw-yellow, at the beginning of Stage III when sensitivity to cracking increases.

Row covers usually form tent-like protective structures of plastic supported by a high wire directly over the tree row and two lower wires at or just beyond the periphery of the tree canopy (Fig. 9.6). These overhead curtains may be fixed from opening until harvest, or they may be devised to slide back and forth to open and close on three wires down the row. This allows them to be retracted manually and closed only during rain events to ensure maximum light interception during the fruit growth period. Row covers can be highly effective at preventing cracking due to rain contact with the fruit, but are relatively costly and can be labor intensive if they must be drawn back again when the sun is shining to prevent excessive heat build-up, encourage airflow and optimize photosynthesis. Automated systems are available, but currently are very expensive.

Wooden structures used in Norway and other European countries are susceptible to breakage in high winds (Fig. 9.7); steel or concrete pole-and-cable structures are less susceptible to wind damage if well engineered and anchored. During significant rainfall events, water running off the row cover may saturate the root zone soil, resulting in fruit cracking due to internal water relations. Planting tree rows on ridges or raised beds (especially effective with dwarfing rootstocks that have limited root zones), and/or installing tile drainage in the tractor alleys, can remove excess water before it infiltrates the root zone. Relative humidity under covers has little effect on fruit cracking; trees grown under covers where humidity levels were high did not develop cracking when soil moisture was kept just below field capacity compared to uncovered control trees (Meland *et al.*, 2014).

Fig. 9.6. Rain covers protect the ripening fruit in this orchard in Chile (courtesy of L.E. Long).

High tunnels can be very effective at preventing cracking due to rain contact with the fruit, but are not without challenges, including a typically higher cost than with individual row covers. Tunnel covers must be removed in fall and reinstalled in the spring after pollination if snowfall is likely to be significant (to prevent damage to the structure) or if their capacity to trap heat would prevent adequate cold acclimatization or chilling during winter; another labor intensive consideration. During significant rainfall events, water running off the tunnel cover may saturate the soils under the tunnels, resulting in as much as 60% or more fruit cracking, despite an absence of free-standing water contact with the fruit surface. A tunnel gutter or soil drainage system in the tunnel leg rows to direct water away from the root zone is one option to address this. Planting tree rows on ridges or raised beds can also be effective. The domed cover geometry means that hot air is more readily trapped in a high tunnel, making heat management during ripening a challenge in warmer, sunnier climates. Yet, this also means that capturing early heat for earlier bloom and ripening early is easier to accomplish in a high tunnel, potentially providing additional value in cost–benefit determinations.

High tunnels are popular in some cherry producing regions of Europe, particularly the UK where Haygrove tunnels originated for use with cherries and small fruits. They have been evaluated at Michigan State University

Fig. 9.7. These structures for rain covers, located in Norway, are made out of wood, but can be susceptible to damage by high winds (courtesy of L.E. Long).

and have shown good potential for cherry production. Their extra cost extends payback time by several years for newly established orchards; however, crop losses to rain cracking will also extend payback time, in proportion to the number of damaging rain events experienced. Although the increase in relative humidity that can develop under the cover does not usually affect fruit cracking, it may result in a higher incidence of powdery mildew and/or brown rot, but also conversely may result in a lower incidence of cherry leaf spot. Typical plastics used for covering tend to reduce ultraviolet light transmission, which together with the protection from rain may increase the residual activity of preharvest fungicide and insecticide applications. However, the altered light transmission can also disorient honeybees, reducing their effectiveness as pollinators. If tunnel covers are in place at flowering, bee hives must be placed at least 3 m (9 ft) from the tunnel, with the hive entrance oriented toward the morning sun, and without barriers between the hive and the sun. Elevate the hives 1 m (3 ft) above the ground on stacked pallets, placing up to six hives per pallet. A 50% glucose solution placed at the far end of each tunnel will encourage honeybees to traverse its full length. Commercial bumblebee hives can be more effective than honeybees for pollination, since bumblebee navigation is not affected by altered light transmission.

9.2.2 Foliar sprays to protect the fruit cuticle

In general, cherry fruit are resistant to cracking through Stages I and II of fruit growth, except when grown in areas which have high rainfall during stage III. In drier areas, cherries only become susceptible at the beginning of Stage III. At this point, the wax concentrations in the fruit cuticle starts decreasing as the fruit begins its exponential phase of expansion, causing the waxy layer to thin and microcracks in the cuticle to form. Consequently, the skin of a cherry fruit becomes a semi-permeable membrane. During ripening, the sugar concentration of the flesh increases rapidly, reaching as much as 24°Brix in cherry juice. By contrast, rainwater is relatively pure. This difference in concentrations between the fruit flesh and the raindrops sets up an osmotic differential. Pure rainwater crosses the membrane of the fruit cuticle and skin to reduce the differential by diluting the sugar concentration inside the cherry. When rainwater is in contact with the skin for extended periods of time, cuticular microcracks that form during normal fruit expansion become larger. Water moves into these cracks more rapidly and when sufficient water has been taken up, the cracks expand to the point of collapse, leading to visible cracking. Some spray surfactants can increase cuticle solubility, further thinning the fruit cuticle and increasing cracking susceptibility.

Various spray formulations of materials that form a thin, protective, water-resistant coating on the cuticle have been used to reduce the type of cracking resulting from prolonged water contact with the fruit skin. The most popular formulas available create a carnauba wax coating (Raingard™) or a copolymer of complex carbohydrates, phospholipids and Ca (SureSeal®, marketed as Parka™). Both of these types of coatings prevent rainwater from moving through the fruit epidermis into the fruit. Such sprays have no effect on cracking incidence due to internal water transported from the root zone; consequently, they work best during light to moderate rains (less than 3.8 mm [1.5 in]) that do not saturate the root zone. These must be applied according to label directions, and must cover the entire fruit surface to form a continuous protective film. To achieve this, tractor speed should not exceed 3 kmph (2 mph); spray nozzle apertures should result in droplets >150 μm in diameter. Airblast sprayer pressure must be high enough to ensure that the spray penetrates the entire tree but does not drift more than one row over. Prevailing winds during the time of application should not exceed 8 kmph (5 mph).

RainGard® is an inelastic hydrophobic protective fruit coating. Over time, fruit growth creates microcracks in the coating, so up to three additional applications are needed to fill these microcracks and protect the fruit from water uptake. The first application should be made approximately 4 weeks prior to harvest (straw color), with repeat sprays every 7–10 days thereafter. Additional applications before anticipated rain events, with at least 2 h drying time, improve protection. No less than 24 h must elapse between reapplications. The first application of RainGard® may be combined with GA_3 (applied to enhance fruit quality); subsequent applications should occur at least 2 h prior to a rainfall event. RainGard® was developed by researchers at Washington State University.

Parka™ is an elastic hydrophobic protective fruit coating that stretches as the fruit enlarges. An application of Parka™ at straw color and again 10 days later is sufficient to protect fruit through harvest. Additional benefits from Parka™ can include up to 1% higher sugar concentrations at harvest and increased stem pull force. GA_3 may be added to the first application of Parka™, but only if there is a 48 h window of dry weather. Parka™ was developed by researchers at Oregon State University.

Vapor Gard® is an inelastic, hydrophobic anti-transpirant coating; it is not recommended as an anti-cracking protective coating, although it is sometimes used that way. Several studies across Washington and Oregon found Vapor Gard® to be ineffective or in some instances, to increase cracking, possibly due to anti-transpiration effects on internal water relations under certain climatic conditions. Vapor Gard® can leave a sticky residue and render fruit unsightly at harvest. Cracked fruit treated with Vapor Gard® cannot be diverted to a brine market, as the coating reacts with the brine.

Ca, usually in the form of a spray solution of calcium chloride (0.5–1%), may increase the integrity of the fruit cuticle to serve as a barrier to water uptake and, more directly, reduce the pureness of rainwater, thereby reducing the osmotic differential that pulls water into the fruit. Calcium chloride can be applied prior to rain events, but residues may damage the leaves and/or accumulate on the fruit, reducing fruit luster and possibly requiring postharvest cleaning. It can be applied during rainfall events by continuous tractor spraying or by injection into over-tree sprinkler irrigation lines, operated automatically by a computerized controller connected to a tipping rain bucket. This automated system is controlled by a software program such that a short duration pulse of Ca is applied to coat the fruit, then only reapplied when a threshold level of additional rainfall is detected that is sufficient to wash off the previous Ca application.

9.2.3   Helicopters, airblast sprayers and wind machines

Mechanized solutions are sometimes used to physically remove rainwater from the surface of the fruit by increasing localized air movement through the tree canopy. Airblast sprayers may be difficult to operate in wet orchard soils and the airblast pattern may blow water on to neighboring trees. Helicopters are effective but are expensive strategies for 'drying' fruit to reduce cracking (Fig. 9.8). However, where rain events are relatively uncommon, paying for a helicopter flight may be an economical option when compared to the cost of a crop loss. Flying approximately 4.5 m (15 ft) above the orchard canopy and traveling at walking speed will remove most

Fig. 9.8. Helicopters are used in the Pacific Northwest (PNW) to dry cherries after a rain event (Oregon) (courtesy of M. Omeg).

surface water while minimizing wind damage to the fruit. A ground-based observer should ensure that water is removed from the stem bowl and that cherries are not being damaged by excessive wind. Long lasting rainfall events may preclude the use of this technique. Generating wind with a helicopter is ineffective in the rain; growers must wait for the rain to end, which is often difficult to predict, and payment for the helicopter may include standby time. In general, helicopters are effective at preventing fruit cracking when a rainfall event lasts less than 4 h. Wind machines generally are spaced within an orchard to move large air masses above the orchard for frost protection, rather than moving localized air adequately through the leaves and fruit of tree canopies to effectively remove water from wet surfaces. Thus, wind machine operation may be effective at reducing fruit cracking to a limited extent around the machine, but not across large sections of orchard.

9.2.4 Soil moisture exclusion techniques

Soil moisture management may be one of the most effective tools for preventing or reducing fruit cracking, particularly in combination with techniques to reduce direct contact of rainwater with the fruit. Water-impermeable soil covers exclude excess rainfall from the root zone and can reduce fruit cracking by up to 25%; however, driplines must be installed beneath the covers. Excess soil drying followed by over-irrigation can also increase fruit susceptibility to cracking. Shortening the irrigation interval to no more than 7 days will keep soils near field capacity. Field capacity without saturation (e.g. by regulating drip irrigation) from straw color through to harvest results in less fruit cracking and larger fruit at harvest. Where rainfall can be extreme, ridging tree rows will improve drainage and reduce saturation around the feeder roots. Mulching and increased soil organic matter (OM) also promote drainage, although repeated applications of OM are required for sustained improvement of the soil profile.

9.3 Harvest Indices

Due to the extremely perishable nature of sweet cherries, harvest is always a time of intense activity and significant stress for the grower. As cherries ripen, there are only a few days when they are at their prime. Harvest too early and cherries will be more susceptible to pitting, smaller in size, less sweet and lighter in color. Harvest too late and fruit will be softer, more susceptible to skin pebbling, lower in stem pull force and darker in skin color. The key is to find the right balance of skin and flesh color and flavor that will provide an attractive, firm cherry of good size. Getting the timing right is critical for shipping and good arrival at the market. There is evidence with Bing and Lapins, to indicate that cherries harvested at a slightly darker color will have less stem browning.

Most cherry growers use a combination of skin color and flavor to determine when to harvest (Fig. 9.9). Although these can be good indicators, fruit flesh color and a refractometer to determine sugar content also can be helpful. Flesh color always develops after skin color, so it is possible for cherries to have similar skin colors but different flesh colors, indicating different levels of ripeness. Most North American, European and Chinese consumers prefer darker colored varieties (mahogany), but in Japan a light skin color (pinkish to light red) is favored with lighter or even yellow-white flesh. To maximize profits, growers should be aware of market preferences to choose varieties that have the preferred skin color.

Sugar levels, as percent soluble solids content (SSC) or degrees Brix, can be read with a hand-held refractometer (Fig. 9.10). At harvest, ripe fruit usually vary between 15–20°Brix, although the best flavor is generally obtained with a minimum of 17–18°Brix. A hand-held refractometer is relatively inexpensive and may be well worth the investment. The measurement is achieved by a random pick of several fruit, placed in a plastic bag and mashed to release the juice. The juice is then poured on to the instrument's glass, where a reading is obtained. Some varieties accumulate sugars earlier during maturation than others, so it is important to also consider skin color.

In some years, environmental conditions may interfere with harvest plans. Forecasted rain or extended heat may force an earlier harvest than

Fig. 9.9. A color card, such as this one developed by the Centre Technique Interprofessionnel des Fruits et Legumes (CTIFL) can help determine the right time to harvest fruit (courtesy of L.E. Long).

is optimal. Some varieties, like Regina, are more resistant to rain than others and may withstand a longer period of wetting than Bing or Skeena. Depending on the predicted amount and duration of rain, a grower may decide to harvest Skeena before the rain, even though it might not be quite at optimal harvest quality, but will leave the Regina until it is fully ripe. Like rain, heat also can severely damage a ripening crop. Several days with temperatures at or above 36°C (97°F) can cause softening or sunburning of the fruit. Some varieties, such as Skeena, are particularly sensitive to heat damage, but all varieties will eventually succumb to extreme heat, resulting in softer fruit after a heat episode. Depending on rain or heat levels, it may be better to harvest fruit a few days early and shift to less valuable markets, than to absorb a loss of all fresh marketing potential due to these environmental conditions.

9.4 Harvest Techniques

Prior to harvest, pickers should be trained so that neither fruit nor trees are damaged by the process. Cherries should only be handled by the stems, which should be lifted and pulled back slightly toward the shoot to detach the fruit from the spur. Pulling outward or downward on the stem may

Fig. 9.10. A hand-held refractometer can help the grower make important harvest decisions (courtesy of L.E. Long).

cause the spur to be removed or the stem to be broken. Directly holding the fruit with fingers during the picking process is likely to cause bruising. Fingernails should be trimmed to avoid accidental damage to the fruit skin. It is also important that harvest buckets should be lined with soft foam pads to reduce the potential for pitting damage. As cherries are being placed in the bucket, they should not be dropped more than 20 cm (8 in).

Cherries should only be harvested in the coolest periods of the day. If possible, harvest should begin at dawn, when the fruit first becomes visible to pickers (Fig. 9.11). This is when the fruit is the firmest and freshest. Once the temperature starts to increase and leaves begin transpiring significantly, water may be pulled out of the fruit to other parts of the tree to maintain metabolic functions. This means that fruit shrinks slightly and gets softer as the day progresses. It is important that harvest ends before temperatures rise above 32°C (90°F). At this point, cherry quality and long-term storage potential will be significantly and negatively affected.

To minimize the potential for fruit damage during harvest, the number of fruit transfers in the orchard should be minimized. In many parts of the world, it is common for fruit to be picked into the same bucket that is transported to the packing house for processing. In this case, the picker carries and places the full bucket into a bin, which when full is then moved with other full bins to the packing house. This has the disadvantage of

Fig. 9.11. In the Pacific Northwest (PNW) harvest begins at sunrise, when the fruit is first visible on the tree, and the temperatures are cool (Oregon) (courtesy of L.E. Long).

needing to use a rectangular or round bucket, which does not conform to the contours of the picker's body, making it uncomfortable over the course of a day of picking. In the PNW, cherries are picked into a kidney-shaped bucket secured to the picker's shoulders and back with straps. These buckets hold approximately 9 kg (20 lb). When full, the picker carefully pours the cherries into a bin which holds approximately 160 kg (350 lb) of cherries (Fig. 9.12). This method has the disadvantage of subjecting the cherries to an extra drop that has the potential for mechanical damage to the fruit.

9.5   Postharvest Fruit Quality Considerations

The quality manifestations of mechanical damage to sweet cherry fruits during and after harvest are bruising and pitting. Bruises appear as large rounded depressed areas on the fruit and are caused by compression damage when cells deep in the flesh are injured or die. This damage often occurs when pickers grab the fruit instead of the stem during harvest. In this case, it is common to see bruises on the shoulders of the fruit. Pits are distinct indentations in the skin of the cherries that are caused when cells in a small area are damaged and die (Fig. 9.13). After a period of storage,

Fig. 9.12. This picker is gently pouring freshly picked cherries from her kidney-shaped bucket into a bin (Oregon) (courtesy of L.E. Long).

the damage appears as a pit on the surface of the fruit. This type of damage is caused by sharp impacts to the fruit such as occurs when cherries are dropped into buckets, or they fall to a lower level on the packing line, or the end of the pedicel is pressed into the side of a fruit as the harvested fruit accumulate in the bucket. Mechanical damage also may include nicks or cuts to the skin or injury to the stem. Too much mechanical damage increases the potential for decay and increases the respiration rate of the fruit, affecting its long-term storage potential.

Presence of the same microcracks that increase rain-cracking susceptibility can also have implications for postharvest performance. The greater the number of microcracks, the more rapidly water may be lost from the fruit postharvest, especially when high humidity levels are not maintained. Microcracks are also sites for potential postharvest infection by opportunistic fungal or bacterial pathogens. However, harvested fruit with microcracks also can take up water more readily under conditions of cool temperatures and high humidity, or when exposed to water on the packing line. Thus, fruit harvested under cool and overcast conditions, particularly following a rain event, may be more susceptible to postharvest cracking. Lowering the osmotic potential of packing line water, as with Ca, can help reduce postharvest cracking of susceptible fruit.

Fig. 9.13. Pitting can show up preharvest (see arrows); however, most pitting appears postharvest (courtesy of G.A. Lang).

Unlike some other fruits, cherries do not continue to ripen after harvest. From the point that they are removed from the tree, deterioration begins. Respiration is a metabolic process that naturally occurs in the fruit, even after harvest. This respiration causes heat, and the warmer the cherry becomes, the faster the rate of respiration. This causes further deterioration and leads to fruit darkening and loss of flavor, especially acidity. As acidity is lost, cherries lose the tangy zest that defines the flavor and they become bland. Warmer fruit temperatures also promote the growth of decay organisms.

9.6   Postharvest Temperature and Humidity Control

Temperature is also important in the loss of moisture from the fruit. When the vapor pressure of the surrounding air is lower than the vapor pressure inside the cherry, moisture will move from the cherry to the air. The warmer the temperature, the faster this occurs. The inside of the cherry has a relative humidity (RH) of 100%, so any time the RH of the surrounding air is less than 100%, moisture will move from the cherry to the surrounding atmosphere. In the warm conditions of the orchard where the RH can be very low, a significant amount of moisture can be lost. Even after harvest in a cold storage room, RH must be controlled to prevent this moisture loss. From harvest to the market, the maximum allowable fruit moisture loss is only 5%. Moisture loss greater than 5% will be noticeable through brown stems, pitting, loss of luster, shriveled skin, etc. If most of this occurs in the orchard due to poor handling, then the fruit will have a very short shelf life. To prevent this moisture loss, the temperature of the fruit needs to be reduced to 0°C (32°F) as soon as possible after harvest.

Protecting the harvested fruit from moisture loss begins in the orchard. Harvested fruit should never be left exposed to the sun. Storing fruit temporarily in the shade of a tree until it can be picked up and moved is fine for a short time, but it is important to remember that shade moves with the sun, and fruit placed in the shade may soon be in full sun if not quickly moved. Once a bin is filled, it should be immediately covered with reflective tarp, such as a silvicool tarp (Bushpro, Vernon, Canada; www.bushpro.ca, accessed 19 June 2020). Tarps come in various sizes and can be custom made to cover individual bins or multiple bins loaded on to a truck for delivery to the warehouse. Research has shown that, when placed over a bin with the white side up, silvicool tarps will maintain a constant temperature and humidity for the fruit from the orchard to the packing house (P. Toivonen, Agriculture and Agri-Food Canada, personal communication). Fruit kept under a tree without cover for 4 h lost 6% of its weight in moisture, whereas fruit covered with the silvicool tarp lost less than 1%. Covered fruit had significantly greener stems and less decay than the fruit stored in the open. In fact, fruit kept in orchard shade for 8 h while covered with the silvicool tarp subsequently maintained high quality in cold storage for up to 4 weeks.

The internal browning, wooliness or bleeding around the pit that is a manifestation of fruit chilling injury in storage can be minimized by storing harvested fruit as close to 0°C (32°F) as possible. Chilling injury develops most rapidly at 4–7°C (39–45°F), which often is the temperature at which cherries are packed. However, additional cooling after packing is necessary, especially when long-term storage is planned. This can be accomplished through forced air cooling. The temperature of packed fruit, placed at 0°C (32°F) in a refrigerated container, will not decrease in storage. In fact, fruit temperatures near the center of the pack will actually increase, due to heat released by the respiration of the fruit. Relieving any form of stress during the growing season, and making sure that fruit is rapidly cooled after harvest to near 0°C (32°F), are the best ways to avoid chilling injury.

9.7 Transport from the Orchard to the Packing House

In many parts of the world, the transport of the cherries to the packing house can severely damage the fruit, due to poor road conditions. Orchard roads need to be graded to remove holes and ruts. A fresh layer of gravel can reduce the bouncing and jarring of the buckets or bins of harvested fruit. Little can be done to rectify unimproved public roads; however, reducing tire pressure and using a torsion bar suspension, where the spring is inside the axle, can improve the ride of trucks and trailers. Moderating speeds will also greatly reduce potential damage to the fruit. To determine the impact of the transport vehicle and orchard-to-packing house route on the potential for fruit damage, marbles can be placed in the ashtray or cup holder of the vehicle and the degree to which they are jostled during transport noted.

Proper harvest timing, careful picking, covering the cherries with re-flective tarps in the orchard and cooling the fruit to near 0°C (32°F) as quickly as possible will help ensure that high fruit quality achieved in the orchard will be maintained by the time it reaches market shelves.

For More Information

Knoche, M. and A. Winkler. (2017) Rain-induced cracking of sweet cherries. In: Quero-Garcia, J., Iezzoni, A., Pulawska, J. and Lang, G. (eds) *Cherries: botany, production, and uses.* CAB International, Wallingford, UK, pp. 140–65.

Wang, Y. and Long, L.E. (2015) Physiological and biochemical changes relating to postharvest splitting of sweet cherries affected by calcium application in hydro-cooling water. *Food Chemistry* 181, 241–247.10.1016/j.foodchem.2015.02.100

Acknowledgements

Portions of this chapter were adapted with permission from Oregon State University. *EM 9227 Understanding and preventing sweet cherry fruit cracking,* copyright 2019. Oregon State University Extension and Experiment Station Communications, Corvallis, Oregon. www.catalog. extension.oregonstate.edu/em9227 (accessed 19 June 2020).
Additional contributions to this chapter were made by M. Pasa, L.J. Brewer and P. Toivonen.

References

Hovland, K.L. and Sekse, L. (2003) The development of cuticular fractures in fruit of sweet cherries (*Prunus avium* L.) can vary with cultivar and rootstock. *Journal of the American Pomological Society* 57, 58–62.

Meland, M., Christensen, J.M. and Kaiser, C. (2014) Physical and chemical methods to avoid fruit cracking in cherry. *Agrolife Scientific Journal* 3(1), 177–183.

Wang, Y., Xie, X. and Long, L.E. (2014) The effect of postharvest calcium application in hydro-cooling water on tissue calcium content, biochemical changes, and quality attributes of sweet cherry fruit. *Food Chemistry* 160, 22–30. DOI: 10.1016/j.foodchem.2014.03.073.

Wang, Y., Xie, X., Einhorn, T. and Long, L.E. (2015) Optimizing preharvest calcium application frequency, timing, rate, and sources to increase tissue calcium content and shipping quality of sweet cherry. *HortScience: a publication of the American Society for Horticultural Science* 50(9), S241.

Managing Orchard Pests **10**

10.1 Insect Pests

10.1.1 Integrated pest management

Integrated pest management (IPM) focuses on the judicious use of multiple orchard practices to decrease pest and disease populations necessary for the economic production of quality fruit with the least adverse effect on the environment and ecosystem. Such practices include biological, physical, mechanical, genetic and chemical methods, as well as cultivation practices (e.g. cover crop management) and the use of technological aids (e.g. pheromone traps and mating disruption). Key factors for success of IPM programs include: (i) knowing the biology of the target pests and potentially beneficial organisms; (ii) applying control measures targeted to specific problems and only when necessary; and (iii) staying ahead of pests and diseases in the orchard by continuous scouting, monitoring and recording of their incidence. Routine scouting can alert growers not only to current problems, but also potential problems before they become significant. Some pests and diseases will first appear in specific areas of the orchard (e.g. borders or edges), or nearby areas where they were found in the past (e.g. riparian areas or adjacent fields of other host crops, such as the migration of stinkbugs coming from recently cut alfalfa fields). The entire orchard should be scouted on a regular basis, looking for telltale damage or the pests themselves. Borders and interior areas should be checked separately. Monitor at least 30 trees per block and inspect both sides of the tree.

Insect scouting can be accomplished with sweep nets, traps, sticky cards, beat sheets or pheromone lures in delta traps, depending on the insect of interest. There are some key things to be aware of when scouting: (i) know when to start scouting, then scout regularly in the most pertinent locations; (ii) know and understand the basic biology of the target pests; (iii) scout for insects at the right time of day or night; (iv) inspect the undersides of leaves and within the canopy; and (v) develop orchard maps with locations affected by pests and diseases, and monitor these more closely in future. Tools for scouting include: a 20× magnification hand lens for monitoring traps; a 60× magnification hand lens for looking for mites and their predators; boxes of zipper lock bags (large and small) for collecting samples; and an ink marking pen to write on the bags.

With the prevalence of global positioning satellite (GPS) technology now available in smartphones, if cell phone reception is available in the orchard, locations for scouting finds can be marked and recorded for precise reference (this technique is called 'dropping a pin' in the smartphone mapping application). If the orchard is in a remote area without cell phone reception (a 'deadzone'), the mapping app should be opened before entering the deadzone, then the entire deadzone can be scanned for future reference to enable 'dropping a pin' even without cell reception. The 'dropped pins' can be used to map the orchard for pests and diseases upon returning to a desktop computer. Also consider marking the affected trees with colored tape/ribbon, different colored paint or plastic tags (and screws for attachment) for writing the scouting find with indelible ink. In lieu of using a smartphone, a clipboard, pen and paper can be used for recording scouting notes.

Many insects and diseases have evolved to attack plants when they are most susceptible, e.g. aphids are most likely to infest young tender leaves in spring; mites tend to infest leaves under hot dry conditions in summer; bacterial canker is most damaging to plants under stresses like cold or insect infestations. All insect life cycles are governed by heat units and these are usually synchronized with tree phenology, which is also governed by heat unit accumulation. Consequently, it is imperative to monitor the different phenological stages of cherry tree and crop development, and understand at which stage the pest or disease is most likely to colonize the plant. This requires an understanding of the target pest or disease phenology as well. It is also critical to identify the pest or disease symptoms correctly. Treating for the wrong pest or disease may incur unnecessary costs, and therefore some treatment measures may be ineffective, e.g. treating with an insecticide when an acaricide (miticide) is needed. It is also crucial to set economic thresholds for each pest or disease, and only treat the orchard once these thresholds have been reached.

Since pesticide registrations and application regulations differ from country to country, region to region within the same country, and year to year, this chapter discusses the major pests of sweet cherries, their life cycle/biology, and types of control measures. Specific chemical control measures should be determined from local pesticide regulatory authorities, and applications made in accordance with each pesticide's label guidelines and restrictions.

10.1.2 Cherry fruit fly and spotted wing *Drosophila*

Among the most damaging insects for sweet cherry production world-wide are a range of fruit fly species that include western (*Rhagoletis indiferens*), eastern (*Rhagoletis cingulata*), European (*Rhagoletis cerasi*) and black (*Rhagoletis fausta*) cherry fruit flies (CFF), and more recently, spotted wing *Drosophila* (SWD, *Drosophila suzukii*). In Europe and Asia, European CFF and SWD predominate, while in North America, western CFF occur in the major production regions of the Pacific coast states and provinces, eastern CFF and black CFF occur in the Midwest and eastern states and provinces, and European CFF have now been detected in Ontario (2016) and western New York near Niagara Falls (2017). Black CFF are more common in sour cherries than in sweet cherries.

10.1.2.1 Cherry fruit fly life cycle and damage

CFF overwinter as pupa in the soil, requiring about 150 days of temperatures above 0°C (32°F) to emerge. Adult CFF begin emerging in orchards in May and can continue through to July (northern hemisphere), or in November (southern hemisphere) continuing through to January, usually coinciding with rapid fruit growth and ripening. Adult CFF are slightly smaller than a housefly, with black bodies about 4–5 mm (~0.2 in) long and transparent wings with several thick, dark brown to black bands, depending on species (Fig. 10.1). For 5–15 days after emergence, they may feed on aphid honeydew, pollen, sap exuded from leaf stipules, and other food sources while they mature sexually, followed by ovipositing just under the fruit skin for up to 3–4 weeks. Each female can lay as many as ten eggs per

Fig. 10.1. Western cherry fruit fly (CFF) adult (courtesy of M. Bush, Washington State Department of Agriculture).

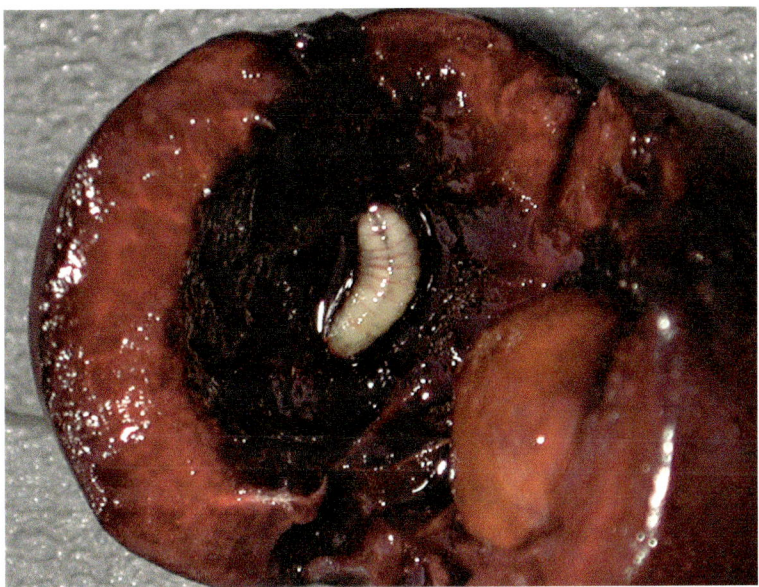

Fig. 10.2. Western cherry fruit fly (CFF) larva (courtesy of M. Bush, Washington State Department of Agriculture).

day and 350 eggs per lifetime. Cherry fruit become susceptible to oviposition once the color changes from green to yellow or red, but ovipositing may occur throughout ripening. Males are extremely territorial and will not tolerate the presence of another fly unless it is a mating female. Females apply a pheromone to fruit as they oviposit which deters other females from ovipositing additional eggs in the fruit.

When the eggs hatch, the young larvae (maggots) are white and can only be seen if the fruit is dissected (Fig. 10.2). Cherries infested with CFF larvae eventually become soft, misshapen, discolored and shrunken due to the larval feeding tunnels in the fruit flesh, and will decay rapidly. The larvae feed for 10–21 days and develop through three instar stages, before mature larvae leave the fruit through an exit hole in the cherry epidermis. Upon exiting, they drop to the ground where they pupate in the soil at about– 8 cm (1– in) deep. Western CFF have only one generation per year. Some Western CFF pupae may survive in the soil for up to 3 years. Drying of the soil due to drought can delay or reduce adult emergence or larval survival during entry into the soil.

10.1.2.2   Cherry fruit fly monitoring and management

Markets generally have zero tolerance for CFF (or SWD) infestations, which makes fruit fly monitoring and control a high priority for growers. Early ripening cultivars tend to be less susceptible to CFF damage due to lower populations of mature females before the fruit is harvested, and conversely, late ripening cultivars tend to have much higher risks of infestation. Degree day (DD) models can be used to determine first emergence of CFF in major cherry production regions. An early model developed in Albany, Oregon used a threshold temperature of 5.0°C (41.0°F) and found that first emergence of western CFF adults occurred at 468 DD (°C) (842 DD [°F]), with peak emergence at 634 DD (°C) (1142 DD [°F]) and last emergence at 975 DD (°C) (1755 DD [°F]) (Aliniazee, 1976). A more recent study in mid-Columbia, Oregon with a higher threshold temperature of 8.3°C (47°F) found first adult emergence at 57.8 DD (°C) (104 DD [°F]) (Song *et al.*, 2004). Another study in Utah and Washington used the lower base temperature of 5.0°C (41.0°F) and found that first emergence occurred at 573 DD (°C) (1031 DD [°F]) in Utah and at 592 DD (°C) (1066 DD [°F]) in Washington (Alston *et al.*, 1991). The initiation and duration of emergence can vary based on orchard topography and slope, soil moisture and other microclimatic factors, as well as variations in locally adapted CFF strains.

Growers can monitor fruit fly emergence by placing yellow card sticky traps in sunny parts of the tree to attract adults. One trap baited with ammonium acetate should be placed every 2 ha (5 acres) and daily monitoring should begin as soon as the model predicts first egg hatch. It is important to note that a single fruit fly caught per week only indicates a potential infestation, as females will initially feed for about a week as their ovaries mature. If weekly catches exceed three flies per trap, a full cover spray must be applied to target ovipositing females, and repeated about 10 days later. Late ripening cultivars may require a third spray application. Spray materials include systemic insecticides, such as organophosphates and neonicotinoids, which provide excellent control but may have long preharvest interval (PHI) and re-entry interval (REI) limitations. Contact insecticides, such as pyrethrum and spinosad materials, have shorter application interval limitations and can be used for organic production, but they may need to be applied more frequently.

Catches of ten or more flies per trap per week indicate 'hot spots' in or near the orchard in which the trap is placed. Obviously, this orchard must be sprayed, but more importantly, the source of infestation must be found and eradicated, including the potential eradication of nearby wild fence row or forest host plants. Note that where infestation has occurred in orchards in previous years, a bait spray program should be implemented once the model predicts first emergence has taken place; and given the zero threshold, eradication spraying should not wait for traps to confirm the presence of fruit flies. A bait spray can be particularly effective against CFF, but droplet size should not be smaller than a pinhead. Bait sprays should be applied in the early morning to avoid spray evaporation during the hottest part of the day. Application of a postharvest orchard spray should also be considered if fruit were not completely removed from all trees, to reduce the potential population that will emerge in the following year. Where monitoring is not conducted, sprays can be applied once fruit becomes susceptible to ovipositing (i.e. when the color begins changing from green) and repeated at 10-day intervals, but this may result in unnecessary expense and environmental impact if CFF are not present.

10.1.2.3   *Spotted wing* Drosophila *life cycle and damage*

SWD are a particularly invasive pest of cherries that also target a wide range of alternative fruits, which increases the potential for difficult-to-control populations on alternative wild hosts such as brambles. SWD are similar in size to common vinegar flies (e.g. *Drosophila melanogaster*) that are attracted to the smell of yeast and fermenting or rotting fruit that has fallen to the ground. However, SWD prefer ripe to overripe fruit that are still attached to the tree, and female SWD have a specialized, serrated ovipositor that allows them to deposit their eggs in ripening fruit. In most cherry producing regions of the world, SWD are considered a very recent, but economically important, pest. SWD are native to South-east Asia, being first identified in Japan in 1916, and subsequently recorded in parts of Thailand, India, China, Korea, Myanmar, Russia and Hawaii. It was not until 2008 that SWD were discovered in North America, on California raspberries and strawberries. As California-grown berries are shipped widely across the USA and Canada, SWD were rapidly identified in Oregon, Washington, Florida and Canada by 2009. By 2014, SWD had become ubiquitous throughout the western and eastern US. A similar rapid progression from discovery to widespread infestation occurred concomitantly across Europe beginning in Spain in 2008, and more recently, in Mexico, Brazil and Chile.

Unlike CFF, SWD can overwinter as an adult in warm refuges within the environment, rather than as pupa in the soil. Adult SWD are typically 2–3 mm (0.07–0.12 in) in length, with red eyes and a yellowish-brown amber colored body. Several dark, continuous bands are visible around the abdomen of both the male and female. This also describes several other vinegar flies, so two key characteristics that easily distinguish SWD from other common vinegar flies include: (i) a black spot (sometimes dark, sometimes faded) near the leading top edge of adult male wings (females do not have wing spots) (Fig. 10.3); and (ii) the prominent saw-like ovipositor (used to insert eggs into fruit) on the female's posterior end (Fig. 10.4). Capture of female SWD in a liquid bait (e.g. apple cider vinegar, yeast) within a monitoring trap tends to cause extrusion of the ovipositor from the body, thus aiding in identification.

Female SWD can lay several hundred eggs each during their 3–4-week reproductive period over the summer months. Eggs have two long respiratory stalks that protrude from the fruit skin. Cherry fruit damage from SWD is similar to that of CFF, resulting from larval feeding on the fruit flesh that causes exudation of liquid from the scar or hole where eggs were

Fig. 10.3. Male spotted wing *Drosophila* (SWD) have a characteristic black mark on their wings (courtesy of M. Bush, Washington State Department of Agriculture).

laid; softening, collapsing and/or bruising of fruit at the damage site; and small white larvae and pupae that can be seen in the fruit flesh. Unlike CFF, multiple eggs may be laid in the same fruit, resulting in the potential for multiple larvae per fruit. Fruit damaged by larvae may then attract common vinegar flies and/or serve as entry points for fungal or bacterial

Fig. 10.4. A female spotted wing *Drosophila* (SWD) showing ovipositor and egg (courtesy of M. Bush, Washington State Department of Agriculture).

disease organisms. The larvae feed inside the fruit for 4–5 days until they are ready to pupate. Pupae are brownish-yellow in color (as opposed to the white CFF pupa) and this is a non-feeding stage which lasts 4–7 days. Pupae can exit the fruit or remain inside fruit with their star-like (6–8 points) respiratory horns sticking out of the fruit until adult fly emergence. Adult flies will then mate and begin a new generation of pests. It has been estimated that three to ten generations per year can occur, depending on environmental conditions.

Food and moisture are extremely important for SWD survival. Dry conditions, wind and hot or freezing weather may limit activity and survival; however, SWD easily find adequately warm refuges for protection during winter. Both SWD activity and male fertility decrease at air temperatures >30°C (>86°F). Dense, shaded, humid plant canopies tend to favor higher SWD populations; populations are lower in portions of the tree canopy exposed to the sun. It remains to be seen whether narrow fruiting wall cherry canopies that minimize shading, such as the upright fruiting offshoots (UFO) training system, are less habitable for SWD. Flight activity tends to be higher from sunrise to mid-morning and from late afternoon to sunset. The lower and upper biological development threshold temperatures for SWD are 10°C (50°F) and 30°C (86°F), respectively. Overwintering females first lay eggs after 150 DD (°C) (270 DD [°F]) using 1 January as a start date in the northern hemisphere. Peak (~50%) ovipositing by overwintering females and emergence of the first generation occurs after 285 DD (°C) (513 DD [°F]), and ovipositing by the first generation females begins at 319 DD (°C) (574 DD [°F]). Adult emergence of the first generation ovipositing peaks at 421 DD (°C) (758 DD [°F]), and peak egg laying by the first generation is complete by 559 DD (°C) (1006 DD [°F]), at which time the second generation have begun emerging.

10.1.2.4   Spotted wing Drosophila monitoring and management

It is recommended to begin monitoring for SWD in spring, when fruits reach the stage at which potential damage might occur. Higher numbers of SWD are captured by liquid reservoir traps with red or yellow color, increased entry area (mesh screen or many small holes) on the trap side, and bait reservoirs having a large surface area and small headspace (space between the bait surface and the exit holes). Trap baits include apple cider vinegar, yeast-sugar, white wine, rice wine (sake) or one of several commercial baits. In spring, traps should be hung where flies are most likely to be intercepted (e.g. field or garden edge with ample shade and coolness; near adjacent earlier fruiting plants; near fence borders shared with other fruit growers). During the growing season, traps must be placed strategically on borders of susceptible fruiting plants, before the fruit starts to ripen and darken in color. Traps should be placed inside the cherry tree canopy at the same height above ground as the fruit. A minimum of 7–8 traps/ha (3 traps/acre) should be placed in the orchard to increase the chances of SWD detection. Traps should be checked, and bait refreshed, weekly. Numerous species of *Drosophila* and other insects will also be attracted to these traps, especially if the yeast bait is used. It is difficult to confirm SWD in a cloudy yeast bait, so contents must be filtered from the traps using a strainer into a pan with a solid white background. Spread the flies out (using a small paintbrush or tweezers), and examine them with a magnifying hand lens or under a microscope.

Unfortunately, to date, trap counts have not been well correlated with infestation pressure and fruit damage. Alternatively, or in addition to canopy trapping, SWD larvae in ripening fruit can be monitored, though a positive identification obviously means an infestation is well under way. Fruit suspected of being infested with SWD can be collected in a clear plastic bag or a shallow white pan and crushed by hand or with a rolling pin. A solution of sugar or salt water (see below) is then poured over the fruit, which will cause any larvae to exit the fruit flesh and float to the surface. The extraction solution is prepared by dissolving plain salt (sodium chloride) or brown sugar in warm water at 60 ml/l (1 cup/gal) or 150 ml/l (2.5 cups/gal), respectively. Solutions must be thoroughly dissolved for larvae to float to the top for easy viewing. Wait for at least 15 min for the majority of SWD to exit the fruit. The sugar solution will keep larvae alive longer than salt; once the larvae die, they sink to the bottom of the solution.

Chemical controls should be coupled with monitoring efforts. Generally, the same pesticides (spinosads, pyrethroids, carbamates, neonicotinoids and organophosphates) that are useful for CFF control are also effective for SWD. However, due to the potential for multiple generations of SWD, more applications may be needed and insecticide chemical families should be rotated to avoid potential resistance. Most of these are contact pesticides that kill SWD adults, but have no effect on larvae that are developing within and protected by the fruit. The most effective management of SWD combines multiple cultural strategies, most of which are also useful for CFF control, in addition to chemical insecticides. Eliminate non-crop SWD alternative host plants from the orchard and its proximity. Ripe fruit should be harvested in a timely manner and at regular intervals, to prevent opportunities for SWD to lay eggs. Remove all fruit by the end of harvest; avoid leaving overripe fruit on the tree and clean up infested fruit on the tree or on the ground. Insect exclusion netting (with mesh <1 mm) over tree rows or the entire orchard has shown some promise, though row netting tends to reduce effectiveness of chemical sprays. Shallow cultivation of the soil under the tree canopy can help reduce pupae that drop to the ground, especially last stage pupae which give rise to the adults that overwinter to initiate subsequent year infestations. Research is under way to determine specific predators and parasitoids (e.g. wasps), and even entomopathogenic fungi that attack SWD larvae and pupae. Field observations suggest that ants, spiders, predacious bugs (e.g. minute pirate bugs, big-eyed bugs), yellow jackets, lacewing larvae and parasitoid wasps may be important biological control agents. Cooling the fruit to <1°C (<34°F) for 4+days after harvest can kill SWD eggs and slow or kill young larvae.

10.1.3   Cherry leaf rollers

There are three major leaf rollers that affect cherries. They are the oblique-banded leaf roller (OBLR, *Choristoneura rosaceana*), the *Pandemis* leaf roller (*Pandemis pyrusana*) and the fruit tree leaf roller (*Archips argyrospilus*). All three are similar in appearance. However, the fruit tree leaf roller overwinters as eggs, which start to hatch at the same time as flowering. The *Pandemis* leaf roller and OBLR overwinter as mature larvae. *Pandemis* leaf roll larvae are green with a green or brown head and are approximately 2.5 cm (1 in) in length when fully grown. OBLR larvae are similar except that the head is dark brown to black. Fruit tree leaf roller larvae are shorter, reaching a maximum of 19 mm (0.75 in) in length and are greenish with a black head.

10.1.3.1   Fruit tree leaf roller

Overwintering egg masses may be found on twigs and branches. Eggs hatch in spring and the larvae feed for 4–6 weeks. Newly hatched fruit tree leaf roller larvae may enter blossoms and damage developing cherries, as well as vegetative buds. They pupate in rolled leaves and emerge as moths in early summer (Fig. 10.5). Overwintering eggs are laid in midsummer and will hatch the following spring. There is only one generation per year. From

Fig. 10.5. Fruit tree leaf roller (courtesy of M. Bush, Washington State Department of Agriculture).

the time of egg laying to hatching is an extremely long period for them to be exposed to predators and parasites, and to freezing temperatures in winter, resulting in low populations. Natural enemies include spiders, parasitic wasps, lacewings and earwigs.

Apart from natural biological control, a dormant season oil spray is recommended, but coverage must be adequate to smother the eggs. A *Bacillus thuringiensis* serotype *kurstaki* (Btk) spray is an effective product for controlling the fruit tree leaf roller. Other more broad spectrum chemistries are registered for controlling larvae; however, these should only be used for the purpose of preventing resistance build-up.

10.1.3.2   Oblique-banded leaf roller and Pandemis *leaf roller*

Overwintering larvae of both these species are fully developed when they begin foraging in spring. Both feed on leaves and enter the fruit, causing contamination. The larvae feed for several weeks, then pupate in rolled leaves. In the northern hemisphere, adult moths emerge in late April and May. Adult females lay eggs for the second generation which hatches in early summer. These second generation larvae also may feed on the fruit, creating small holes for entry. Rain-induced cracks on the ripening fruit also serve as an easy point of entry.

Early leaf feeding damage in spring is a clear sign of infestation, as are rolled leaves in mid-spring and early summer. OBLR adults may be trapped with pheromones from late April through to May. Dormant oils, kaolin clay and biocontrol products such as Btk and *Beauvaria bassiana* have been shown to be very effective against OBLR and *Pandemis* leaf rollers. However, a comprehensive list of more potent chemicals is available

Fig. 10.6. A black cherry aphid (BCA) infestation showing the protected colony in curled, distorted leaves (courtesy of C. Kaiser).

for use in situations when these softer approaches fail to work. It should be noted that much of the need for more potent chemistries is due to the eradication of natural enemies by broad spectrum chemicals, and when these natural controls are lost, growers must adjust their control measures accordingly.

10.1.4 Black cherry aphid

The black cherry aphid (BCA, *Myzus cerasi*) is found throughout the world and will attack cherry trees of all ages, but damage is mainly significant on young trees due to stunting of early season canopy development. BCA feed on plant sap from the underside of leaves, causing leaf and even shoot growing tip distortion, premature defoliation of leaves, and sooty mold which can further reduce photosynthesis (Fig. 10.6). BCA populations can increase very quickly. Since BCA emerge just before bloom from overwintering eggs laid at the base of buds, on spurs, and shoot apices, the best control is achieved with dormant oil applications that kill the eggs. Several natural enemies, such as lady beetles and lacewings, can help control BCA populations weakened by the dormant oil treatments, and populations rarely persist in being problematic by early summer. If natural enemies fail to control populations from reaching damaging levels, a single contact or systemic insecticide application is usually sufficient for control, providing good coverage to reach the undersides of leaves where BCA feed.

10.1.5 Plum curculio

Plum curculio (PC, *Conotrachelus nenuphar*) are a serious insect pest of cherries and many other tree fruits in the eastern and midwestern US. The adults are long-snouted beetles about 6 mm (0.2 in) long that migrate from their protected overwintering sites, such as brush piles and leaf litter, into orchards following several days of warm spring temperatures, usually coincident with bloom. They feed on buds, blossoms and young fruits, and females mate and can begin laying eggs in the young fruit as soon as set occurs. This can continue for 4–6 weeks, with females capable of laying from 100 to 500 eggs during the season. Evidence of an infestation is readily identified by the crescent-shaped scar left on the fruit during ovipositing. Fruit damage occurs from larvae feeding in the fruit flesh. Unless the infested fruit drops first, as the larvae mature, they exit from the fruit, and drop to the ground where they pupate at 7–8 cm (3 in) deep in the soil. The adults emerge during summer, and in northern locations more typical of cherry production (northern hemisphere), only one generation per year is typical.

PC tend to cause the most damage to early ripening cultivars, from both feeding and ovipositing. PC are best controlled by insecticide applications beginning around petal fall and repeated about 10 days later. Traps for PC facilitate monitoring of their movement into the orchard, particularly from their protective cover in hedgerows and forests, to better time insecticide sprays for control. Intensive scouting of orchards and DD models are also successful IPM strategies for pesticide application timing. Reducing potential overwintering sites near the orchard is also recommended.

10.1.6 Japanese beetle

Japanese beetle (JB, *Popilla japonica*) are another serious insect pest of cherries in the eastern and midwestern US (Fig. 10.7). Adult beetles, about 12 mm (0.5 in), emerge throughout the summer, beginning in late June in Michigan, and each wave of emerging beetles can feed voraciously on young leaves during their month-long lifespan. JB are very mobile and move readily from forest edges and wild brambles to cherry orchards. Total foliar damage, and hence the need for repeated control measures, may extend 8–10 weeks, into late summer. Defoliation of the youngest, most sun-exposed leaves can reduce photosynthesis and nitrogen remobilization after cherry harvest, a time when storage reserves are being built up (which improves the tree's ability to cold acclimatize) and the period of flower bud differentiation for the next season's cropping potential.

Fig. 10.7. Japanese beetle (JB) (courtesy of M. Bush, Washington State Department of Agriculture).

The JB life cycle begins with mating during midsummer; grassy meadows and sodded orchard tractor alleys are preferred habitats for JB egg laying, providing the larval grubs with abundant roots on which to feed. Moist soil conditions are preferred, so dry years can reduce the natural survival and/or emergence of JB. Treating grassy areas in or near the orchards with soil-applied insecticides to kill the grubs prior to emergence can reduce early season damage. However, as the season progresses, JB may migrate into the orchard from adjoining areas, so contact and stomach poison insecticides applied to the trees may be needed to control these foraging populations. The use of foliar applications of white kaolin clay after harvest may deter some JB foraging, though little work has been reported on its effectiveness in cherries. Kaolin should not be applied prior to cherry fruit harvest, as it is difficult to remove the unsightly residue for marketing of the fruit. Some plastic row covering materials can reduce JB foraging, as they navigate best or prefer to feed in direct sunlight. Since JB are mobile, the use of traps with attractants are not recommended, as they may draw distant JB into the orchard, increasing the potential for damage and larval populations in the soil.

10.1.7 Borers

10.1.7.1 American plum borer

American plum borer (*Euzophera semifuneralis*) are the larvae of a gray-brown moth with a wingspan of up to 28 mm (1.1 in) (Fig. 10.8). The larvae typically bore into the trees at wound sites, using scaly bark for protection and concealment. The larval tunnels are shallow and usually filled with

Fig. 10.8. American plum borer (courtesy of M. Bush, Washington State Department of Agriculture).

frass. Larvae feed exclusively on the cambium for up to 6.5 weeks and since larvae feed horizontally, the damage can result in girdling if the pest is left uncontrolled over time. Pupation takes place in loosely spun silk cocoons up to 12 mm (0.5 in) long in the larval tunnels surrounded by red frass. During summer, the pupal stage may be as short as 10 days, but overwintering generations may take up to 33 days. The number of generations per season depends on the climate, with up to five generations occurring in the southern US. In Michigan, only two generations typically complete each growing season. Emergence of first generation adult moths usually begins just after full bloom. Most first generation females lay eggs by petal fall, although adult emergence continues for another 3 weeks. In the northern hemisphere, adult emergence of the second generation begins in June, peaks in mid-July, and may continue to September. Peak emergence and egg laying coincide with harvest; where cherries are harvested mechanically, an abundance of fresh cracks and wounds on the bark of the trees is attractive to the female moths, which lay single small pink eggs in the wounds. In severely infested trees, up to ten larvae per tree are common.

Monitoring should be performed by trapping adult moths using sex pheromones to assess male flight. One trap per 2 ha (5 acres) is recommended and should be placed in the center of the block. An average weekly catch exceeding six moths per trap constitutes an economic threshold, and chemical control should be implemented. Alternatively, peeling away bark near wound sites in early spring before the white tip flower bud stage, or in summer shortly before harvest, may reveal white hibernaculae with red frass. The presence of a live larva or pupae is considered a 'hit'. Up to three 'hits' per tree in several trees is considered a threshold for treatment.

Fig. 10.9. Evidence of damage by the peachtree borer includes these blackened gum masses containing sawdust and frass (courtesy of M. Bush, Washington State Department of Agriculture).

10.1.7.2   Peachtree borer

Peachtree borer (*Synanthedon exitiosa*) damage is caused by the larvae, which are white with a brown head and may be up to 38 mm (1.5 in) long at maturity. There is only one generation per year, but larvae overwinter in a wide range of larval stages and may burrow into the heartwood of the tree, though typically they overwinter below ground. Larvae become active in early spring and resume feeding through early to midsummer. Most adults emerge in mid- to late summer. Adults resemble wasps and the female moths are dark blue with a single orange band around the body. Male moths are smaller with three or more narrow yellow bands across the body. Female moths live for up to 7 days and may lay up to 500 eggs in trunk cracks and under scaly bark. Eggs hatch in 10 days and larvae bore into the bark at the base of the trunk. Once beneath the bark, the larvae feed on the cambium layer. Damage is usually confined to the trunk, a few inches above to a few inches below the soil level. Young trees are most susceptible, as feeding can result in complete girdling of the bark. Evidence of damage at the base of the tree is usually accompanied by a gum mass containing small wood chips, sawdust and frass (Fig. 10.9).

Monitoring usually uses one pheromone trap per 2 ha (5 acres), and catching more than ten adults per week is considered the economic threshold. Older trees with more than one larva per tree should be treated. Several species of parasitic *Hymenoptera* have been reared from larvae or pupae of the peachtree borer. They include *Bracon sanninoideae* and *Macrocentrus marginator* (Braconidae); *Syntomophyrum clisiocampae* (Eulophidae); *Cryptus rufovinctus, Phaeogenes ater,* and *Venturia nigricoxalis* (Ichneumonidae); and *Telenomus quaintacei* (Scelonidae). Mating disruption with pheromones is considered to be highly effective at controlling this pest. As a last resort, chemical pesticides can be applied to the trunk and ground around the base of the tree using a handgun in midsummer and again one month later, but the PHI should be noted if the fruit have not yet been harvested.

10.1.7.3   Pacific flatheaded borer

Pacific flatheaded borer (*Chrysobothris mali*) are reddish-bronze beetles
with copper spots that attack many different tree and shrub species and
may reach 12 mm (0.5 in) in length. Larvae are white to pale yellow and
also reach 12 mm (0.5 in) in length when fully developed. Immediately
behind the head is a broad flat enlargement, hence the 'flathead' name.
Larvae feed beneath the bark and may girdle trunks and limbs. Young trees
and those that are stressed are most susceptible. Insects overwinter in host
plants as mature larvae. In spring, they pupate and then bore holes into
the hardwood. Emergence in the northern hemisphere usually begins in
May to June. Females lay eggs from June through to July in bark crevices
and usually on the sunny side of the tree trunks below the lowest branches.
Eggs hatch and larvae bore into the vascular cambium where they feed
until fully grown. Burrows are irregular and characteristically broad. When
monitoring, look for depressions in the bark or cracks, through which frass
can be seen. Female beetles are attracted to weak, sunburned or injured
parts of the trunk and lay eggs in the bark exposed to the sun. Protect
newly planted trees from sunburn by painting them with 100% water-based
white latex paint. Chemical control should begin with a dormant spray and
focus on a trunk spray in summer. These beetles are extremely difficult to
kill and current recommendations are for use of a registered pyrethrin in
rotation with other registered neonicotinoids.

10.1.7.4   Shothole borer and ambrosia beetle

Shothole borer (*Scolytus rugulosus*) and the closely related ambrosia beetle (*Xyleborini genera*) are very small black beetles that attack a wide range
of fruit trees, ornamental and forest trees and shrubs (Fig. 10.10). Apple,
pear, cherry, prune and plum trees are all susceptible. These borers are
primarily found in weakened and stressed trees, and population explosions often occur in years following severe winter freeze damage. Adults are
brownish-black winged beetles about 5 mm (0.2 in) in length and ambrosia
beetle adults may be covered with fine yellow hairs. Larvae of the shothole
borer are white and legless and up to 4 mm (0.17 in) in length when fully
mature. Larvae of the ambrosia beetle are up to 5 mm (0.2 in) in length
and are pinkish-white, legless and cylindrical when mature. Larvae of both
species bore into the vasculature and bark cambium of trees, causing wilting and dieback of individual stems and branches. Trunks and branches
can be thoroughly riddled with galleries.

Shothole borer overwinter as larvae in the burrows beneath the bark
of infested trees. In spring, they pupate and emerge, and in early summer
when they mate, will fly to susceptible trees where they feed at the base of
leaves and small twigs. Adult shothole borer then tunnel into the tree, exca-
vating galleries parallel to the wood grain. Females lay eggs in the galleries.

Fig. 10.10. Shothole borer beetle on cherry (courtesy of M. Bush, Washington State Department of Agriculture).

Once the eggs hatch, larvae burrow into the wood perpendicular to the burrow, which results in characteristic damage in the trunk and branches. Burrows are filled with reddish-brown frass and tunnels become wider as the larvae increase in size and burrow further away from the original tunnel. Larvae pupate after 6–8 weeks at the ends of the galleries and emerge as adults starting in late summer, creating multiple round exit holes ('shotholes'). There are two generations per year.

Ambrosia beetles overwinter as mature larvae, pupae or adults in the galleries within the tree. When temperatures exceed 18°C (65°F), adult females become active. They emerge from the galleries and fly to susceptible hosts where they bore into the trees and lay eggs in the gallery. As the larvae develop, the female beetle carefully tends them and cultivates *Ambrosiella* fungal gardens along the walls of the gallery. The larvae feed on the fungus conidia and conidiphores. Monitoring should begin in late spring by looking for shotholes 3 mm (0.12 in) in diameter on the trunk and branches. Often these are disguised by the presence of oozing sap and sawdust. In some cases, a gummy congealed mass also may be seen. To confirm the presence of shothole borer, slice the gummy mass off under the bark and a shothole will be revealed. Check weakened and diseased trees first. Destroy all prunings and burn infested prunings the same day they are

Fig. 10.11. Peach twig borer larva (courtesy of M. Bush, Washington State Department of Agriculture).

removed from the trees. Leaving them in piles in the orchards overnight allows the beetles to fly out of this wood into trees nearby. Paint trunks white, especially of young trees, and keep them well irrigated to prevent stress. Chemical control should focus on dormant sprays and during the growing season, trunk and limb applications should be made using a registered pyrethrin or esfenvalerate at the label recommended rates.

10.1.7.5   Peach twig borer

Peach twig borers (*Anarsia lineatella*) are native to Europe but also found throughout North America (Fig. 10.11). They attack a wide range of stone fruits including, apricots, peaches, nectarines, plums and prunes, but on rare occasions peach twig borer also have been found in mature cherry orchards. The larva can kill young succulent twigs and disfigure or infest fruit. Wasps parasitize the eggs at shuck fall. If damage is noticed, mined shoots can be cut out below the wilted area to eliminate the larvae. Do not use pesticides toxic to bees during flowering; a softer alternative, such as Btk, may be considered for spring use.

10.1.8   Cherry slug

Cherry slug (*Caliroa cerasi*) also known as the pear slug, cherry sawfly or pear sawfly, is a slimy, olive green larva with a head that is wider than the rest of the body. Mature larvae reach 10 mm (0.39 in) in length. Larvae feed on the upper leaf surface (Fig. 10.12). After 3–4 weeks, they drop to the soil and pupate. There are two generations per year. Cherry slugs overwinter at 5–7.5 cm (2–3 in) deep in the soil as a cocoon-protected pupa. In the

Fig. 10.12. Cherry slug showing the typical rasping damage caused by the insect (courtesy of C. Kaiser).

northern hemisphere, adults emerge from late April into May. Fertilized females insert their eggs into leaf tissue. The eggs hatch in 10–15 days and the larvae begin feeding on the leaves immediately, causing skeletonization. The second generation emerges in midsummer and feeds through late summer. Evidence for monitoring includes leaf skeletonization and noticeable olive green slimy slugs in late summer when populations build. For control, insecticidal soaps work well, as does spinosad and kaolin-based clay applications.

10.1.9   Oriental fruit moth

Oriental fruit moth (OFM, *Grapholita molesta*) originated in China, were first found in the USA in 1913, and now are common throughout North America, Europe, Asia, Australia and South America. The larvae bore into shoots and feed on the fruit. Although the primary hosts are peach and nectarine, it will also attack cherry, apple, pear, plum, prune, apricot and quince (*Cydonia oblonga*). OFM overwinter as a fully grown larva or hibernacula of silk webbing, which may be found in crevices of the bark or in the ground cover. Pupation takes place in spring and moths of the first generation appear when trees are in full bloom. Eggs, which are white, flat, oval discs, are laid on the foliage, typically on the upper surface of

Fig. 10.13. An adult oriental fruit moth (OFM) caught in a pheromone trap (courtesy of M. Bush, Washington State Department of Agriculture).

leaves on terminal shoots. Females lay up to 200 eggs and incubation may take between 5–21 days, depending on the weather during flowering. First generation larvae bore into newly emerging shoots where they feed and mature, hatching in early through midsummer. There may be three to four generations per year in the Pacific Northwest (PNW). The second and third generation larvae can damage cherry fruit.

OFM larvae bore into the succulent terminals of rapidly growing shoots during spring and summer, advancing 5–15 cm (2–6 in) towards the trunk. Once the larva matures, it no longer requires protection from the twig and exits it. Infested twigs have wilted leaves and shriveled apical meristems. If the twig is dark colored with dry leaves and gummosis, the larva has exited the twig. In young trees, damage results in a loss of apical dominance in these twigs and axillary buds below the damaged twig will begin to grow, often resulting in bushy growth.

OFM do not normally feed on cherry fruit. Adult OFM populations can be monitored with pheromone traps, which should be placed in orchards by late winter (15 February in the PNW) (Fig. 10.13). Traps are most effective when placed about 2 m (6–7 ft) high in the tree where direct or western sun exposure is avoided. For orchards or varietal blocks less than 12 ha (30 acres) in size, use three traps. For larger orchards, use one trap per 4 ha (10 acres) for 12–32 ha (30–80 acre) orchards, and one trap per 8 ha (20 acres) for orchards larger than 32 ha (80 acres). Check traps weekly and replace pheromone lures according to the manufacturer's directions. Replace trap liners when dirty, or after counting and removing an accumulated total of 150 moths. Mating disruption has proved highly effective in controlling OFM, but dispensers must be placed in the orchard before the pre-bud break flight. In warm years, a second application of dispensers may be needed since they typically last only 2 months.

Biological control is often very effective if the braconid wasp, *Macrocentrus ancylivorus*, is present. Female wasps lay eggs in the OFM larvae, which then hatch, parasitize the larvae, and mature as the host cocoons. If insecticides are needed, the target of control sprays is the young larvae before they have bored into shoots. To determine optimum time to spray, accumulate DDs beginning with the first male moth trapped from the second flight, which usually occurs in May in the northern hemisphere (November in the southern hemisphere). Use a lower threshold of 7.2°C (45°F) and an upper threshold of 36.2°C (90°F). The optimum time to treat for OFM is 500–600 DD after biofix (i.e. the first trapped male of the growing season). One generation requires 1500–1600 DD after biofix. Insecticides that are highly effective against OFM while being less harmful to beneficial insects include chlorantraniliprole, spinetoram and spinosad.

10.1.10 Mites

10.1.10.1 *Two-spotted spider mite*

Two-spotted spider mite (*Tetranychus urticae*) is ubiquitous and feeds on all fruit trees, as well as vegetables, ornamentals and many weed species. It is in the same family as the European red mite. Eggs are spherical and about 0.14 mm (0.0055 in) in diameter, and when first laid, they are translucent, taking on the greenish tinge of the leaf (Fig. 10.14). As the eggs mature, they become more opaque and finally turn a pale yellow. Red eyespots of the embryo are visible just prior to hatching. The larva is round, similar in size to the egg, with three pairs of legs. Initially the larva is also translucent (except the red eyespots), but once feeding commences, it varies from straw color to pale green, and characteristic black spots begin to appear on the dorsum (back). The protonymph is more oval, larger than the first instar, and has four pairs of legs, as do all other successive instars. The two dorsal spots are more pronounced, and the green color becomes darker. Each immature stage goes through three phases: active feeding, a quiescent (or resting) period and a molt. The integument (the outer covering of the body) usually takes on a silvery appearance in the quiescent stage as it separates from the body in preparation for the molt.

Adult males are smaller than the females and are characterized by a distinctly pointed abdomen. Males may have an orange-brown tinge and are more active than females. The female is about 0.42 mm (0.0165 in), has

Fig. 10.14. A female two-spotted spider mite after laying an egg on this leaf (courtesy of H. Riedl).

a more oval shape, and color can vary greatly, including pale leaf-green and yellow, brown or orange margins. There are normally two distinct spots on the front half of the dorsum behind the eyes. Fertilized females are orange with no spots and they overwinter at the base of trees and in sheltered sites under scaly bark. In late spring, females emerge and begin feeding, progressively regaining their greenish hue and dorsal spots. After 2–5 days, egg laying starts, primarily on the bottom surface of newly expanded leaves, and later on fruitlets. Overwintering females lay approximately 40 eggs and live for about 23 days, a considerably shorter time than the summer morphs. Depending on the temperature, eggs may take up to 3 weeks to hatch and from this point forward, generations begin to overlap. Summer morph females can lay up to 100 eggs over a period of 30 days. These eggs take only 1–2 days to hatch during the warmer part of the summer, and one summer generation (oviposition to adult) may take as little as 10 days to complete.

When leaf quality begins to deteriorate due to excessive mite feeding, or in fall when cool temperatures and shorter day lengths occur, the orange overwintering forms will again begin to appear. Monitoring should be ongoing and ten trees should be sampled per block. From each of these trees, sample 25 leaves and record the number of leaves that have more than two mites. This can be used to calculate the average number of mites per leaf and develop a local economic threshold. Look for natural enemies while counting too, as this will help determine whether chemical sprays are warranted. Predatory mites most commonly used in biological control of spider mites are in the family Phytoseiidae. Some Phytoseiid mites are spe-cialists like *Phytoseiulus persimilis*, which feeds exclusively on spider mites. Others, such as *Amblyseius fallacis*, are adaptable feeders, utilizing alterna-tive prey and food such as pollen. Both types are beneficial. *P. persimilis* can increase its population rapidly in response to surging spider mite popula-tions, but is susceptible to population crashes as its food source becomes scarce. *A. fallacis* only does well when spider mite populations are low, and it can survive on alternate food, allowing growers to use it early in a crop cycle to suppress mite outbreaks and avoid damage. When populations explode and chemical control is warranted, most mite populations can be controlled with insecticidal/miticidal oil and soap sprays. A dormant oil spray is a critical tool and miticides that do not affect predatory mites should be applied only as a last resort. Ensure that the miticide is labeled for the specific mite to be controlled, as pesticides that are listed as 'mite suppressants' are generally weak and usually require follow-up applica-tions. Avoid using pyrethrins and carbamates, as these are highly toxic to predatory mites.

10.1.10.2 European red mite

European red mite (*Panonychus ulmi*) is a pest of many crops and ornamen-tals, and is especially problematic in cherries and apples. Overwintering eggs are small, bright red, oval shaped and up to 0.15mm (0.0059 in) in length, with a short stipe or stalk. They are usually laid on the woody parts of the tree, especially at the base of spurs, in crevices of the bark and even on fruit at the pedicel or opposite end if left unharvested. Overwintering eggs hatch in spring during flowering. The six-legged larvae are yellow to light orange, but change to reddish-brown before moulting to become an eight-legged nymph, which is green. The adult stage is bright red. The female has three characteristic pairs of hairs arising from white spots on its back.

Eggs laid during the growing season are slightly smaller than overwin-tering eggs. These are laid on the leaves and, if populations are high, also on the fruit. Where eggs occur singly, they are difficult to see; however, in groups, they are clearly visible due to their distinct red color. Egg laying is triggered by day length, low temperatures and lack of availability of food. One female may lay between ten and 90 eggs, depending on conditions. The life cycle of the European red mite takes between 2–6 weeks, depend-ing on temperature, and between four and eight generations may be com-pleted during a single growing season.

Damage caused by the European red mite results in bronzing of leaves, but during the early stages, this is localized on the upper leaf surface and first appears as mottling. Feeding damage is cumulative and is usually ex-pressed as 'mite days' (Beers *et al.*, 2007). Monitoring involves collecting ten leaves from ten trees, counting the number of mites per leaf (using a microscope), calculating the per cent infested leaves, and converting that to numbers of mites per leaf using a conversion table. Setting an economic threshold may vary by time of year, by site and by orchard. Experience will help refine orchard thresholds, but a starting point would be two to five mites per leaf. European red mite populations usually peak in late July or August. When populations increase to high levels, they disperse by a technique known as 'ballooning'. Often in great masses, they crawl to a high point (e.g. the tip of a branch) and raise the anterior part of their body. They then spin a silken thread, which can catch air currents, lifting the mites and taking them away, much like a hot air balloon. This is one of the primary means of mite dispersal to nearby trees where canopies are not overlapping. Females lay overwintering eggs in late summer.

Biological control is preferred to chemical control, but monitoring for predatory Phytoseiid mites (e.g. *Typhlodromus occidentalis*) is crucial. In the PNW, a Stigmaeid mite (*Zetzellia mali*) is another great predator of European red mites. Cultural practices also affect populations, and irrigation type and frequency have a major impact on mite populations by modifying temperature and humidity. Dormant oil sprays are critical and specific miticides that do not affect predatory mites are recommended in fall to prevent overwintering female morphs from developing. Avoid using pyrethrins and carbamates, as these are highly toxic to predatory mites.

10.1.11   Western flower thrips

Western flower thrips (*Frankliniella occidentalis*) are slender, 1 mm (0.04 in) long insects with four long, thin membranous wings. Adults are mostly yellow but males are lighter colored and smaller than females. The female lays its eggs in fruit, leaves and buds. Larvae are white to whitish-yellow with small dark eyes. There are two feeding life stages (first and second instar larvae), followed by two immobile non-feeding stages (the propupa and pupa) that both occur in the soil. Damage is mostly superficial and affects the aesthetics of the fruit (Fig. 10.15). Adults and feeding larvae rasp the fruit surface, resulting in russeting, scar formation and distorted growth, which in mature fruit is sometimes confused with wind damage. Later in the season, thrips damage can appear as a ring or halo on the side of the fruit that is touching another fruit (Fig. 10.16). Where cherries are growing in close proximity to alfalfa, thrips will migrate to the cherries in large numbers when the alfalfa dries out in summer, as adults can move large distances by floating on the wind. In the PNW, thrips can move out of sagebrush plants into orchards as the season progresses. When the weather is warm, the life cycle from egg to adult may be completed in as little as 2 weeks.

Monitoring for thrip adults and larvae can be done by branch beating or gently shaking foliage or flowers on to a light colored sheet of paper, beating tray, or small cloth. For thrips that feed in buds or unexpanded

Fig. 10.15. Feeding by thrips early in the season causes a dimple on the surface of the fruit (courtesy of H. Riedl).

Fig. 10.16. Late season damage by thrips appears as a circular ring, or halo in the location adjacent to two touching fruit (courtesy of G.A. Lang).

shoot tips, prune off several plant parts suspected of harboring thrips, submerge them in a jar half-filled with 70% ethanol (alcohol), and swirl vigorously to dislodge the thrips. Strain the solution through filter paper or fine cloth so that the thrips can be better observed. Adult thrips can also be trapped by hanging bright yellow sticky cards in or near the tree.

If management is necessary, use an integrated program that combines the use of good cultural practices, natural enemies and the most selective or least toxic insecticides that are effective in that situation. Biological control is the most economical option and predatory thrips (e.g. banded wing thrips black hunter thrips, six-spotted thrips, and vespiform thrips), green lacewings, minature pirate bugs and some parasitic wasps are all excellent natural enemies of thrips. Cultural practices also can help prevent populations from exploding. Avoid planting plant species that are preferred hosts (e.g. alfalfa) and control weeds that are alternative hosts in close proximity to cherries. If a catch crop is implemented, keep plants well irrigated and fertilized, but avoid excess nitrogen applications as this will result in a thrip population explosion. Prune and destroy injured and infested terminals in spring and summer if the problem is identified at an early stage.

Thrips can be difficult to control effectively with insecticides, partly because of their mobility, feeding behavior, protection of the egg and pupal stages, improper timing of application, failure to treat the proper plant parts, and inadequate spray coverage when using contact materials. Contact insecticides (e.g. azadirachtin, insecticidal soaps [M-pede], neem oil and pyrethrins) do not leave persistent residues, have low toxicity to people, pets and pollinators and relatively little adverse impact on biological pest control. For maximum efficacy, contact sprays must be applied to thoroughly cover buds, shoot tips, and other susceptible plant parts where thrips are present. In orchards with a history of unacceptable damage, treatment should start early when thrips or their damage is first observed. Where damage is severe, repeat applications of contact insecticides in rotation may be needed since eggs and pupae are protected. Spinosad is another very effective insecticide against thrips, having week-long translaminar activity. Adding horticultural oil to the spray mix can increase spinosad persistence within plant tissue. However, spinosad is toxic to bees and certain natural enemies (e.g. predatory mites, syrphid fly larvae) for approximately 1 day after application, so do not apply spinosad to plants that are flowering.

10.1.12   San Jose scale

San Jose scale (*Quadraspidiotus perniciosus*) are mostly problematic on twigs and leaves of cherry trees, but can also damage fruit. Adult females are visible as small reddish-brown bumps on the bark of stems and branches and on the underside of leaves. The bumps are 5 mm (0.2 in) in diameter. San Jose scale is distinct from other scales as the shell that covers the female is

Fig. 10.17. Trees infested with San Jose scale fail to drop leaves in the autumn. Close inspection of young twigs will reveal the cone-shaped gray to black shell covering the female scale (courtesy of M. Bush, Washington State Department of Agriculture).

hard, gray to black in color and cone-shaped (Fig. 10.17). There is also a tiny white knob in the center of the cone with concentric rings emanating outwards from it. Damage is caused mostly by the females, which suck the sap from twigs and leaves and inject a toxin, resulting in vigor reduction and lower productivity. Untreated infestations can kill a tree in 1–2 years and predispose it to bacterial canker infection. Damage on fruit results in dimples and an unsightly, red halo around the feeding site.

San Jose scale overwinter as immature stages on twigs and branches and are black. In spring, males emerge and mate with wingless females. About 1 month later, females give birth and the resultant crawlers are flat, oval and pinkish-brown in color. Crawlers move around on the bark and foliage before settling down to feed. Young crawlers can be spread around by wind, rain, irrigation, mechanical equipment and on laborers' clothing. In the northern hemisphere, crawlers are present in June through to July (first generation) and again in August through to September (second generation). There are only two generations per year. Pest monitoring is easy as infested cherry leaves do not drop in fall, indicating a need for closer inspection. Observe young bark for reddish-purple halos. During June/July, close inspection of nearby trees using a 10× magnifying glass will reveal crawlers on the bark and twigs. Another useful tip is to wrap the bark of a shoot with black electrical tape with the sticky side out as this will trap crawlers.

For biological control, green lacewings and lady beetles are voracious feeders of San Jose scale. In addition, *Encarsia* and *Aphytis* parasitic wasps can lay an egg under the scale cover. The parasite larva consumes the scale body, and the new adult parasite cuts a circular hole in the scale cover to emerge. Broad spectrum pesticides applied during the summer may destroy natural enemy populations and result in increased scale infestations. San Jose scale is easily controlled using dormant winter oil sprays and insecticides should only be added to these when scale insect populations are severe. If an insecticide is necessary, consider using an insect growth regulator (e.g. pyriproxyfen) that will have less effect on natural enemy populations compared to broad spectrum alternative insecticides.

10.1.13   Army tentworm/army tent caterpillar

Army tentworm/army tent caterpillar (*Malacosoma* spp.) hatch in the spring from eggs laid on branches the previous year and emerge as leaves and new shoots are expanding and elongating. They typically progress through five to six larval instars, developing over a period of 7–8 weeks, and often are not noticed until the later instars, as their web 'tents' become apparent. The most significant foliar damage due to defoliation occurs during the final instar. While defoliation of individual branches can be extensive, damage to the entire canopy is less common, though proportionally greater on younger trees when the canopy is still filling its orchard space. Outbreaks significant enough to be noticeable generally are spot-treated since their occurrence rarely extends across an entire orchard, often due to natural control by parasitoids and diseases.

10.2   Vertebrate Pests

10.2.1   Birds

Bird damage to cherry fruit is a common, persistent and serious problem, causing economic losses that vary from season to season and orchard block to orchard block (Fig. 10.18). Usually, bird predation pressure is greater for early ripening cultivars since cherries are one of their earliest fruit food sources. The earliest ripening cultivars incur proportionally higher damage than later ripening blocks when overall fruits are more abundant. Similarly, damage tends to be proportionally greater in smaller orchards and in areas where cherry orchards are somewhat isolated, compared to large orchards and areas with a high density of cherry orchards, since the latter situations spread out damage from a finite number of birds over a greater number of trees. However, even within areas with a large number of orchards, flocks of birds may disproportionately damage some orchards and not others. In some areas, potential bird damage can occur not only to ripening fruits, but also to flower buds as they swell in the spring or even open flowers during bloom. Orchards that are close to roosting sites, e.g. power lines, tree lines and other vantage points, are usually the first to suffer damage. Orchards that are close to urban environments tend to be very

Fig. 10.18. Bird damage in cherry orchards, especially in isolated areas, can be extensive in the absence of some kind of mitigating strategy (courtesy of G.A. Lang).

susceptible since these settings provide refuge for a number of fruit-eating birds.

Bird damage also can vary from year to year, since birds are opportunistic feeders and many pest birds are migratory. Research funded by the US Department of Agriculture (USDA) has estimated that, in general, if cherry growers make no attempt to manage bird damage, production losses would approach 37% on average (with higher damage levels possible in some orchards). Consequently, mitigation strategies save the cherry industries in Washington, Oregon, California, Michigan and New York up to $143 million worth of crop annually. Despite these strategies, it is estimated that US cherry growers still lose ~13% of their crop overall to birds. Australian stone fruit growers estimate a crop loss of ~16% to birds (Tracey *et al.*, 2007).

The species of birds that may damage sweet cherries is extensive, varying widely around the world. Studies of birds responsible for cherry damage in the USA found that the European starling (*Sternus vulgaris*), American robin (*Turdus migratorius*) and American crow (*Corvus brachyrynchos*) were the three most destructive species, though in some areas cedar waxwings (*Bombycilla cedrorum*) can be very destructive as well. In North America and Europe, house finches (*Haemorhous mexicanus*) and Eurasian bullfinches (*Pyrrhula pyrrhula*), respectively, may eat flower buds or open flowers. In Australia, rosella parrot (*Platycercus* spp.) damage to cherry flower buds of up to 95% has been reported in some orchards (Tracey *et al.*, 2007). This bud damage even varied by cherry cultivar, with six to seven times the damage in Black Douglas and Williams Favorite as compared to Lustre and Makings. Nearly two dozen species of birds in Australia, mostly native (but some common introduced pest species like starlings), have been identified as being damaging to cherries, including cockatoos (*Cacatua galerita*) and lorikeets (*Trichoglossus moluccanus* and *T. chlorolepidotus*). As a rule, effective control methods depend on the legal protection status of the bird species. Consequently, it is imperative to identify the bird causing the problem, e.g. the American robin and the house finch are on the US Migratory Bird Treaty Act List and therefore are protected species, which means they can only be removed with a depredation permit from the US Fish and Wildlife Service or local county agricultural commissioner (Stetson and Baldwin, 2010). A similar regulatory situation exists for control measures needed for most native birds in Australia (Tracey *et al.*, 2007).

The most consistent and effective technique to reduce bird damage in cherry orchards is with exclusion netting, though this is also the most costly (Fig. 10.19). Netting mesh should be 6–12 mm (0.25–0.5 in), with no gaps between netting panels, sides, doorways for equipment access into the orchard, or where the netting contacts the orchard floor, as birds can find small gaps to get into a covered orchard. The pole-and-wire or -cable framework to support the netting should be robustly engineered and anchored to withstand strong winds, and it should be removed seasonally in locations where snow occurs. Thus, the capital costs can be significant, and the annual labor costs for installation, removal, storage, and repair or replacement add additional expense. Drapenetting, which is applied directly over the tree canopy without a support structure, can reduce the cost of exclusion netting, but this can lead to more wear and tear on the netting. When a drapenet rests directly on fruit clusters, damage can still occur from birds pecking a fruit through the netting, and if the drapenet is not tightly closed at its base, many birds will find the openings to get in.

The most effective alternative strategies to exclusion netting usually employ a variety of techniques, and a diversified control program should begin before birds become a problem. Once problematic birds are accustomed to feeding on cherries, it becomes even more difficult to discourage or dissuade them from their orchard food source. Methods to mitigate bird damage include: shooting, trapping, acoustic scare devices, visual scare devices, predator or competitor nesting boxes, and chemical repellents (Fig. 10.20). Success with each of these individual strategies may be limited

Fig. 10.19. Bird netting protects these cherries in New Zealand (courtesy of L.E. Long).

Fig. 10.20. This New Zealand grower uses a more lethal method to control bird damage (courtesy of L.E. Long).

or fleeting, as bird populations shift in and out of the orchard or become habituated. Trapping and shooting are labor intensive and may be limited by local legalities. The effectiveness of any deterrent strategy may be influenced significantly by the availability of alternative food sources.

Where disruptive and scare-type techniques are employed, it has been found that birds become conditioned quickly to repetitive behavior. Consequently, it is imperative that acoustic or visual devices and deterrents be moved every 2–3 days to new locations. Metallic ribbons, Mylar® streamers, spinners, shiny objects such as compact disks and reflective tin foil, 'scare-eye' balloons, plastic hawk silhouettes, plastic owls on poles and plastic snakes all have limited efficacy over time and must be moved regularly. A more recent scare tactic that has shown some success is that of 'air dancers' (plastic tubes through which air is blown for release out the top and smaller attached side tubes, making them rise up and down); however, birds have been known to approach orchards from different directions to avoid the air dancers, so placement around the entire orchard becomes necessary. Research has found that two air dancers per border are needed for a 4 ha (10 acre) block. Another recent scare tactic is the installation of a rotating laser beam that sweeps back and forth above the orchard in different programmable patterns, which can protect up to 6 ha (15 acres) or more, depending on the strength of the laser and the topography of the orchard. Acoustic devices that scare birds by generating noise are also available. Propane-powered cannons and exploders, and electronic sirens and bird distress calls, are all employed by growers with varying results. Cannons can be disturbing to nearby neighbors. Avoiding conditioning of the birds to the noise is a key factor for success, so varying the interval of the noise abatement, as well as timing, intensity, and variation of the type of noise or distress call, is crucial. Habituation of birds to scare devices can actually increase damage in the orchard, as foraging birds that peck at fruit (rather than eating it whole) may leave the fruit they are eating when the scare device is initiated, then quickly return to a previously undamaged fruit to resume foraging (Tracey et al., 2007). Visual and acoustic deterrents used in combination tend to be more effective than each type alone.

Regarding chemical repellents for birds, methyl anthranilate (MA), is currently registered in the US for use on fruits, but results have been mixed, and the fruit must be washed to remove the typical 'grape' smell of MA residue on the fruit. Spraying of cane sugar (sucrose) as a repellent is based on ingestion by birds which are unable to digest this form of sugar (rather than fructose and glucose), thereby causing non-lethal digestive distress. This also is reported to have mixed results, since birds must ingest it while eating fruit, and the sucrose may attract insects. Another deterrent method is that of employing a falconer or establishing boxes for falcons, such as American kestrels (Falco sparverius), to nest in the orchard. However, falconers are rather uncommon, the work is labor intensive, and fruit-eating birds may become conditioned to falconers as well, returning upon the falconer's departure. Nesting boxes only work when occupied, obviously, and the wide hunting range for most falcons means nesting boxes may need to be placed no closer than several kilometers from each other. In some situations, it may be necessary to hire a professional wildlife pest control operator.

10.2.2 Rodents

Rodents can cause significant damage to cherry trees by eating roots and bark, and the tunneling habits of some rodents can damage the soil structure in the orchard as well as may interfere with irrigation equipment and distribution of water. Rodent holes in the soil may even cause injury to orchard workers who may inadvertently step into the holes; these holes can also serve as nesting sites for snakes. Generally, although rodents are considered pests, different localities may specify certain limitations to mitigation and control efforts.

10.2.2.1 Voles

For the most part, voles (*Microtus pennsylvannicus*) are controlled by natural enemies, including snakes, coyotes, foxes, hawks, owls and domestic cats (Fig. 10.21). However, if the habitat provides suitable protection and there is an abundance of high protein food, vole populations can quickly increase to damaging levels. Population explosions typically occur once in 5 years and usually only last for a year. Once in 12 years, populations explosions can become epidemic. Damage to cherry trees usually occurs by girdling and removal of bark from the trunk near the base of the tree, and normally occurs in winter when the voles are foraging under snow cover. Voles do not hibernate and are most active at night. They prefer feeding under cover in brush piles and vegetation more than 15 cm (6 in) in height. They do not like crossing bare ground and their presence is easily recognized from runways in the vegetation or shallow 2.5–5 cm (1–2 in) holes leading to tunnels in the ground.

Habitat modification to reduce vole nesting and activity may be achieved from early spring through late fall. Reducing vegetation by mowing or herbicides will remove the cover that voles prefer and make them susceptible to predation by nocturnal animals and birds. Voles can be excluded from the susceptible portion of the trunk of young trees by wrapping the tree with 9–10 mm (0.35–0.39 in) wire netting or with plastic spiral trunk guards, extending the protection down to the soil level and up at least 15 cm (6 in) above the desired vegetation. Plastic spiral tree guards may serve a double purpose to protect trunks from winter injury in cold climates (see Chapter 7, 'Sweet Cherry Training Systems').

Trapping should begin in spring and traps should be placed perpendicular to vole runways on the soil surface. Use of the pesticide zinc phosphide

Fig. 10.21. This grower in the Pacific Northwest (PNW) encourages owls and other predatory birds to nest in his orchard by erecting bird houses, so they can actively control rodent populations (Oregon) (courtesy of L.E. Long).

bait is restricted and should only be applied in early spring and late fall if conditions permit; do not apply when moisture is expected. There are also some additional restricted-use anticoagulants that require multiple applications and should only be applied in early spring and late fall. Note that anticoagulants can be toxic to other animals. In Washington state, the only legal vole or gopher trapping devices are common mouse and rat trap devices that do not grip or hold the body in any way.

10.2.2.2   Pocket gophers

Pocket gophers, comprising three genera (*Thomomys*, *Geomys* and *Pappogeomys*) and up to 18 species, are burrowing rodents that are active all year and live for up to 5 years. Pocket gophers feed on tree roots and bark, and some have shown preferences for certain cherry rootstocks over others; in a research plot of Chelan sweet cherry on various rootstocks at Washington State University's Roza experimental orchards, gophers killed all young trees on Mahaleb rootstock, while damage was minimal on Mazzard, Colt, Tabel Edabriz, Gisela 3, Gisela 5, Gisela 6 and Gi.195/20 (G. Lang, personal communication). They may also chew on plastic irrigation tubing. Pocket gophers are adapted to a wide range of soil types, but tend to avoid heavy clay soils and those that do not allow free gaseous exchange. Burrows are a network of tunnels that may reach a depth of 2 m (6 ft) and are made up of a main tunnel with several lateral tunnels, including blind tunnels and storage areas. Mating can occur throughout the growing season and gophers can produce up to three litters per year, averaging three to six young per litter.

It is important to understand pocket gopher habits, especially the burrowing system, to control them effectively. Management methods include cultivation of burrows and mounds, hand or mechanical baiting, trapping, fumigation, and gas combustion within burrows (Gunn *et al.*, 2011). For most effective control, use a combination of methods. In cases of heavy infestations, drag or harrow the field to eliminate mounds of soil and subsequently identify active burrows. Cultivation is most effective when done in the late fall when gopher populations are lowest and they are most active preparing for winter. Good weed control removes potential food sources for burrowing gophers.

10.2.2.3   Rabbits and hares

Numerous species of rabbits (*Sylvilagus* spp.) and hares (*Lepus* spp.) may become pests in cherry orchards. Damage to cherry trees usually occurs by girdling and removal of bark from the trunk near the base of the tree, but may also occur further up the trunk during the winter when snow levels provide higher access.

Management methods include cultivation and weed control to make the orchard habitat a poor source of food and brush for cover for rabbits/ hares; hand or mechanical baiting, trapping and shooting are also possible. For most effective control, use a combination of methods. Rabbits tend to be easier to trap than hares. Rabbits and hares can be excluded from the susceptible portion of the trunk of young trees by wrapping the tree loosely with 9–10 mm (0.35–0.39 in) wire netting or with plastic spiral trunk guards, extending the protection down to the soil level and up at least 15 cm (6 in) above the potential snow line in cold climates, since accumulated snowfall can provide rabbit/hare access to higher portions of the trunk compared to voles which burrow under the snow. Plastic spiral tree guards may serve a double purpose to protect trunks from winter injury in cold climates (see Chapter 7). Rabbits and hares can be excluded from the orchard by perimeter fencing in which the wire mesh on the lowest section is closely spaced and extends at least 0.5–0.75 m (2–3 ft) high.

10.2.2.4   Groundhogs, muskrats and beavers

Damage to cherry orchards by large rodents like groundhogs, beavers and muskrats is rare, but not unheard of. Ground hogs (*Marmota monax*), also known as woodchucks or marmots, dig significant burrows that can damage the root zone and create dangerous holes for orchard workers. For orchards near water or wetlands, beavers (*Castor canadensis*) can cut down young trees for building their dams and lodges or for their food storage piles. Muskrats (*Ondatra zibethicus*) can girdle young trees similarly to rabbits and hares. Muskrats and beavers can be excluded from the tree by wrapping the trunks loosely with 9–10 mm (0.35–0.39 in) wire netting or hardware cloth, 0.5–0.75 m (2–3 ft) high, or by bounding the orchard with fencing in which at least the wire mesh on the lowest section is closely spaced and extends to at least 0.5–0.75 m (2–3 ft) high. Trapping or shooting are the most common ways of dealing with these large rodents if they become orchard pests, subject to local laws.

10.2.3   Other mammals

10.2.3.1   Raccoons

Raccoons (*Prodyonn lotor*) can be voracious foragers on ripe cherry fruits, not only consuming fruit just as it ripens, but also damaging fruiting spurs and branches in tree canopies from climbing to feed on the fruits. If the orchard is fenced (as for deer or rabbits), an electrical wire about 10–12 cm (4–5 in) above the ground will be necessary to exclude raccoons since they are adept climbers. Trapping *or shooting are the most common ways of dealing with raccoons, subject to local laws.*

10.2.3.2   Deer

Deer (several species, such as the white-tailed deer, *Odocoileus virginianus*), and other large ruminant mammals such as elk (e.g. *Cervus canadensis*) and moose (*Alces alces*), are native to North America and roam freely in many locations where agricultural fields and forests are in close proximity, leading to the potential for minor to severe damage from foraging. Shoots and leaves of sweet cherry trees are among the most favored foods for browsing deer, and the development and training of young orchards can be set back significantly in a single night of grazing by deer populations of only a half dozen or more. Additional damage can occur in the fall during the rutting season, when male deer rub the velvet off their antlers and scrape the bark off tree trunks during mating rituals. Consequently, where deer population densities are consequential, major economic losses can occur in sweet cherry production, especially with the increasing trend of high density orchards comprised of small statured trees for which development of fruiting wood low in the canopy is critical.

Fencing the entire orchard is the most effective method for minimizing orchard damage from deer and other large wild ruminants, but it can be expensive for large orchards. Suitable fencing consists of welded or woven wire, fixed-knot wire, chain-link, or electric fencing, and the height of the fence is critical for effective control. The minimum height for a deer fence should be 2.5 m (8 ft), with support posts spaced 3–6 m (10–20 ft) apart, closer for larger deer or elk.

Eradication of deer falls under the various local or regional governmental jurisdictions that manage natural resources, which often have the goal of maintaining healthy deer populations for hunting and wildlife viewing, while limiting the effect that the population will have on property damage. Permitted recreational hunting may or may not control deer populations, depending on the number of hunters and the type and duration of hunting season permits. Where an orchard is subject to damage from high deer populations not controlled adequately by recreational hunting, it is usually possible to apply for an out-of-season control permit, which may require a minimum economic loss threshold to be documented. Growers who experience excessive losses should contact the local or regional wildlife biology office, which can help develop a control plan.

For More Information

Edge, W.D. and Loegering, J.P. (1998) Controlling pocket gopher damage to agricultural crops. Oregon State University EC 1117. www.smallfarms.oregonstate.edu/sites/default/files/pocket-gophers-ec1117.pdf (accessed 19 June 2020).

Dreves, A. *et al.* (2019) Non-crop host plants of spotted wing *Drosophila* in North America. Oregon State University Extension publication EM9113. www.catalog.extension.oregonstate.edu/em9113 (accessed 19 June 2020).

van Steenwyck, R. (2014) Spotted wing *Drosophila* (SWD) recommendations for sweet cherry. www.ipm.ucanr.edu/PDF/MISC/2014_Cherry_Spotted_Wing_Drosophila.pdf (accessed 19 June 2020).

University of California Pest Management Guidelines – Cherry. www.ipm.ucanr.edu/PMG/selectnewpest.cherries.html (accessed 19 June 2020).

References

AliNiazee, M.T. (1976) Thermal unit requirements for determining adult emergence of the Western cherry fruit fly (Diptera: Tephritidae) in the Willamette Valley of Oregon 1. *Environmental Entomology* 5(3), 397–402.

Alston, D.G., Alston, D.G., Brunner, J.F., Davis, D.W. and Shelton, M.D. (1991) Phenology of the Western cherry fruit fly (Diptera: Tephritidae) in Utah and Washington. *Annals of the Entomological Society of America* 84(5), 488–492.

Beers, E.H., Hoyt, S.C., Alston, D.G., Alston, D.G., Brunner, J.F. *et al.* (2007) European red mite. In: Tree Fruit Research & Extension Center. Orchard Pest Management Online. WSU. Available at: www.treefruit.wsu.edu/crop-protection/opm/spider-mites (accessed 19 June 2020).

Gunn, D., Hirnyck, R., Shewmaker, G., Takatori, S. and Ellis, L. (2011) Meadow voles and pocket gophers: management in lawns, gardens and cropland. PNW publication 627, 16. Available at: www.extension.uidaho.edu/publishing/pdf/PNW/PNW0627.pdf (accessed 19 June 2020).

Song, Y.-H., Riedl, H., Coop, L., Omeg, M., Castagnoli, S. *et al.* (2004) Development and validation of phenology models for predicting cherry fruit fly emergence and oviposition in the mid-Columbia area. Poster presented at Portland Pest and Disease Conference, January 2004: Available www.uspest.org/ipm/cff_poster_portland.jpg (accessed 19 June 2020),Jan 2004.

Stetson, D.I. and Baldwin, R.A. (2010) Birds on tree fruits and vines. University of California publication 74152. Available at: www.ipm.ucanr.edu/PDF/PESTNOTES/pnbirdstreefruitsvines.pdf (accessed 19 June 2020).

Tracey, J., Bomford, M., Hart, Q., Saunders, G. and Sinclair, R. (2007) *Managing bird damage to fruit and other horticultural crops.* Bureau of Rural Sciences, Canberra, pp. 268.

Managing Orchard Pathogens and Disorders   **11**

11.1   Understanding Disease Phenology

There are a number of infectious diseases that affect cherries around the world. Common bacterial diseases include bacterial canker (*Pseudomonas syringae* pathovars *syringae* and *morsprunorum*) and crown gall (*Agrobacterium tumefaciens*). The most serious fungal disease of cherries in the Pacific Northwest (PNW) production areas of the USA is powdery mildew (*Podosphaera clandestina*), but where spring and summer rains are more prevalent, cherry leaf spot (*Blumeriella jaapii*), brown rot (primarily *Monilinia fructicola*, but also *M. fructigena*, *M. laxa* and *M. polystroma*) and silver leaf (*Chondrostereum purpureum*) can cause significant losses. Viruses like Prunus necrotic ringspot virus (PNRSV) and little cherry disease (caused by little cherry virus 1 and 2 [LChV1, LChV2] and western X phytoplasma) can cause the loss of entire orchard blocks. Infectious diseases can be transmitted from one tree to the next and, depending on the disease, may infect a few or all of the trees in a block.

In order for an infection to take place, the following conditions must be met: the tree must be susceptible to the disease; the pathogen must be present; and the environment must be conducive for infection to occur. For example, Chelan is resistant to powdery mildew; therefore, Chelan trees will not succumb to the disease even if the pathogen is present and the environment is favorable for infection. Unfortunately, most sweet cherry breeding programs around the world have not selected for disease resistance, so there are few examples of disease-resistant cultivars.

Although *P. syringae* is present in all cherry production areas of the world, a favorable environment is needed for the population to reach infectious levels and cause bacterial canker. In this case, free water and a wound (such as a pruning cut or frost-damaged flower) or natural opening (such as a leaf scar in the fall) are needed for infection. The tree is susceptible to infection for several days after the wound is made if free water, in the form of rain or dew, is present on that wound. Shifting the time of pruning to a period of predicted dry weather when precipitation or dew is less likely to occur, can significantly reduce infection. In this case, the environment is no longer conducive to infection and the disease cycle is stopped.

In the case of viruses, another factor is introduced into the cycle. Most viruses are transmitted from plant to plant by insect vectors such as aphids, leafhoppers or thrips. If the vector is not present, then it may be possible to slow or stop the transmission of the disease. An example of this can be seen with LChV2. Both apple and grape mealybugs serve as vectors for this disease. Managing the disease is a three-fold process that includes the planting of disease free trees, destruction of symptomatic trees, and controlling the insect vectors. PNRSV (and prune dwarf virus, PDV) are pollen-borne viruses, so tree removal is critical since pollinating insects are necessary for cropping.

Non-infectious plant diseases cannot be transmitted from a diseased plant to a healthy plant. Examples of these situations include nutrient deficiencies (see Chapter 8, section 8.4.2, Nutrient Disorders) and physiological disorders. Wind, hail and rain can also wreak havoc on cherry tree health and technically fall into the category of non-infectious diseases; however, they are most often thought of as plant injury rather than disease.

11.2 Bacterial Diseases

The most current accounting of bacterial pathogens found in sweet cherries can be found in Pulawska *et al.* (2017). The following sections describe the most significant widespread bacterial diseases of economic importance.

11.2.1 Bacterial canker

Bacterial canker is perhaps the most important disease of sweet cherries worldwide. It is found everywhere that sweet cherries are grown, and under conditions favorable to the disease, reports of up to 75% mortality have been recorded. More commonly, losses between 10–20% are documented in young orchards (Spotts *et al.*, 2010b).

11.2.1.1   Symptoms and disease cycle

Pseudomonas syringae can cause three distinct symptoms on cherry trees, depending on the tissue that has been infected. The most common symptom is a canker that forms as a slightly sunken, darker area of bark on the trunk or branches. Removing the bark exposes orange to brown streaks that run vertically along the wood, eventually girdling the branch. Often a light to dark amber colored ooze will seep from the canker (Fig. 11.1). This ooze is filled with bacteria that can be moved to new infection sites by insects,

Fig. 11.1. Ooze emanates from this canker as a result of a *Pseudomonas syringae* infection (courtesy of J.W. Pscheidt, © Oregon State University).

rain splashing, or pruning shearers. As temperatures increase, the branch or tree portion above the canker may begin to wilt, leaves turn light green to yellow or orange, and this portion of the tree may collapse entirely in hot weather.

The second symptom is known as dead bud, blossom blight, or spur blast. Frost, just prior to or during bloom, can damage buds, allowing an entry point for the bacterium. When leaf scars become infected in the fall, the infection may move systemically into leaf or flower buds in the spring. Infected flowers and buds may fail to open, ultimately dying. The dead flowers may remain attached to the buds, and cankers then form at the base of the buds, expanding to twigs and entire spurs. Alternatively, infected buds may open normally but later collapse, resulting in shriveled leaves and fruit.

The third symptom, leaf or fruit spots, is seen primarily in areas with high spring or summer rainfall. Leaf spots generally occur on young leaves following a heavy rain event, and initially are small and water soaked, but become dry and brown as the leaves mature (Fig. 11.2.). These necrotic spots often separate from the leaf, leaving shot holes later in the season. Fruit infections are similar in size, causing a dark pitted area to form in the surface of the fruit.

P. syringae colonizes many plant species, including grasses and weeds commonly found in orchards, without causing these plants any harm. However, this means that there is always a ready source of inoculum in the orchard, especially during bloom and leaf fall, two periods when the trees are most susceptible to infection. *P. syringae* populations in spring can increase 10–100-fold during bloom. As temperatures increase during the summer, *P. syringae* populations decline and active cankers go dormant.

Fig. 11.2. Note the angular nature of these leaf spots caused by *Pseudomonas syringae*, a common indicator of a bacterial spot, compared to the more circular spot of a fungal leaf spot (courtesy of G.A. Lang).

In the fall when cooler temperatures and rain return, populations of the bacterium increase once again.

Young trees are at the greatest risk and mortality is highest in the first 2 years after planting. Infection and mortality are rare in plants that are 8 years or older. Spring frost and winter freeze damage, pruning cuts, insect injury and newly formed leaf scars are most commonly associated with infection. Infection also can occur during warmer periods of winter as trees are being pruned. Additionally, any situation that causes stress or weakens the tree can lead to infection, such as drought or infestations by piercing/ sucking insects such as San Jose scale (*Quadraspidiotus perniciosus*).

11.2.1.2 Management

Good cultural practices to maintain stress free, healthy trees can reduce the potential for bacterial canker infection. This includes planting in sites that avoid frost pockets, use of frost mitigation measures like wind machines (and avoiding over-tree sprinkler frost protection), avoiding low or high pH soils or soils with high nematode infestations (especially ring nematodes), and providing adequate irrigation and fertilization to avoid water stress and nutrient deficiencies. Orchard rain covers have been found to significantly reduce the incidence of bacterial canker and facilitate improved healing and reduced spread of cankers that were initiated prior to covering.

The choice of plant materials can influence the risk of canker infection. There is evidence, both from the scientific literature and grower experience, that cultivars vary in degree of susceptibility to bacterial canker. In a study in Oregon, Spotts *et al.* (2010a) reported that Bing and Sweetheart were the most susceptible cultivars, while Regina and Rainier appeared to be more resistant. Bing trees had the highest mortality of any cultivar tested, with 70% dead at the end of the 3-year study. Growers in Oregon and Germany have reported similar observations for Sweetheart and Regina.

Growers have long believed that rootstocks noticeably impart varying levels of resistance or susceptibility to bacterial canker. The F12-1 clone of Mazzard is thought to have some resistance to bacterial canker. In the Willamette Valley of Oregon, where disease pressure is very high, nurseries high-bud scions on to F12-1 to prevent total tree mortality if limbs become infected. Spotts *et al.* (2010a) reported that Bing on Gisela 6 and on Colt had the highest (90%) and lowest (0%) incidence of dead trees after inoculation, respectively. Spotts *et al.* (2010b) stated that Bing on Krymsk 5 had smaller heading cut cankers than trees on Mazzard or Gisela 6, and 43% of trees on Krymsk 5 died after inoculation compared with 50% on Mazzard. High levels of bacterial canker on trees grown on Gisela 6 have been reported by growers in Oregon's Hood River Valley as well as by growers in Chile, where Colt has been the rootstock of choice due to observed resistance to bacterial canker. MaxMa 14 has also gained popularity in Chile for the same reason, while CAB-6P has recently lost favor due to perceived high levels of bacterial canker infection.

Pruning trees during or within several days of a rain event significantly increases the potential for infection to take place. In fact, Spotts *et al.* (2010a) found that pruning cuts remained susceptible to infection for up to 3 weeks in the winter and up to 1 week in the summer. Due to the sizeable period of susceptibility in the winter, pruning during winter can incur risks where canker is prevalent. In the highest risk areas, or where significant infection has already taken place within a block, summer pruning during a rain free period may be the best option until the trees become old enough for reduced susceptibility. Unfortunately, it may not be possible, or advantageous, to prune all trees in the summer, especially those on dwarfing rootstocks. A reasonable alternative to this timing is during the colder, dry periods of winter, where temperatures are near freezing and the bacterium is less active.

Notches (scoring) made in the bark for the purpose of branch initiation can also provide sites for infection. Therefore, bark scoring should be completed before spring conditions promote a large population increase of *P. syringae*. Alternatively, application of Promalin® with a surfactant such as Pentra-Bark™ at a 2% rate can promote branch formation on one-year-old wood without wounding the tree.

Copper is the primary chemical used for controlling bacterial canker in the USA. The first application is typically made prior to the first fall rains, as leaves begin to fall, to reduce bacterial populations as the leaf scars become susceptible. Additional applications of copper are also made prior to, and/or just after, pruning as well as to protect newly planted trees. Reduced rates are often applied during budbreak to help control symptoms associated with dead bud. Unfortunately, *P. syringae* has developed resistance to copper in some parts of Oregon, Michigan, and probably other cherry production areas where it has been the primary control measure over the years. Copper applications in these areas are no longer effective and applying copper to trees where resistance is prevalent may have the adverse effect of increasing the disease level due to reduction in the populations of other potentially competitive bacteria.

Surface disinfectants containing hydrogen peroxide are registered but have very short residual activity. Multiple applications per week are recommended, timed to temporarily reduce populations just before periods of susceptibility like pruning or bloom. Although used by and sometimes recommended by growers, the efficacy of such surface disinfectants has not yet been tested extensively.

Successful control of bacterial canker must be accomplished through an integrated approach. This is true not only in copper-resistant areas, but wherever cherries are grown. The following suggestions have been made as an integrated approach to bacterial canker control (Spotts *et al.*, 2010b):

1. Do not interplant new trees with old trees, which are major sources of *P. syringae*.
2. Keep irrigation water off the part of the trees above ground as much as possible for the first 2 or 3 years after planting.
3. Avoid all types of injury – mechanical, insect, frost. Paint all trunks white with latex paint to prevent winter injury. Adding copper to the paint may be of little benefit.
4. Less disease occurs when summer pruning is used. Prune only during dry weather. Pruning at any time under row or orchard covers can greatly reduce the incidence of bacterial canker.
5. Remove and destroy branches and trees killed by *P. syringae*.
6. For recommendations on less susceptible cultivars and rootstocks, see discussion in text.
7. Locate the orchard in an area less likely to be affected by frost and slow drying conditions. Wind machines may be useful for reduction of blossom blast.

8. Test for and control plant pathogenic nematodes before planting. High populations of ring nematodes have been associated with more bacterial canker.
9. In high infection areas, plant trees later in spring to avoid cool, wet conditions.
10. Provide optimal soil conditions for growth, including attention to pH and nutrition. Application of excess nitrogen, especially late in the growing season, may promote late season growth that is susceptible to low temperature injury in early winter, followed by bacterial infection.
11. Control weeds, especially grasses, which often support large populations of *P. syringae*. Clover and vetch ground covers support lower populations. Consider clean cultivation of row middles for the first 3 years of orchard establishment.
12. Strains of *P. syringae* resistant to copper are widespread in some traditional fruit production areas of Oregon, Michigan and elsewhere. In such areas, copper sprays may result in more bacterial canker than no sprays and should not be used.

11.2.2   Crown gall

The crown gall bacterium enters plants through wounds in the roots or crown, causing a gall on the affected plant part. Common forms of injury that cause entry points include the natural growth of roots through the soil or the formation of lateral roots, pruning roots prior to planting, damage to roots from cultivation or root pruning, insect damage and soil heaving caused by freezing soil water. Once infection occurs, the bacterium transfers a strand of its DNA into the plant, causing tumor cells to grow as undifferentiated masses of tissue that form the gall. This undifferentiated tissue impedes the flow of water and nutrients through the plant's vascular tissues, leading to plant stress and potential mortality.

11.2.2.1   Symptoms

Agrobacterium tumefaciens can produce galls that vary in size from that of a pea to 5 cm (2 in) or more across (Fig. 11.3). Young galls can be soft, spongy, or wart-like, but as they mature, they harden into a very dense, rough, fissured surface. Infected plants will be less vigorous than their non-infected neighbors, and mature trees may have significant dieback in the tree top.

11.2.2.2   Management

Although it is possible to reduce the population of the bacterium with common soil fumigants, fumigation does not result in satisfactory control of the disease. It is possible to protect trees from infection by dipping or spraying the roots and crown of nursery stock in a non-pathogenic strain

Fig. 11.3. Galls reduce tree vigor and can cause dieback, usually in the top of the tree (from Oregon State University Plant Clinic collection, © Oregon State University).

of *Agrobacterium radiobacter* (strain 84, marketed as 'Galltrol-A') which competes with the pathogenic strains, preventing infection. It is only a preventative treatment and will not cure infections. Existing galls can be treated with a mixture of chemicals that specifically attack the gall tissue, but are safe for uninfected tissue. An example is marketed as 'Gallex', which can be painted on visible galls. Unfortunately, many galls are untreatable due to their location below ground.

Consequently, prevention and avoidance of situations favorable for crown gall formation are best management practices. These include:

- Inspect all nursery plants for galls on roots or crown. Destroy any infected trees.
- Avoid planting into soils with a recent history of crown gall.
- Symptoms may not develop when temperatures are below 15°C (59°F). If possible, plant when soil temperatures are below 10°C (50°F).
- Plant in only well drained soils.
- Avoid injuring the roots or crowns of plants while planting, or once established.
- Destroy seriously diseased mature trees with visible galls, removing as much of the root system as possible.

11.3   Fungal Diseases

The most current accounting of fungal pathogens found in sweet cherries can be found in Borve *et al.* (2017). The following sections describe the most important widespread fungal diseases of economic importance.

Fig. 11.4. Brown rot has completely engulfed many of the fruit in these clusters (courtesy of G.A. Lang).

11.3.1   Brown rot

11.3.1.1   Symptoms

In most parts of the world, brown rot is a serious disease of cherries, causing blossom blight, twig and branch dieback, and fruit rot. When climatic conditions are favorable for fruit infection, the entire crop may be at risk of infection nearly overnight. At 10°C (50°F), a wetting period of 18h is needed for infection, while at 25°C (77°F), only 5h of wetting is needed. Fruit are susceptible at all stages of development, but infections are most common on mature fruit, which are subject to microcracks as the fruit swell, as well as rain-induced macrocracks. Infection can also be initiated where fruits have been damaged by birds, insects or hail. Fruit infections begin with small light brown spots that can grow rapidly in concentric rings, covering the fruit and infecting adjacent fruit (Fig. 11.4). Rotted fruit may abscise or persist as mummies into the next season, creating

Fig. 11.5. Brown rot blossom blight is a widespread problem in areas where rain is common during bloom (courtesy of J.W. Pscheidt, © Oregon State University).

future sources of inoculum. Dispersion of spores can be driven by both rainfall (conidial spores) or wind following rain (ascospores). Production of conidial spores is inhibited by long periods of dry weather; conversely, several days of warm, humid conditions can create rampant and catastrophic infections.

Rainfall during bloom can trigger a blossom blight infection (Fig. 11.5). Infected flowers turn light brown followed by the production of buff colored (*M. fructicola*) or gray (*M. laxa*) spores. Infected flowers remain attached to the tree and provide the inoculum for later infections to take place on fruit or flowers the following year. Infected flowers also act as the entry way for infection to move into small twigs or spurs. Buff or gray spores can also form at these sites, along with ooze. In dry, arid cherry producing regions, only the fruit rot symptoms are common, and then only when fruit is damaged by rain.

Fig. 11.6. A spur infected with brown rot can serve as a source of inoculum as flowers emerge in the spring (courtesy of G.A. Lang).

11.3.1.2 Management

Good orchard sanitation is important for brown rot control in commercial orchards, and generally must be supplemented with chemical controls in humid production regions. Infected twigs and branches should be pruned out and destroyed, along with all mummified fruit (Fig. 11.6). Fruit damaging insects, such as leaf rollers and oriental fruit moth (OFM), should be controlled, and proper picking techniques should be practiced to ensure that fruit is not damaged during harvest. Fruit should be cooled rapidly after harvest.

Where the blossom blight stage of this disease is a problem, up to three applications of fungicides should be applied, such as at 'popcorn' bloom stage, full bloom, and/or petal fall. In areas with less pressure, one or two sprays are usually sufficient if a systemic product is used. For control of the fruit rot stage, fungicide applications should be made just prior to expected rain events, or immediately after damage has occurred. To reduce the risk of developing chemical resistance, fungicides with different modes of action should be alternated or tank mixed. No more than two applications should be made from any one group per year. Biological control (e.g. with certain epiphytic fungi such as *Aureobasidium pullulans*) is reported to be possible but has yet to be adopted at a commercial level.

Measures to control brown rot during postharvest include prevention of injuries during harvest, rapid cooling of harvested fruit, and the use of chlorine- or peracetic acid-based disinfectants in hydrocooling water and packing lines. Orchard rain covers have not been found to reduce brown rot incidence.

11.3.2 Gray mold/fruit rot

Gray mold, also called blossom rot or green fruit rot, is caused by *Botrytis cinerea*, a ubiquitous fungal pathogen found around the world. In cool wet or very humid conditions, it can attack flowers, including the peduncle, and in some cases infection can even extend into the twig. Flower infections can subsequently infect developing fruit.

11.3.2.1 Symptoms

Infected flowers may die and become colonized with gray fungal spores, which can infect the rest of the flower cluster. Where young fruit come in contact with infected flower parts, smooth brown lesions may form: the infection may be latent (invisible), or quiescent (visible as small red halos around a small tan lesion). Under favorable (wet) climatic conditions, fruit infections can progress rapidly, leading to a significant loss of young developing fruit. Conversely, latent or quiescent infections may remain barely noticeable until postharvest storage, and then rot fruit quickly. New infections also can occur near harvest if spores are present and fruit wounds or microcracks provide openings for spore germination.

11.3.2.2 Management

Application of fungicides is necessitated primarily if climatic conditions at bloom are very rainy. However, fungicides applied to control brown rot will generally control gray mold as well. To reduce the risk of developing chemical resistance, fungicides with different modes of action should be alternated or tank mixed. No more than two applications should be made from any one group per year. Inoculum can be reduced by clean cultivation of the orchard floor. Unlike the situation with brown rot, orchard rain covers have been shown to significantly reduce the incidence of gray mold fruit rot.

11.3.3 Anthracnose

11.3.3.1 Symptoms

Anthracnose, caused by the fungus *Colletotrichum acutatum*, causes sunken, brown concentric circular lesions on the fruit, as though it had been pressed against a hot surface and partially melted. Often, orange fungal fruiting bodies can be found in the lesion rings. The fruit lesions become most evident on ripening fruit near harvest.

11.3.3.2 Management

Anthracnose spores are spread by rain splash and infections are promoted by warm moist weather, so orchard covers can reduce infection incidence. Removal of mummified fruit at the end of the season helps to reduce inoculum. Two applications of fungicides at the green fruit stage have been reported to provide satisfactory control of anthracnose in sweet cherry.

11.3.4 Cherry leaf spot

11.3.4.1 Symptoms

Cherry leaf spot is prevalent in moist and humid production areas around the world (Fig. 11.7.). Although it tends to be most severe in sour cherries, it can also cause significant foliar damage and defoliation in sweet cherries when climatic conditions are highly favorable. Such damage can lead to poor winter survival. Initial infections cause small, reddish-purple spots on the upper surface of leaves. The spots may enlarge, turn brown, and coalesce, and the leaf may lose chlorophyll and become cholorotic before abscising. In wet weather, whitish patches of conidial spores appear on the underside of leaves. Premature defoliation can significantly weaken the tree, predisposing trees to winter injury and/or subsequently affecting flower bud development, fruit set, spur leaf size, and fruit size. Cherry leaf spot has been reported occasionally in some arid production regions, but is not common and usually does not need treatment in such climates.

11.3.4.2 Management

In high pressure growing regions, it is necessary to protect newly emerging leaves with a fungicide application at petal fall, shuck fall and on a regular basis, every 7–10 days, until harvest. In wet years, at least one or two postharvest applications also should be considered, beginning 2–3 weeks after harvest. To reduce the risk of developing chemical resistance, fungicides with different modes of action should be alternated or tank mixed. No more than two applications should be made from any one group per year. To reduce inoculum, fallen leaves should be mowed throughout the

Fig. 11.7. Cherry leaf spot can be a serious problem in wetter climate areas such as Michigan (courtesy of G.A. Lang).

growing season so that they will decompose quickly. In fall, apply urea to the fallen leaves to promote more rapid decomposition.

11.3.5 Powdery mildew

In most cherry producing regions of the world, powdery mildew is of minor economic consequence compared to brown rot and bacterial canker. However, in the arid production areas of the PNW, it is the most severe disease to be managed.

Fig. 11.8. White patches of powdery mildew conidia spores colonize both the leaves and the fruit (courtesy of J.W. Pscheidt, © Oregon State University).

11.3.5.1   Symptoms and disease cycle

Powdery mildew infection is characterized by the presence of a white, powdery coating (comprised of conidia) often covering the underside of leaves, causing distortion and puckering (Fig. 11.8.). Newly formed leaves are most susceptible. Often, fast growing suckers in the center of the tree canopy are the first leaves infected. Later infection will move to newly formed leaves on the perimeter of the tree. Finally, ripening fruit will show initially subtle signs of infection. Careful examination will reveal small patches of white spores that are often difficult to see. As the disease progresses, the fruit tissue becomes sunken and spores become more visible. Green fruit are not as susceptible as ripening fruit, but spores can remain quiescent on the fruit for some time before fruit infection takes place.

In orchards where powdery mildew is present, specialized fruiting bodies known as 'chasmothecia' (formerly known as cleistothecia) form on infected leaves in fall and winter. Chasmothecia are small, round and black, and will wash off the leaves with rain in the fall. Chasmothecia overwinter in cracks and bark crevices, and begin releasing ascospores in spring when there have been three consecutive days each with 6 h consecutively of temperatures between 21.1–29.4°C (70–85°F) in the presence of free-standing water. This typically occurs during or shortly after budbreak, as temperatures begin to warm, and rain or irrigation can trigger the release of ascospores that infect wet leaves. These infection sites can be found on either leaf surface and appear as light green, circular spots. Typically, these primary infections first become visible around shuck fall.

Warm temperatures (21–27°C or 70–80°F) and high humidity are required for the secondary spores, conidia, to infect newly developing leaves. Under these conditions, conidia can infect and develop new sporulating colonies in as little as 4–5 days. This is the secondary infection phase that can grow quickly on leaves and fruit, causing significant loss. Wet leaves are not needed for the secondary infection to occur.

11.3.5.2   Management

As primary infection typically occurs around shuck fall, chemical control measures should begin at this time. Applications may need to be continued at 7–10-day intervals or when degree day (DD) models determine timing based on temperature and humidity. To reduce the risk of developing chemical resistance, fungicides with different modes of action should be alternated or tank mixed. In addition, in any given year, avoid applying more than two applications from any chemical group.

Highly susceptible cultivars, such as Bing, Rainier, and especially those that are late ripening such as Sweetheart and Staccato, should be avoided if feasible. Chelan is resistant and Regina shows some tolerance to the powdery mildew. Early ripening cultivars tend to avoid economic infection levels since fruit is harvested before mildew populations begin to build. Excessive tree vigor should be avoided, such as by using dwarfing rootstocks and judicial use of nitrogen fertilizers. Thinning cuts to open up tree canopies for good air movement and removal of suckers in canopy centers can reduce the potential for infection, as can training to narrow planar 'fruiting wall' type canopies. Any heavily infected shoots should be pruned out and destroyed.

11.3.6   Silver leaf

Although silver leaf is a common disease on many species of plants, especially *Prunus* spp., it is rarely of economic importance in US cherry production areas, but can be of consequence in Chile, New Zealand and some other production regions. In these areas, extreme care must be taken to avoid infection, especially in young trees.

11.3.6.1   Symptoms and disease cycle

The silver leaf fungus is a saprophyte, growing on dead trees and branches, but can be an aggressive parasite when it gains entry into a live plant through pruning cuts, winter injury or insect damage. Pruning cuts can remain susceptible for as long as 1 week after wounding. Once infection takes place, the fungus grows systemically in the tree's water-conducting xylem tissue. As it grows, a toxin is produced that is transported through the xylem to leaves, causing them to turn silvery-gray and eventually to abscise. Leaf symptoms are most often seen in the spring on newer leaves. Young trees, 3–5 years old, are most susceptible to the disease, although the fungus can infect trees of any age. Often the disease is fatal, leading to a progressive decline of the tree beginning with individual branches and ultimately death of the entire tree. In dead tissue, the fungus grows to the exterior of the wood, forming 2.5–8 cm (1–3 in) fungal fruiting bodies (basidiospores) that sporulate and spread the disease.

11.3.6.2   Management

There is no chemical control for silver leaf. The best prevention is through maintenance of good tree health, avoidance of large pruning cuts as much as possible, and avoidance of pruning trees in the rain. Where disease pressure is strong, pruning in the dry season is critical. All pruning cuts should be treated immediately with a sealant; usually this task is assigned to a worker in tandem with a worker making the pruning cuts. All pruned wood should be removed from the orchard and destroyed. Pruned wood should not be piled near the orchard, as the fungus will grow and sporulate on dead wood, serving as a source of inoculum.

11.3.7   Shot hole

Sweet cherry leaf shot holes may form due to a number of causes, including: (i) bacterial infection by *Pseudomonas syringae* or *Xanthomonas pruni*; (ii) fungal infection by *Thyrostroma carpophilum* (previously known as *Wilsonomyces*; also *Coryneum*, *Clasterosporium* and *Stigmina*) or (iii) viral infection by PNRSV.

11.3.7.1   Symptoms and disease cycle

Distinguishing features of shot hole include the color and shape of the lesion spots that initially form on the leaves, the centers of which often die and drop out, leaving a shot hole appearance (unlike cherry leaf spot). Initially, fungal leaf spots are usually reddish-purple and round (Fig. 11.9.), whereas bacterial leaf spots are brownish-black and angular (Fig. 11.10.). Bacterial shot holes are most common following heavy rain events when leaves are still young and tender and easily infected. These usually do not

Fig. 11.9. Shot hole of cherry leaves caused by *Thyrostroma carpophilum* with close-up of the red/purple leaf spot. Note the circular and bullseye characteristics of these spots, both symptoms indicative of a fungal infection (courtesy of C. Kaiser).

Fig. 11.10. Shot hole of cherry leaves caused by the bacterium *Xanthomonas pruni*. Unlike fungal spots, bacterial spots are generally angular in nature as they are limited by the mature leaf veins (courtesy of C. Kaiser).

lead to any further infections or significant loss of physiological function (unlike cherry leaf spot, which usually leads to leaf abscission). PNRSV (see Section 11.4.3) tends to cause necrotic blotches as well as irregular spots on leaves; when the blotches dry up in warm weather and drop out, the leaves take on a tattered appearance rather than a shot hole appearance.

The first signs of fungal shot hole infection are slightly raised reddish pinhead specks on the young shoots. With time, these increase in size and show canker-like lengthwise splits, and gum may be exuded. If the cankers are large and numerous, they may result in bark girdling and subsequent shoot tip dieback. On the leaves, the lesions begin as reddish or purplish spots, surrounded by a light green to yellowish halo, typically up to 8 mm (0.25 in) in diameter. Fruit infections begin as purple dots that eventually turn into grayish-white lesions, then sunken brown to black necrotic areas on the fruit. Infected buds take on a dark appearance.

11.3.7.2   *Management*

Fungal shot hole diseases must be carefully monitored in late winter and early spring, particularly in windy, rainy conditions. For fungal shot hole, a fungicide should be applied at petal fall, shuck fall, and 2 weeks after shuck fall. To reduce the risk of developing chemical resistance, fungicides with different modes of action should be alternated or tank mixed. In addition, in any given year, avoid applying more than two applications from any chemical group. Avoid over-tree watering, as leaves must be moist for infection to occur. Prune out and destroy dead buds and cankered twigs if present. Rake and destroy infected leaves to reduce inoculum. If inoculum is allowed to build up within the orchard, control can become quite difficult.

11.3.8 *Leucostoma (Cytospora)* canker

11.3.8.1 *Symptoms*

Fungal cankers caused by *Leucostoma* species occur around the world, marked by sunken areas of bark on branches or the tree trunk, often with gumming at the margin. Black pinhead-sized fruiting bodies may appear on the dead bark of the canker. Leaves on infected branches begin to wilt, turn yellow, and die (Fig. 11.11).

11.3.8.2 *Management*

Leucostoma species are opportunistic and can infect a wide host range, including stone fruits, apples, pears, mountain ash, Russian olive, serviceberry and willow. They are dispersed by rain and wind. Trees become more susceptible to infection when under stress, including poorly drained soils, low temperature injury, sunburn or insect attack (e.g. borers). There are no chemical treatments for *Leucostoma* canker, so maintaining good tree health, and rapid removal and destruction (e.g. burning) of any wood exhibiting cankers or that is damaged in any way, including mechanical damage or earlier infection by brown rot, are key strategies for management.

11.3.9 *Armillaria* root rot

Armillaria root rot is common in most, if not all, cherry production areas of the world. There are many species of the fungus that attack a wide host range. It can be particularly devastating when orchards are planted on land where forest trees or scrub oak have been recently cleared, as the fungus survives on many wild species of trees and shrubs including Douglas fir, yellow pine, maple, oak, willow and aspen.

11.3.9.1 *Symptoms*

Infection occurs when cherry roots come in contact with partially decayed roots of infected plants or rhizomorphs (black, intertwining fungal threads that develop in the soil around infected trees). Trees with *Armillaria* infections may become weak, exhibiting low vigor and leaves that change to a

Fig. 11.11. A *Leucostoma* (*Cytospora*) canker on the trunk has caused branches above that point to die, leaving dead, brown leaves attached to the tree (courtesy of J.W. Pscheidt, © Oregon State University).

reddish-purple color earlier in fall than healthy trees. Typical wilt symptoms may affect all or part of the tree, or seemingly normal trees may suddenly collapse in hot weather. Leaves often remain attached to the tree after it has died. The crown and roots of infected plants exhibit characteristic mycelial fans just under the bark. Mycelial fans are white layers of the fungus body that spread out in a fan shape wherever the tree is infected.

Soil fumigation is not effective at controlling *Armillaria* root rot and the fungal propagules can remain viable in the soil for many years. Consequently, there is no known treatment for this disease, and to date, no known *Armillaria*-resistant rootstocks, although trees on Mazzard tend to survive longer than trees on Mahaleb. Thus, any potential orchard site with a history of *Armillaria* should be avoided for sweet cherry production.

Fig. 11.12. *Phytophthora* root rot can be a common problem, especially when cherry trees are grown on heavier soils (from Oregon State University Plant Clinic collection, © Oregon State University).

11.3.10　*Phytophthora* root and crown rots

11.3.10.1　Symptoms

Phytophthora root and crown rots (also known as 'collar rots') are caused by many species of *Phytophthora* and are widespread, destructive diseases of cherry trees throughout the world. *Phytophthora* infections are more prevalent where soils are heavy and drainage is poor, such as in low-lying areas, or where hardpan layers prevent good subsoil drainage (Fig. 11.12). Affected trees exhibit poor vigor and sparse growth, and may show wilting that includes sagging yellow or reddish leaves. Tree decline occurs over time, and in some cases, trees may grow normally while temperatures are mild, but then suddenly collapse in the heat of summer. The cambium region of the crown will be reddish-brown on infected trees and roots will show distinct signs of rot.

11.3.10.2   Management

Soil fumigation often is ineffective since some *Phytophthora* species survive fumigation treatments. The main practices for management include site selection and preparation. Orchard site soils should be well drained. In poorly drained conditions, trees should be planted on raised berms at least 30 cm (12 in) high, and marginal sites should be amended before planting by installing drainage tiles or breaking up subsoil hardpans that prevent good drainage. Mahaleb or MaxMa 14 rootstocks are particularly sensitive and should not be planted on poorly drained sites.

11.3.11   *Verticillium* wilt

Caused by the soil fungal organism *Verticillium dahliae*, *Verticillium* wilt can be deadly to sweet cherry trees. It is prevalent around the world and can infect other stone fruits and a wide range of other crops, such as potatoes, tomatoes, peppers, raspberries, strawberries, as well as many weed species (e.g. geranium, lambsquarters, nightshade, phlox, shepherds purse). It is long-lived in the soil, so prior history of the orchard site (in terms of previous crops and weed species) should be considered when planning a new sweet cherry orchard.

11.3.11.1   Symptoms

Initial symptoms of *Verticillium* wilt infection include sudden wilting of new extension shoot growth in early summer with dull-greenish flagged leaves retained, or progressive yellowish-reddish leaf coloration beginning at the base of extension growth that moves to the shoot terminal by midsummer, with eventual defoliation. Wilting is caused by partial blockage by the fungus of the water-conducting vascular system as well as by toxins produced by the fungus. By the end of the growing season, the entire branch or scaffold may be defoliated, except possibly the most terminal (youngest) leaves. Only one side or a portion of the canopy may be affected in the first season of symptoms (Fig. 11.13). The affected branches will reveal a brownish-red discoloration of the vascular tissue, and if present, fruit will be small.

11.3.11.2   Management

In addition to careful site selection based on prior cropping history, orchard management to avoid vigorous, succulent tree growth (such as excessive irrigation, nitrogen fertilization or pruning) or plant stress (such as nutrient deficiencies) will reduce plant susceptibility and soil conditions conducive to *Verticillium* infection. Infections are often fatal, especially for younger trees, though older trees with only a portion of the canopy affected initially may recover under good orchard management. There are no

Fig. 11.13. In the initial stages of the *Verticillium* wilt disease, symptoms of wilting may appear on only one side of the tree (from Oregon State University Extension Plant Pathology collection, © Oregon State University).

chemical controls, though preplant fumigation of sites with a prior history of infection may reduce inoculum or the parasitic nematode populations that can increase infection incidence and severity.

11.4   Virus and Virus-Like Diseases

Sweet cherry trees are at risk of infection by a wide diversity of viruses, some of which, like cherry virus A (CVA), do not appear to cause major negative symptoms. However, some viruses can significantly impact economic returns or even cause tree mortality. Furthermore, some viruses may cause no obvious symptoms in certain cultivar/rootstock/environment

Fig. 11.14. The lower branch is showing classic symptoms of little cherry virus (LChV), with small, poorly colored fruit (from Oregon State University Extension Plant Pathology collection, © Oregon State University).

combinations, while eliciting virulent reactions in other combinations, depending not only on the plant genotypes but also the virus strain, the climate, tree age and the combination of viruses present in the tree. Virus-infected trees may be more susceptible to winter injury. All viruses can be transmitted by grafting, hence the use of budwood that is free from known viruses is always important. Also, there are no cures for virus-infected trees, so identification of virus infections and expedient removal of infected trees is often the best strategy for minimizing the spread and impact of detrimental viruses. The most current accounting of viruses and virus-like pathogens found in sweet cherries can be found in James *et al.* (2017). The following sections describe the most important widespread viruses of economic importance.

11.4.1   Little cherry virus and western X disease

Little cherry virus (LChV) and western X disease have been associated with several fluctuating cycles of economic importance in US sweet cherry production areas, as they result in small fruit size with poor flavor and color, making the fruit unmarketable (Fig. 11.14). LChV1 (less common) and LChV2 (most common) were of importance in Washington state in the 1950s, and British Columbia through the 1970s, while western X significantly impacted California cherry production through the 1970s, but meticulous attention to budwood sources had largely reduced incidence to inconsequential levels by the 1990s. However, the rapid expansion of sweet cherry production in the Pacific western regions of the USA since 2000 included rapid grower adoption of many new rootstocks and varieties, some of which were not adequately tested for virus status. Consequently, LChV and western X disease have recently shown a significant resurgence in these areas, particularly in Washington and Oregon, with major losses of orchards currently ongoing.

Infection symptoms of these diseases generally become evident about 2 weeks before harvest, as fruit fail to size and color, with red-fruited varieties remaining very light colored with a bland to bitter flavor. Some varieties (such as Bing) exhibit more pronounced small fruit symptoms, while others (Van and Sam) may achieve almost normal size but still have poor flavor. Some cultivars (e.g. Lambert) may also exhibit a pointed fruit tip. Shoots may grow normally or may exhibit reduced extension, sometimes resulting in leaf rosetting. Symptoms tend to begin on a portion of a single branch, but extend throughout the tree over time and spread to neighboring trees. The levels of virus or phytoplasma in the branches can fluctuate seasonally, so some branches may appear to recover, only to revert again later. Eventually, branches may begin to die back. Western X disease is known to infect alternative hosts (including some weeds) as well as most *Prunus* species, and is transmitted readily by leafhoppers as well as grafting. LChV2 is transmitted by apple and grape mealybugs, while the vector for LChV1 is currently unknown. Both LChV1 and LChV2 can be transmitted by grafting.

There are no control measures for diseases caused by viruses and phytoplasmas. Thus, management begins with the planting of pathogen free nursery trees, which itself begins with budwood and rootstocks certified to be free of known pathogens. When infected trees have been identified in the orchard, they must be removed. Tree removal is often conducted just after harvest, since infected trees are more readily identified by the fruit symptoms. Such trees should be treated to control leafhoppers before removal, to prevent resident leafhoppers from moving from the infected trees to non-infected trees. It may be desirable to also remove trees adjacent to the infected tree, since they may be infected by root grafting or leafhopper movement but are not yet exhibiting symptoms. Treatment of the trees just before removal, or the stumps just after removal, with a systemic herbicide can help prevent spread by assuring root death. In addition to routine control of leafhoppers and mealybugs throughout the growing season, weeds that can serve as alternative hosts for leafhoppers should also be eliminated or minimized. These include dandelions, clovers, curly dock and chokecherries. Mealybug control measures may be chemical or biological, with a good selection of beneficial insects that prey on mealybugs, including lacewings, spiders, ladybird beetles and parasitic wasps.

Fig. 11.15. The infected tree on the right exhibits the narrow leaf symptom common with prune dwarf virus (PDV) (from Oregon State University Extension Plant Pathology collection, © Oregon State University).

11.4.2   Prune dwarf virus

Prune dwarf virus (PDV) is usually a sublethal virus in sweet cherry trees, except when trees are planted on sensitive or hypersensitive rootstocks. Trees infected with PDV may exhibit somewhat narrow, rough leaves and reduced flower bud formation on branches, particularly in older trees (Fig. 11.15). Fruit traits are generally unaffected, or may even have a firmer texture than normal. Tree growth may be slightly reduced, and success rates for nursery budding are usually lower when using PDV-infected budwood. Trees grafted on sensitive rootstocks may survive for several years, declining but flowering before eventually dying. Trees grafted on hypersensitive rootstocks tend to exude gum at the graft union and die quickly upon infection, generally before the next pollination season. PDV infection can

Fig. 11.16. Note the shot hole symptoms causing a tattered appearance on infected Prunus necrotic ringspot virus (PNRSV) leaves (courtesy of J.W. Pscheidt, © Oregon State University).

make other virus infections more damaging, such as PNRSV or cherry leaf roll virus (CLRV). PDV transmission is primarily by grafting, infected seeds (as when used for Mazzard rootstocks), or especially infected pollen.

11.4.3 Prunus necrotic ringspot virus/rugose mosaic virus

Like PDV, Prunus necrotic ringspot virus (PNRSV) is usually a sublethal virus in sweet cherry trees, except when trees are planted on sensitive or hypersensitive rootstocks, such as Krymsk 5 and 6 (see Chapter 3, 'Sweet Cherry Rootstocks') or when the PNRSV strain is particularly virulent, in which case it is referred to as 'rugose mosaic virus'. PNRSV symptoms include leaf shot holes that may eventually cause a tattered or lacy leaf appearance and leaf chlorosis (Fig. 11.16). Trees infected with PNRSV may exhibit uneven fruit ripening and show some dead buds, though symptoms that initially appear severe may become less apparent in subsequent seasons. Trees infected with the rugose mosaic strains may exhibit small outgrowths of leaf tissue, called 'enations', on the underside of leaves along the midrib. More virulent strains may cause deformed leaves, blossom blast, reduced pollen germination, delayed pollen tube growth, reduced tree vigor and/ or delays in fruit ripening of 10 days or more (up to 3 weeks), thereby having more negative effects on yields.

Fig. 11.17. This tree was positive for cherry leaf roll virus (CLRV), prune dwarf virus (PDV) and Prunus necrotic ringspot virus (PNRSV) (courtesy of J.W. Pscheidt, © Oregon State University).

PNRSV transmission is primarily by grafting, infected seeds (as when used for Mazzard rootstocks), or especially infected pollen. Therefore, infected trees should be removed before bloom to reduce the potential for transmission to neighboring trees. Trees grafted on sensitive rootstocks may survive for several years, declining but flowering before eventually dying. Trees grafted on hypersensitive rootstocks tend to exude gum at the graft union and die quickly upon infection, generally before the next pollination season.

11.4.4 Cherry leaf roll virus

Cherry leaf roll virus (CLRV) is particularly important in combination with infections by PDV or PNRSV. Incidence of CLRV infection in the PNW cherry production regions has increased significantly since the early 2000s. Infection of Bing trees on Colt or Gisela 6 rootstock has led to hypersensitive reactions and tree death, although with other cultivars, Colt and Krymsk 5 are reported to be tolerant or resistant to CLRV. In general, symptoms include delayed bloom and rolling of leaves that may have a chlorotic appearance. When co-infection includes PDV, leaf enations may be seen. When co-infection with PNRSV occurs, trees may decline rapidly and eventually die (Fig. 11.17). Fruit on CLRV-infected trees set heavily, are small, ripen late and generally are not marketable. Premature leaf

Fig. 11.18. Note the extensive enations on the underside of the leaf common to cherry rasp leaf virus (CRLV) (courtesy of J.W. Pscheidt, © Oregon State University).

abscission may begin in spring, buds may abort causing blind wood and shoots often die over winter.

CLRV transmission is primarily by grafting, root grafting, infected seeds (as when used for Mazzard rootstocks) and presumably infected pollen. Therefore, infected trees should be removed before bloom to reduce the potential for transmission to neighboring trees. It has a relatively wide host range, including woody and herbaceous plants, so pollen transmission can occur due to nearby plants other than sweet cherry.

11.4.5 Cherry rasp leaf virus

Cherry rasp leaf virus (CRLV) causes extremely deformed leaves, with deep marginal serrations, a rough surface texture, and often with many enations on the underside along the midrib (Fig. 11.18). Buds and branches may die, leading to bare, open sections of the tree canopy and reduced yields.

CRLV is transmitted by grafting and by dagger nematodes, and occurs in alternate hosts such as apples, many other *Prunus* species, potatoes, dandelions and red raspberries. Infected trees should be removed and alternate host plants should be eliminated within the orchard. If dagger nematodes are present, nematicides should be applied, particularly before planting a new orchard.

Fig. 11.19. This leaf, infected with cherry mottle leaf virus (ChMLV) shows the typical patches of yellow and dark green colors (from Oregon State University Plant Clinic collection, © Oregon State University).

11.4.6   Cherry mottle leaf virus

Cherry mottle leaf virus (ChMLV) causes distorted leaves with mottled patches of yellow or light green on otherwise dark green leaves (Fig. 11.19). Terminal leaves may become distorted. Shoot internode lengths decrease, resulting in a somewhat rosetted appearance. Fruit from infected trees are small, delayed in ripening, and have poor flavor. Symptoms may be suppressed in warmer climates. ChMLV infections can be severe in some production areas.

ChMLV is transmitted by grafting as well as by the scale mite (*Eriophyes inaequalis*). Some cultivars (e.g. Van, Rainier, Sam, Black Tartarian) may be symptomless carriers, while others (e.g. Bing, Sweetheart, Lapins, Royal Ann, Celeste) exhibit severe symptoms. Wild cherries (*Prunus emarginata*) are symptomless alternate hosts, so nearby wild cherry trees and infected orchard trees should be removed in a timely manner to minimize spread by mites. ChMLV also may infect peaches, apricots and ornamental flowering cherries.

11.4.7   Cherry necrotic rusty mottle virus/cherry rusty mottle virus

Cherry necrotic rusty mottle virus (CNRMV) causes angular dead spots on leaves, with the dead portions creating cracks and open angular holes in the leaf, usually appearing in spring. Leaves with significant necrotic sections may abscise early, and non-necrotic leaves tend to turn a mottled yellow-green color in fall (Fig. 11.20). Buds on infected young shoots may fail to swell and open, eventually abscising, and entire twigs or branches may eventually die, reducing yields. Bark may exhibit pockets or blisters of

Fig. 11.20. The cherry necrotic rusty mottle virus (CNRMV) caused severe necrotic spots to form on the leaves of this Sweetheart tree (courtesy of J.W. Pscheidt, © Oregon State University).

gum. Symptoms tend to be more severe during cool, wet springs, and may become less obvious during the season when temperatures are high.

Cherry rusty mottle virus (CRMV) causes similar, but less severe, symptoms which appear in late spring as a chlorotic mottling of older leaves. Leaves may exhibit yellow rings and progress to red colors, with a majority abscising near fruit harvest. Fruit ripening may be delayed. The mottling on non-abscised leaves eventually turns rusty brown. Fall coloring of leaves tends to occur early and rapidly. Infected trees decline in vigor and begin exhibiting limb death.

CNRMV and CRMV can be transmitted by grafting, but how they are transmitted naturally is unknown. CRMV may be spread by leafhoppers. Spread from tree to tree is slow, so tree removal upon identification of infected trees is worthwhile.

11.4.8 Prunus stem pitting/tomato ringspot virus

Prunus stem pitting is caused by the tomato ringspot virus (ToRSV). Infected trees lack vigor, leaves may be somewhat chlorotic and droop or cup upwards later in the season, developing fall colors and abscising earlier than normal (Fig. 11.21). The root systems of infected trees develop poorly. Consequently, tree nutrition and health suffer due to the poor root system and inefficient vascular conducting tissues. Trees can be diagnosed by

Fig. 11.21. The enations on the back of this leaf are due to infection from Prunus stem pitting virus and prune dwarf virus (PDV) (courtesy of J.W. Pscheidt, © Oregon State University).

removing a patch of bark (which may be thicker than normal and spongy) from around and below the graft union. Pitting (elongated indentations) of the wood is usually evident, though expression can vary depending on cultivar and rootstock. In trees of Royal Ann, pitting only occurs in the rootstock.

ToRSV is transmitted primarily by dagger nematodes, and weeds in the orchard, such as dandelions, can serve as a viral source for infection. ToRSV infects a wide range of alternate hosts, including most Prunus species, apples, and many berries, vegetables and ornamentals. Consequently, control of broadleaf weeds can reduce potential stem pitting risk. Colt rootstock is reported to exhibit a hypersensitive response, thereby isolating the virus to the nematode transmission site of the root and preventing systemic infection. Thus, Colt may be useful for producing sweet cherries on sites where dagger nematodes and ToRSV are known to occur.

11.5   Non-Infectious Physiological Disorders

11.5.1   Cherry crinkle leaf and fruit deep suture

Cherry crinkle leaf and fruit deep suture are thought to be physiological disorders since no causal organisms have yet been identified. Cherry crinkle leaf symptoms include distorted leaves with irregular margins and a mottled green color, small and abnormally developed flowers, and small pointed fruit, often with a raised suture (Fig. 11.22). Fruit deep suture symptoms include a significant fruit depression along the suture, and leaves are often long and narrow (Fig. 11.23). Both disorders reduce marketable yields. Branches exhibiting such defects can be pruned out. Wood from such trees should not be used for propagation.

For information on nutrient deficiencies, see Section 8.4.2: 'Nutrient disorders'.

Fig. 11.22. The Bing leaf and two cherries on the left are free of cherry crinkle leaf, whereas all the other fruit and the leaf on the right are showing symptoms of the disorder (from Oregon State University Extension Plant Pathology collection, © Oregon State University).

Fig. 11.23. The leaf on the right and the two top rows of cherries are all showing evidence of deep suture (from Oregon State University Extension Plant Pathology collection, © Oregon State University).

References

Borve, J., Ippolito, A., Tanovic, B., Michalecka, M., Sanzani, S.M. *et al.* (2017) Fungal diseases. In: Quero-Garcia, J., Iezzoni, A., Pulawska, J. and Lang, G. (eds) *Cherries: Botany, Production and Uses.* CAB International, Wallingford, UK, pp. 338–364.

James, D., Cieslinska, M., Pallas, V., Flores, R., Candresse, T. *et al.* (2017) Viruses, viroids, phytoplasmas and genetic disorders of cherry. In: Quero-Garcia, J., Iezzoni, A., Pulawska, J. and Lang, G. (eds) *Cherries: Botany, Production and Uses.* CAB International, Wallingford, UK, pp. 386–419.

Pulawska, J., Getaz, M., Kaluzna, M., Kuzmanovic, N., Obradovic, A. *et al.* (2017) Bacterial diseases. In: Quero-Garcia, J., Iezzoni, A., Pulawska, J. and Lang, G. (eds) *Cherries: Botany, Production and Uses.* CAB International, Wallingford, UK, pp. 365–385.

Spotts, R.A., Wallis, K.M., Serdani, M. and Azarenko, A.N. (2010a) Bacterial canker of sweet cherry in Oregon– infection of horticultural and natural wounds, and resistance of cultivar and rootstock combinations. *Plant Disease* 94(3), 345–350. DOI: 10.1094/PDIS-94-3-0345.

Spotts, R.A., Olsen, J., Long, L.E. and Pscheidt, J.W. (2010b) Bacterial canker of sweet cherry in Oregon – disease symptoms, cycle, and management. EM 9007. Oregon State University, Corvallis, OR.

The Future of Cherry Production

12

Sweet cherry production has increased consistently and significantly worldwide over the past 20 years, driven by improvements in fruit quality, consistency in yields, high economic returns, packing line sorting technologies and expanding middle-class populations that will pay well for this highly desirable fruit. Sweet cherries remain challenging to grow and subject to many production risks, but as long as market demand continues to increase, dedicated growers will find ways to overcome the risks. Rapid advances in horticultural production technologies, including electronic sensing, autonomous orchard equipment, machine learning and artificial intelligence, and robotics will contribute significantly to future production trends, as will new challenges due to invasive species, climate change and the ever unpredictable geopolitical landscape.

12.1   Orchard Technologies

The state of recent cherry production advances was summarized in Chapter 1, 'Trends in Sweet Cherry Production'. Vigor-controlling, precocious rootstocks have led to much more efficient, high density orchards, and simplified planar 'fruiting wall' canopy architectures are increasingly being adopted around the world. These radical changes in tree architecture are designed to optimize tree physiology through light interception and growth habit. Such changes enable labor efficiency through fruiting structure simplification and precision, and will likely be key to increased mechanization: initially tasks such as general tractor-based hedging and eventually some level of autonomous task-oriented robotic pruning. Simplification and repetition of uniform fruiting units within the canopy will facilitate increased utilization of non-destructive sensors of plant stress, canopy leaf area, canopy crop load and/or fruit quality during maturation. These technologies will generate increasing amounts of data that growers will be able to use to make better decisions regarding management of optimized leaf area-to-fruit (LA:F) ratios, selective applications of fertilizer and protective sprays, and variable irrigation to avoid transient stresses. The need for minimally skilled labor will decrease (though not disappear), while the need for labor with skills to operate technologically advanced equipment and understand data management and analysis programs will increase. While autonomous robotic pruning is likely to appear on the horizon, sweet cherries will probably be among the last tree fruits to be successfully harvested by robots, due to the fruit's relatively small size among large leaves, its delicate texture at maturity, and the current market demand that fruit be picked with the stem attached.

While climate change is clearly an increasing challenge, with ramifications for fundamental production factors such as adequate winter chilling, high and low temperature extremes, irrigation water availability, and the incidence of catastrophic climatic events such as hail, torrential rains or droughts, there is already a clear trend worldwide for increasing adoption of climate modification technologies such as wind machines, rain covers, high tunnels and greenhouses for production. The revolution in tree canopy architecture has facilitated the feasibility of installing protective orchard structures, including those for protection from birds or even insects. On a related tangent, recent innovative research has also been undertaken to develop mechanical pollination technologies, for the application of pollen at bloom via electrostatic sprayers. This technology could supplement or even replace honeybee pollinators in order to achieve more consistent flower fertilization and fruit set, in light of declining honeybee colonies and the impact of increasing spring climatic variation on pollination. Some commercial orchards are now planted simply for the harvest of pollen for such services.

Other advances in electronic technologies that are just now emerging or in advanced stages of research include light-emitting diode (LED) and other lighting sources that can provide specific physiologically active wavelengths of light to supplement natural sunshine. For example, equipment to provide specific 'doses' of ultraviolet (UV) light that disrupt the life cycle of important pathogens is currently under way for powdery mildew in grapes and gray mold in strawberries. Two new hand-held devices that bring the analytical capabilities of near-infrared (NIR) spectroscopy directly into the orchard for non-destructive assessments of fruit soluble solids, dry matter, acidity, color and possibly other quality traits, have the potential to greatly improve harvest decisions based on optimal fruit quality or factors for long-distance shipping to export markets. As noted in Chapter 8, 'Managing the Orchard Environment', soil moisture sensors and irrigation valving can now be automated or controlled remotely by smartphone applications, improving the monitoring of orchard water use and the timeliness of applying water and/or fertilizer via fertigation.

12.2 Genetic Improvement

In terms of genetic technologies for cultivar improvement, major strides have been made in the past 10 years in the development, through traditional breeding strategies, of cultivars with low chilling requirements for adapting to mild production climates. The generation of new improved low chill cultivars will likely accelerate as the next generations build upon the foundational generations of the past decade. Advances in greater disease resistance are likely to be sporadic, based on the difficulty of identifying sources of resistance for cross-breeding with germplasm having high fruit quality traits. Breeding populations with powdery mildew resistance are quite close to commercial quality, while those for bacterial canker resistance are nowhere in sight. Breeding for resistance to important insects, such as spotted wing *Drosophila* (SWD), is currently a similar distant dream, as is breeding for resistance to fruit cracking (given the multiple factors involved in cracking).

In terms of technologies for non-traditional genetic improvement, the greatest challenge that currently remains is consumer acceptance of any genetically modified temperate zone fruit crop. A major sustained effort has been made to commercialize a number of apple cultivars that utilize gene silencing, the targeted genetic modification to alter a single natural enzyme responsible for flesh browning. The Arctic® apple series down-regulates the gene for polyphenol oxidase (PPO) to only 10% of normal activity, which prevents the typical browning reaction of cut fruit, and this genetic modification has been introduced into industry standard varieties such as Golden Delicious, Granny Smith, Fuji and Gala. The company, Okanagan Specialty Fruits (now owned by Intrexon), recently announced that work is under way to create a similar line of PPO-downregulated sweet cherries that presumably would not 'bruise'. In the case of Rainier, these fruits would be easier to handle without suffering from brown marking of the yellow flesh.

Similar gene silencing has been applied successfully to modify a cherry rootstock for resistance to Prunus necrotic ringspot virus (PNRSV), and follow-up studies have shown that this resistance can be transferred to any scion cultivar grafted to the resistant rootstock. Consequently, this technology can confer virus resistance to any industry standard cultivar, such as Bing, Lapins or Regina, without any genetic modification of the cultivar itself. This may be a promising technological approach to solving some of the more difficult virus challenges in sweet cherry production, like little cherry virus. Sweet cherry genome mapping has long promised more efficient traditional breeding cycles as the list of molecular markers for important traits increases. For now, that list remains very limited and to date, no new cultivars have yet been developed from breeding populations that utilize molecular marker selection, but this will likely change in the very near future. Other advances in genetic improvement

technologies, such as clustered regularly interspaced short palindromic re-
peats (CRISPR)/CRISPR-associated protein 9 (Cas9) gene editing, have
not yet been applied to sweet cherry improvement, but efforts are surely
not far off. A major hurdle to the use of genetic engineering technologies
is the difficulty with which sweet cherries can be both transformed and
regenerated from transformed cell lines.

The 21st century continues to be an exciting time for sweet cherry
producers and all those involved in the industry. The production of high
quality fruit from tree to market will ensure continued demand and new
opportunities for cherry growers around the world.

Index

Note: The locators in bold and italics represents tables and figures.